AF595455

Artificial Intelligence Applied to Medical Imaging and Computational Biology

Artificial Intelligence Applied to Medical Imaging and Computational Biology

Editors

Leonardo Rundo
Carmelo Militello
Andrea Tangherloni

MDPI • Basel • Beijing • Wuhan • Barcelona • Belgrade • Manchester • Tokyo • Cluj • Tianjin

Editors
Leonardo Rundo
University of Salerno
Italy

Carmelo Militello
National Research Council
(ICAR-CNR)
Italy

Andrea Tangherloni
University of Bergamo
Italy

Editorial Office
MDPI
St. Alban-Anlage 66
4052 Basel, Switzerland

This is a reprint of articles from the Special Issue published online in the open access journal *Applied Sciences* (ISSN 2076-3417) (available at: https://www.mdpi.com/journal/applsci/special_issues/AI_medical_imaging_biology).

For citation purposes, cite each article independently as indicated on the article page online and as indicated below:

LastName, A.A.; LastName, B.B.; LastName, C.C. Article Title. *Journal Name* **Year**, *Volume Number*, Page Range.

ISBN 978-3-0365-6487-6 (Hbk)
ISBN 978-3-0365-6488-3 (PDF)

Contents

Editorial

Artificial Intelligence Applied to Medical Imaging and Computational Biology

Leonardo Rundo [1], Andrea Tangherloni [2] and Carmelo Militello [3,*]

[1] Department of Information and Electrical Engineering and Applied Mathematics (DIEM), University of Salerno, 84084 Fisciano, Italy
[2] Department of Human and Social Sciences, University of Bergamo, 24129 Bergamo, Italy
[3] Institute of Molecular Bioimaging and Physiology, Italian National Research Council (IBFM-CNR), 90015 Cefalù, Italy
* Correspondence: carmelo.militello@cnr.it

Citation: Rundo, L.; Tangherloni, A.; Militello, C. Artificial Intelligence Applied to Medical Imaging and Computational Biology. *Appl. Sci.* **2022**, *12*, 9052. https://doi.org/10.3390/app12189052

Received: 5 September 2022
Accepted: 6 September 2022
Published: 8 September 2022

The Special Issue "*Artificial Intelligence Applied to Medical Imaging and Computational Biology*" of the Applied Sciences Journal has been curated from February 2021 to May 2022, which covered the state-of-the-art and novel algorithms and applications of Artificial Intelligence methods for biomedical data analysis, ranging from classic Machine Learning to Deep Learning.

Medical imaging and computational biology continuously pose new fundamental medical and biological questions that often give rise to novel challenges in Artificial Intelligence. Moreover, the amount of biomedical data is constantly increasing due to the different image acquisition modalities and high-throughput technologies [1,2]. In these research fields, there is thus an increasing need for the application of cutting-edge computational approaches that generally involve Machine Learning or Computational Intelligence techniques, able to provide high-performance and specialized services in medical contexts [3]. Machine Learning and Computational Intelligence techniques can effectively perform image processing operations (such as segmentation [4–10], classification [11–14], and quantification [15–18]), in the fields of neuroimaging and oncological imaging. Although manual approaches often remain the golden standard in several tasks, Machine Learning can be exploited to automate and facilitate the work of researchers and clinicians. In addition, these fields often present new clustering and classification challenges, as well as combinatorial problems, which can be effectively addressed using novel strategies based on Machine Learning and Computational Intelligence techniques.

More recently, Deep Learning approaches [4,5,7,11,14,19] were shown to be very successful in computer vision and bioinformatics tasks owing to their ability to automatically extract hierarchical descriptive features from input images or gene expression data. They have also been used in the oncological, neuroimaging, and microscopy imaging domains for automatic disease diagnosis [12,13], tissue segmentation [16,20], and even synthetic image generation. However, the main issue remains the relative sample paucity of the typical datasets that leads to a poor generalization of the employed deep Artificial Neural Networks, considering the high number of required parameters. Consequently, parameter-efficient design paradigms specifically tailored to biomedical applications ought to be devised, also by exploiting Computational Intelligence based techniques (e.g., Evolutionary Computation, Swarm Intelligence, and neuroevolution).

In this context, advanced Machine Learning techniques were suitably exploited to combine heterogeneous sources of information, allowing for multiomics data integration [21,22]. Such kinds of analyses represent a significant step towards personalized medicine.

Author Contributions: Conceptualization, L.R., A.T. and C.M.; writing—original draft preparation, L.R., A.T. and C.M.; writing—review and editing, L.R., A.T. and C.M.; visualization, L.R. and A.T.; supervision, C.M. All authors have read and agreed to the published version of the manuscript.

Funding: This research received no external funding.

Conflicts of Interest: The authors declare no conflict of interest.

References

1. Rundo, L.; Militello, C.; Vitabile, S.; Russo, G.; Sala, E.; Gilardi, M.C. A Survey on Nature-Inspired Medical Image Analysis: A Step Further in Biomedical Data Integration. *Fund. Inform.* **2019**, *171*, 345–365. [CrossRef]
2. Castiglioni, I.; Rundo, L.; Codari, M.; Di Leo, G.; Salvatore, C.; Interlenghi, M.; Gallivanone, F.; Cozzi, A.; D'Amico, N.C.; Sardanelli, F. AI Applications to Medical Images: From Machine Learning to Deep Learning. *Phys. Med.* **2021**, *83*, 9–24. [CrossRef] [PubMed]
3. Conti, V.; Militello, C.; Rundo, L.; Vitabile, S. A Novel Bio-Inspired Approach for High-Performance Management in Service-Oriented Networks. *IEEE Trans. Emerg. Top. Comput.* **2021**, *9*, 1709–1722. [CrossRef]
4. Weis, C.-A.; Weihrauch, K.R.; Kriegsmann, K.; Kriegsmann, M. Unsupervised Segmentation in NSCLC: How to Map the Output of Unsupervised Segmentation to Meaningful Histological Labels by Linear Combination? *Appl. Sci.* **2022**, *12*, 3718. [CrossRef]
5. Park, S.; Kim, H.; Shim, E.; Hwang, B.-Y.; Kim, Y.; Lee, J.-W.; Seo, H. Deep Learning-Based Automatic Segmentation of Mandible and Maxilla in Multi-Center CT Images. *Appl. Sci.* **2022**, *12*, 1358. [CrossRef]
6. Militello, C.; Ranieri, A.; Rundo, L.; D'Angelo, I.; Marinozzi, F.; Bartolotta, T.V.; Bini, F.; Russo, G. On Unsupervised Methods for Medical Image Segmentation: Investigating Classic Approaches in Breast Cancer DCE-MRI. *Appl. Sci.* **2021**, *12*, 162. [CrossRef]
7. Wu, S.; Wu, Y.; Chang, H.; Su, F.T.; Liao, H.; Tseng, W.; Liao, C.; Lai, F.; Hsu, F.; Xiao, F. Deep Learning-Based Segmentation of Various Brain Lesions for Radiosurgery. *Appl. Sci.* **2021**, *11*, 9180. [CrossRef]
8. Militello, C.; Rundo, L.; Dimarco, M.; Orlando, A.; Conti, V.; Woitek, R.; D'Angelo, I.; Bartolotta, T.V.; Russo, G. Semi-Automated and Interactive Segmentation of Contrast-Enhancing Masses on Breast DCE-MRI Using Spatial Fuzzy Clustering. *Biomed. Signal Process. Control.* **2022**, *71*, 103113. [CrossRef]
9. Militello, C.; Vitabile, S.; Rundo, L.; Russo, G.; Midiri, M.; Gilardi, M.C. A Fully Automatic 2D Segmentation Method for Uterine Fibroid in MRgFUS Treatment Evaluation. *Comput. Biol. Med.* **2015**, *62*, 277–292. [CrossRef]
10. Isensee, F.; Jaeger, P.F.; Kohl, S.A.A.; Petersen, J.; Maier-Hein, K.H. nnU-Net: A Self-Configuring Method for Deep Learning-Based Biomedical Image Segmentation. *Nat. Methods* **2021**, *18*, 203–211. [CrossRef]
11. Asami, Y.; Yoshimura, T.; Manabe, K.; Yamada, T.; Sugimori, H. Development of Detection and Volumetric Methods for the Triceps of the Lower Leg Using Magnetic Resonance Images with Deep Learning. *Appl. Sci.* **2021**, *11*, 12006. [CrossRef]
12. Baazaoui, H.; Hubertus, S.; Maros, M.E.; Mohamed, S.A.; Förster, A.; Schad, L.R.; Wenz, H. Artificial Neural Network-Derived Cerebral Metabolic Rate of Oxygen for Differentiating Glioblastoma and Brain Metastasis in MRI: A Feasibility Study. *Appl. Sci.* **2021**, *11*, 9928. [CrossRef]
13. Taibouni, K.; Miere, A.; Samake, A.; Souied, E.; Petit, E.; Chenoune, Y. Choroidal Neovascularization Screening on OCT-Angiography Choriocapillaris Images by Convolutional Neural Networks. *Appl. Sci.* **2021**, *11*, 9313. [CrossRef]
14. Karhade, J.; Ghosh, S.K.; Gajbhiye, P.; Tripathy, R.K.; Rajendra Acharya, U. Multichannel Multiscale Two-Stage Convolutional Neural Network for the Detection and Localization of Myocardial Infarction Using Vectorcardiogram Signal. *Appl. Sci.* **2021**, *11*, 7920. [CrossRef]
15. Zhang, J.; Huang, Y.; Ye, F.; Yang, B.; Li, Z.; Hu, X. Evaluation of Post-Stroke Impairment in Fine Tactile Sensation by Electroencephalography (EEG)-Based Machine Learning. *Appl. Sci.* **2022**, *12*, 4796. [CrossRef]
16. Sharma, M.; Goudar, V.S.; Koduri, M.P.; Tseng, F.G.; Bhattacharya, M. Quantitative and Qualitative Image Analysis of In Vitro Co-Culture 3D Tumor Spheroid Model by Employing Image-Processing Techniques. *Appl. Sci.* **2021**, *11*, 4636. [CrossRef]
17. Rundo, L.; Tangherloni, A.; Militello, C.; Gilardi, M.C.; Mauri, G. Multimodal Medical Image Registration Using Particle Swarm Optimization: A Review. In Proceedings of the 2016 IEEE Symposium Series on Computational Intelligence (SSCI), Athens, Greece, 6–9 December 2016; IEEE: Piscataway, NJ, USA, 2016; pp. 1–8.
18. Rundo, L.; Militello, C.; Vitabile, S.; Casarino, C.; Russo, G.; Midiri, M.; Gilardi, M.C. Combining Split-and-Merge and Multi-Seed Region Growing Algorithms for Uterine Fibroid Segmentation in MRgFUS Treatments. *Med. Biol. Eng. Comput.* **2016**, *54*, 1071–1084. [CrossRef]
19. Lee, J.; Chung, S.W. Deep Learning for Orthopedic Disease Based on Medical Image Analysis: Present and Future. *Appl. Sci.* **2022**, *12*, 681. [CrossRef]
20. Fasoula, A.; Duchesne, L.; Cano, J.D.G.; Moloney, B.M.; Abd Elwahab, S.M.; Kerin, M.J. Automated Breast Lesion Detection and Characterization with the Wavelia Microwave Breast Imaging System: Methodological Proof-of-Concept on First-in-Human Patient Data. *Appl. Sci.* **2021**, *11*, 9998. [CrossRef]
21. Simidjievski, N.; Bodnar, C.; Tariq, I.; Scherer, P.; Andres Terre, H.; Shams, Z.; Jamnik, M.; Liò, P. Variational Autoencoders for Cancer Data Integration: Design Principles and Computational Practice. *Front. Genet.* **2019**, *10*, 1205. [CrossRef]
22. Tangherloni, A.; Ricciuti, F.; Besozzi, D.; Liò, P.; Cvejic, A. Analysis of Single-Cell RNA Sequencing Data Based on Autoencoders. *BMC Bioinform.* **2021**, *22*, 309. [CrossRef] [PubMed]

Article

Evaluation of Post-Stroke Impairment in Fine Tactile Sensation by Electroencephalography (EEG)-Based Machine Learning

Jianing Zhang [1], Yanhuan Huang [1], Fuqiang Ye [1], Bibo Yang [1], Zengyong Li [2] and Xiaoling Hu [1,3,4,5,*]

1 Department of Biomedical Engineering, The Hong Kong Polytechnic University, Hong Kong 999077, China; jayce.zhang@connect.polyu.hk (J.Z.); yanhuan.huang@connect.polyu.hk (Y.H.); 19073922r@connect.polyu.hk (F.Y.); bibo.yang@polyu.edu.hk (B.Y.)
2 National Research Centre for Rehabilitation Technical Aids Beijing, Beijing Key Laboratory of Rehabilitation Technical Aids for Old-Age Disability, Beijing 100176, China; lizengyong@nrcrta.cn
3 University Research Facility in Behavioural and Systems Neuroscience (UBSN), The Hong Kong Polytechnic University, Hong Kong 999077, China
4 Shenzhen Research Institute, The Hong Kong Polytechnic University, Shenzhen 518000, China
5 Research Institute for Smart Ageing (RISA), The Hong Kong Polytechnic University, Hong Kong 999077, China
* Correspondence: xiaoling.hu@polyu.edu.hk; Tel.: +852-3400-3206

Abstract: Electroencephalography (EEG)-based measurements of fine tactile sensation produce large amounts of data, with high costs for manual evaluation. In this study, an EEG-based machine-learning (ML) model with support vector machine (SVM) was established to automatically evaluate post-stroke impairments in fine tactile sensation. Stroke survivors (n = 12, stroke group) and unimpaired participants (n = 15, control group) received stimulations with cotton, nylon, and wool fabrics to the different upper limbs of a stroke participant and the dominant side of the control. The average and maximal values of relative spectral power (RSP) of EEG in the stimulations were used as the inputs to the SVM-ML model, which was first optimized for classification accuracies for different limb sides through hyperparameter selection (γ, C) in radial basis function (RBF) kernel and cross-validation during cotton stimulation. Model generalization was investigated by comparing accuracies during stimulations with different fabrics to different limbs. The highest accuracies were achieved with ($\gamma = 2^1$, $C = 2^3$) for the RBF kernel (76.8%) and six-fold cross-validation (75.4%), respectively, in the gamma band for cotton stimulation; these were selected as optimal parameters for the SVM-ML model. In model generalization, significant differences in the post-stroke fabric stimulation accuracies were shifted to higher (beta/gamma) bands. The EEG-based SVM-ML model generated results similar to manual evaluation of cortical responses to fabric stimulations; this may aid automatic assessments of post-stroke fine tactile sensations.

Keywords: stroke; fine tactile sensation; electroencephalography; machine learning; evaluation

Citation: Zhang, J.; Huang, Y.; Ye, F.; Yang, B.; Li, Z.; Hu, X. Evaluation of Post-Stroke Impairment in Fine Tactile Sensation by Electroencephalography (EEG)-Based Machine Learning. *Appl. Sci.* **2022**, *12*, 4796. https://doi.org/10.3390/app12094796

Academic Editors: Leonardo Rundo, Carmelo Militello and Andrea Tangherloni

Received: 25 March 2022
Accepted: 6 May 2022
Published: 9 May 2022

Publisher's Note: MDPI stays neutral with regard to jurisdictional claims in published maps and institutional affiliations.

1. Introduction

Approximately 50% of stroke survivors have reported persistent sensory deficiencies for both somatosensation and proprioception [1,2]. For example, they often have difficulties in perceiving pain, temperature, pressure, posture, and light touch [3]. Sensory deficiencies have profound negative impacts on the functional ability and independency in daily living, which further affect motor recovery after stroke [4,5]. Fine tactile sensation is an elementary somatosensory function for obtaining external information through touch [6]. Previous studies have shown that fine tactile sensation also provide valid spatial references for body positions to reduce postural sway [7], and it may act as an indicator to enhance sensory feedback in position control [8,9]. However, rehabilitation for sensory functions has been overlooked in the traditional practices, when compared with efforts for motor restoration; this is attributed to the lack of effective evaluation measures for sensory impairments [10].

Objective and efficient assessments of sensory impairments are important for long-term post-stroke rehabilitation with repeated measurements during follow up [11]. However, subjective and manual measurements have been used traditionally for sensory impairment assessments [12]. For example, the Fugl–Meyer assessment (FMA) [13] and Semmes–Weinstein monofilament test [14] are commonly used in current evaluations of fine tactile sensations because of the ease of interpretation of the assessment results. Additionally, the measurement process highly relies on the personal experiences of the assessor, where achieving consistency in measurements is challenging when the stroke population increases during long-term service [15].

Neuroimaging techniques have been introduced to provide objective data for sensory impairment assessments [16]. The common neuroimaging techniques include functional magnetic resonance imaging (fMRI), positron emission tomography (PET), neuromolecular imaging, and electroencephalography (EEG), among others [17–23]. These approaches characterize neural circuitry changes during post-stroke sensorimotor recovery; however, such medical equipment is expensive and the preparations before neuroimaging-based examinations are complicated compared to the traditional clinical assessments [24]. Among these techniques, owing to the advantage of high temporal resolution, EEG has been applied to detect transient sensory neural responses during fine tactile stimulations [25,26]. For example, Ahn et al. compared the effects of different tactile exploration tasks, i.e., passively or actively moving a tactile board, on post-stroke brain activation using EEG [27]. The sensory motor rhythm indicated by the EEG relative powers from the right prefrontal and parietal lobes during active tactile perceptions were significantly greater than those in the damaged left hemisphere during passive tactile perception [27]. In our previous work [28], post-stroke sensory impairment of fine tactile sensation was measured quantitatively via EEG during textile fabric stimulation, i.e., simulation of the common fabric–skin touch. We observed EEG relative spectral power (RSP) differences after stroke, i.e., RSP intensities in different frequency bands between unimpaired and stroke populations [28]. However, neuroimaging-based measurements usually generate large amounts of data, whose interpretations still heavily rely on human professionals, which is time consuming and labor demanding [29,30].

Neuroimaging data interpretation by machine-learning (ML) techniques has been a promising approach to reduce manpower workload in data interpretations [31]. ML is a technique that can help develop an automatic predictive model by learning the relationships between features and targets from a given set of historical data before application to repeated analyses on massive data [32]. Various ML algorithms, e.g., linear discriminant analysis (LDA), artificial neural network (ANN), and support vector machine (SVM), are being explored for the detection, classification, and characterization of neuroimaging data, e.g., EEG [31]. For instance, Jochumsen et al. classified single-trial movement intentions associated with different hand grasp types using the EEG spectra as input features to an LDA model [33]. Usama et al. distinguished correct/error feedbacks during hand and foot movements by feeding the EEG waveform features into an ANN model [34]. Limited classification accuracies were obtained in both studies: 41–86% [33,34]. This may be attributed to insufficiencies in feature mapping by simple linear transformation of the LDA, leading to inefficient construction of the optimal decision function (classification boundary) for multichannel EEG [35,36]. Although ANN-based models offer nonlinear feature mapping abilities during classifications, overfitting often occurs when there are several hyperparameters, e.g., numbers of hidden layers and nodes, to be determined during network optimization [36]. In contrast to the ANN, SVM-based models reduce the disadvantages of overfitting of the classification results with the help of kernel functions [37]. SVM with kernel functions effectively minimize model complexities via implicitly realizing nonlinear transformations of the feature spaces without explicit mathematical expressions, so that only specific hyperparameters related to the kernel functions of the SVM need to be optimized during model development [38]. In the SVM-ML models, several kernel functions are commonly used, namely linear, polynomial, and radial basis function (RBF) kernels. For example, Liu et al. extracted the spectrum features from subject-related EEG frequency bands and channels, and the SVM with linear kernel was applied to each subject's EEG-based motor imagery classification [39]. Ghumman et al. investigated the classification performance of SVM with a polynomial kernel in multiclass motor imagery EEG [40]. Bousseta et al. used SVM with RBF kernel to classify the EEGs of imagined hand movements [41]. These studies reported a classification range of 67–92.8% [39–41]. Among the practical applications of the kernel functions mentioned above, the RBF kernel is a common choice in SVM-ML models because of its better performance on the nonlinearities in feature mapping capabilities with less hyperparameters compared to the other two types of kernel functions [42,43].

Automatic evaluations of neuroimaging data by SVM-ML techniques have not been fully explored in literature, e.g., fine tactile sensation. Kim et al. extracted the powers of the alpha and gamma bands as features representing EEG during touch with different objects, i.e., fabric, glass, and paper [44]. However, they only evaluated the tactile perception of unimpaired persons and obtained limited classification performance (68.1%) with the LDA model [44]. The purpose of this study was to automatically evaluate and assess post-stroke impairments in fine tactile sensation using a new EEG-based SVM-ML model.

2. Methodology

In this study, an SVM-ML model was established based on EEG measurements of cortical responses to fine tactile stimulations to the upper limbs in persons who have experienced stroke and in unimpaired participants via stimulations with different types of fabrics (cotton, nylon, and wool). The SVM-ML model was first developed and optimized using EEG RSP features with cotton fabric stimulation as the baseline input for classifying the responses from multiple upper limb groups, i.e., stimulation to the (1) affected sides of persons after stroke (SA), (2) unaffected sides of persons after stroke (SU), and (3) dominant sides of unimpaired participants (UD). Then, the generalization performance of the model was evaluated using the EEG RSP features during stimulations with different fabrics with and without considering arm differences.

2.1. EEG Acquisitions during Fabric Stimulations

After obtaining ethical approval from the Human Subjects Ethics Sub-committee (HSESC) of the Hong Kong Polytechnic University, twelve survivors of chronic stroke were recruited as the "stroke group", and fifteen unimpaired participants were recruited as the "control group", whose demographic details are listed in Table 1. The inclusion criteria of stroke group were: (1) individuals must be at least six months after the singular and unilateral brain lesion due to stroke; (2) the lesions occurring due to stroke were experienced in the subcortical area, to ensure the detectable EEG from the cortical area. All unimpaired participants were right-handed. No significant difference was found in age between the stroke and control groups ($p > 0.05$) by the independent *t*-test after verifying normality evaluations via the Shapiro–Wilk test [45].

Table 1. Demographic characteristics and clinical scores of the stroke and control groups [28].

Measure	Stroke Group (n = 12)	Control Group (n = 15)
Age in years	55.1 ± 16.0	46.4 ± 17.4
Gender (male/female)	11/1	5/10
Stroke type (ischemic/hemorrhagic)	10/2	-
Affected side (right/left)	6/6	-
Years since stroke	14.9 ± 5.8	-
FMA (upper extremity)	42.5 ± 15.2	-
FMA (light touch on forearm)	1 ± 0	-
MAS (elbow)	1.1 ± 0.7	-

Note: Data are given as mean ± standard deviation. MAS: Modified Ashworth scale; FMA: Fugl–Meyer assessment [28].

The experimental setup and protocol for the fabric fine tactile stimulation are shown in Figure 1. The three types of fabrics, i.e., cotton, nylon, and wool of the same size and different textural properties, were alternatively placed on the skin surface of the ventral forearm of the upper limb, i.e., a single stimulation trial (Figure 1c). Each trial consisted of a 30 s baseline measurement, i.e., no fabric stimulation to the skin, followed by alternative stimuli with the three different fabrics in a random sequence for 13 s stimulation with each fabric and 60 s gaps in between. The stimulation trial was repeated thrice for each target forearm. The whole brain EEG with 64 channels (BP-01830, Brain Products Inc., Gilching, Germany) based on the 10–20 system [46] was captured during the stimulation trials at a sampling frequency of 1000 Hz. Each subject was required to stay awake and calm during the EEG measurements while wearing ear plugs and an eye mask, whose purpose was to minimize visual and audio disturbances from the environment. The detailed experimental procedure is described in [28].

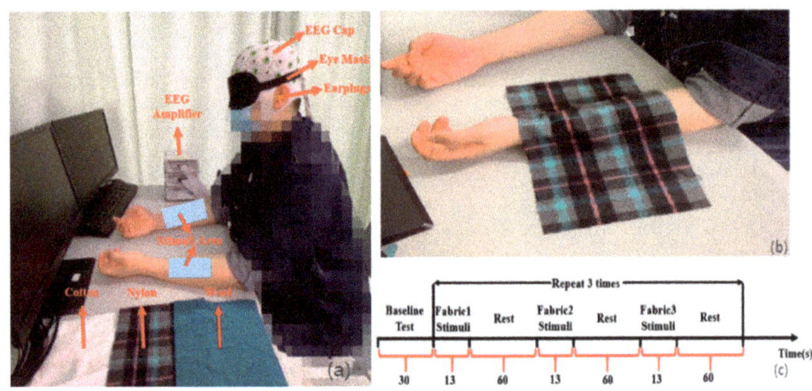

Figure 1. Electroencephalography (EEG) experimental setup and protocol. (**a**) Participant wearing the EEG cap, ear plugs, and eye mask, is seated in a chair and tested with their hands comfortably placed on the table. The areas of each fabric stimulation are the ventral forearms. The three fabric samples, namely cotton, nylon, and wool, are placed on the table. (**b**) Illustration of fabric stimulation. The nylon fabric is placed on the ventral forearm. (**c**) Protocol for fabric stimulation.

2.2. EEG-Specific Feature Extraction for the SVM-ML Model

In this work, the RSP_{mean} and RSP_{max} values in different frequency bands were selected as the EEG features for input to the SVM-ML model as they represent the average and maximal cortical changes, respectively, during fabric stimulations based on manual recognition from the previous study [28]. During the real-time EEG recording, the sampling frequency was 1000 Hz. In the preparation of the EEG RSP features, a Butterworth bandpass filter from 0.1 to 100 Hz was first applied to the EEG to eliminate irrelevant high-frequency components. Then, an additional Butterworth notch filter from 49 Hz to 51 Hz was applied to eliminate the 50 Hz noise from the environment. Following this, the filtered EEG was segmented into different epochs, i.e., 30 s pre-stimulus baseline and three 13 s stimuli with different fabrics. The numbers of EEG samples after segmentation were 108 from the SA group (12 participants × 3 trials × 3 fabric stimuli), 108 for the SU group (12 participants × 3 trials × 3 fabric stimuli), and 135 for the UD group (15 participants × 3 trials × 3 fabric stimuli). Next, the EEG samples were transformed into their power spectra by Pwelch estimation [47], and the entire frequency band (0.1–100 Hz) of each segmented EEG epoch was decomposed into five frequency bands, i.e., delta (0.5–4 Hz), theta (4–8 Hz), alpha

(8–12 Hz), beta (12–30 Hz), and gamma (30–100 Hz) [48]. Finally, the RSP [49] of each frequency band for each fabric stimulus was calculated using the following equations:

$$P(f_1, f_2) = \int_{f_1}^{f_2} p(f)df \quad (1)$$

$$RSP(f_1, f_2) = \frac{P(f_1, f_2)}{P(0.1, 100)} - \frac{P_{baseline}(f_1, f_2)}{P(0.1, 100)}, \quad (2)$$

where $p(f)$ is the power spectral density; f_1 and f_2 are the low and high cutoff frequencies of a given EEG frequency band, respectively; $P(f_1, f_2)$ is the power spectrum from f_1 to f_2; and $P_{baseline}$ is the power spectrum of the EEG segments during the baseline test in each trial. The above spectral analysis of the raw EEG signals was implemented offline with the EEGLAB v12 toolbox in MATLAB (The MathWorks Inc., Natick, MA, USA).

After obtaining the RSP value from each EEG channel, the RSP_{mean} and RSP_{max} values were acquired to represent the RSP features of the multichannel EEG, where RSP_{mean} is the average value of the RSPs of all the channels in a given frequency band of a signal epoch, and RSP_{max} is the highest value among all the EEG channels. Then, the RSP_{mean} and RSP_{max} of the 62-channel EEG (ground and reference channels were neglected), which covered the entire cortical area, were calculated for each frequency band. To minimize the diversity of the ranges for the RSP_{mean} and RSP_{max}, the original RSP_{mean} and RSP_{max} were further normalized as in the following equation according to z-score normalization, which scales all the RSP_{mean} and RSP_{max} values in varying ranges with a zero mean and unit standard deviation [50]:

$$RSP_i' = \frac{RSP_i - \mu_{RSP}}{\sigma_{RSP}}, \quad (3)$$

where RSP_i is the original spectral feature, i.e., RSP_{mean} or RSP_{max}; μ_{RSP} is the mean of RSP_i; σ_{RSP} is the standard deviation of RSP_i; and RSP_i' is the normalized spectral feature. The normalized features were then used as the inputs to the SVM-ML model.

2.3. SVM-ML Model Configuration

Figure 2 shows the configuration of the SVM-ML model, including optimization of the SVM RBF kernel function and k-fold cross-validation (CV) strategy. The normalized EEG features (i.e., RSP_{mean} and RSP_{max}) during stimulation with cotton fabric were adopted as the baseline inputs for model establishment. This is because cotton is the most widely used fabric that is in intimate contact with skin in daily living and provides minimum stimulation intensity with a comfortable feeling compared to other fabrics [51]. In addition, compared to nylon and wool, the textile physical properties of cotton fabric as quantitatively measured by the fabric touch tester (FTT) [52] were neutral with equivalent distances in the aspects of smoothness, thickness, etc. [28]. Therefore, the EEG RSP features evoked by the cotton fabric were used as the baseline inputs to configure the SVM-ML model.

The RBF kernel function of the SVM-ML model was determined by optimizing the classification boundaries that achieved the best accuracy on the RSP features related to cotton stimulation. For an RBF kernel, two hyperparameters, namely the kernel scaling parameter γ and regularization parameter C [53], are optimized in the SVM-ML model development to classify the different upper-limb groups. The search for optimal (γ, C) was conducted by a "grid search" approach [54]. The candidate values of (γ, C) were first defined as exponentially increasing sequences ($\gamma = 2^{-15}, 2^{-13}, \ldots, 2^{9}; C = 2^{-5}, 2^{-3}, \ldots, 2^{15}$), which were the ranges adopted by most EEG-based SVM-ML studies to identify the optimal (γ, C) values [42,55,56]. Following this, different pairs of γ and C ($13 \times 11 = 143$ pairs) values were generated, and each pair was used to construct the RBF kernel of the SVM. The classification accuracies with the different hyperparameter pairs were evaluated by three-fold CV according to the greatest common divisor of the number of stroke patients (i.e., $n = 12$) and unimpaired controls (i.e., $n = 15$); this is a common pilot estimation approach used in previous studies [57]. The value pair that achieved the best classification accuracy was then adopted as the optimal hyperparameters for model configuration. The above SVM algorithm was implemented using the Scikit-learn toolbox, an open-source ML toolbox in Python [58].

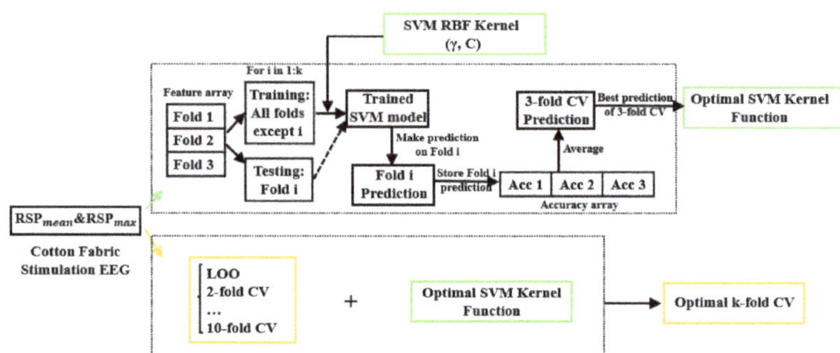

Figure 2. Configuration flowchart for the SVM-ML model. RSP: relative spectral power; SVM: support vector machine; RBF: radial basis function; Acc: accuracy; LOO: leave-one-out; CV: cross validation.

The grid search results of γ and C are displayed in Figure 3, where Figure 3a–e present the accuracies of different (γ, C) pairs for distinguishing the UD, SA, and SU groups with the RSP_{mean} and RSP_{max} of the 62-channel EEG in the delta, theta, alpha, beta, and gamma bands, respectively, as input features. The coordinates and values of the highest accuracy for each frequency band are indicated by the red dots in Figure 3a–e. Among all the accuracies, the model with the highest accuracy of 67.4% ($\gamma = 2^3$, $C = 2^9$) was achieved in the gamma band.

As the sensorimotor cortex is the main response area for sensory stimulations [59], the classification performance achieved by including only the EEG channels covering the sensorimotor cortex was evaluated in the SVM-ML model configuration. The RSP_{mean} and RSP_{max} of the corresponding 21-channel EEG (i.e., FC1–FC6, FCZ, C1–C6, CZ, CP1–CP6, CPZ), which cover the sensorimotor area [60], were used as the inputs to the model. Figure 4 shows the accuracies with the RSP features for the 21-channel EEG, and the highest accuracy of 76.8% ($\gamma = 2^1$, $C = 2^3$) was obtained for the gamma band as well.

The accuracies of the SVM-ML model for classifying the UD, SA, and SU groups with the RBF kernel hyperparameter pairs in the different bands are summarized in Table 2. Compared to other frequency bands, the gamma band has the best average accuracy performance for both channel set selections. The average and peak accuracies of the gamma band of the 21-channel EEG were better than those of the 62-channel EEG. Therefore, the hyperparameter pair ($\gamma = 2^1$, $C = 2^3$) from the 21-channel EEG was selected as the optimal RBF kernel hyperparameters.

After the RBF kernel function was determined, the k-fold CV was also configured using the RSP features of the 21-channel EEG as inputs to improve generalization of the SVM-ML model. Compared to the simple train/test split, the k-fold CV ensures that each sample from the original dataset has the chance of appearing in the training and testing set, which results in less biased evaluations [61]. Since the partition of k folds is random, the k-fold CV was performed 10 times to calculate the mean estimate to decrease the variance of accuracy estimations of the one-shot k-fold CV [62–64]. Typically, the configuration of k is 5 or 10, as these values have been shown to be the bias-variance trade-off for model evaluation [61,65]. In our experiment, different selections of k from 2 to 10 were employed to compare the influence of k on model performance. In addition, the leave-one-out CV, where k is the number of samples in the dataset, was used as a complementary comparison to different k-fold CV. Although the leave-one-out CV is more computationally expensive compared to the above strategies, i.e., five-fold and ten-fold CV, it offers an unbiased evaluation of the model performance as each sample is given the opportunity to represent the entirety of the test dataset [61].

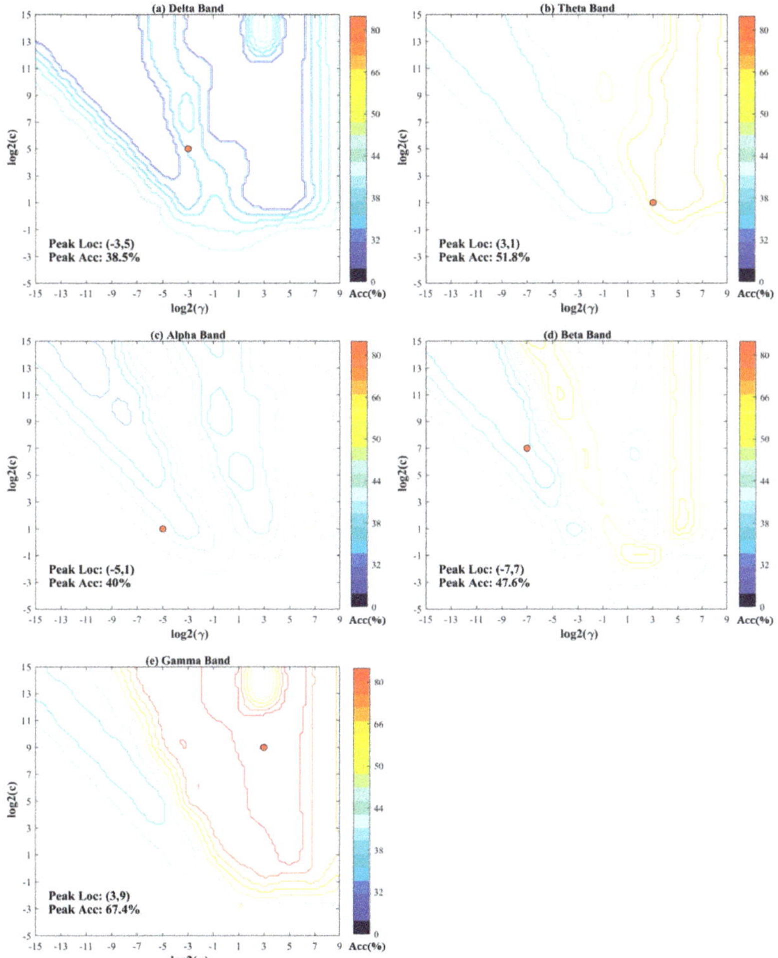

Figure 3. Grid search results of γ and C in the SVM-ML model with RBF kernel using the 62-channel EEG RSP features of the delta, theta, alpha, beta, and gamma bands. Acc: accuracy. Peak Acc: highest classification accuracy of the SVM-ML model in the predefined range of (γ, C); Peak Loc: location (γ, C) corresponding to the highest classification accuracy of the SVM-ML model.

The accuracies of the SVM-ML model for distinguishing between the UD, SA, and SU groups with different k-fold CV strategies in the different frequency bands are shown in Table 3. The model achieved the highest accuracy of 75.4% in the gamma band by six-fold CV. For the leave-one-out CV, the model obtained the highest classification accuracy of 74.4% in the gamma band as well. Therefore, the six-fold CV was selected as the optimal evaluation strategy for the model when using the RSP features of the 21-channel EEG as inputs.

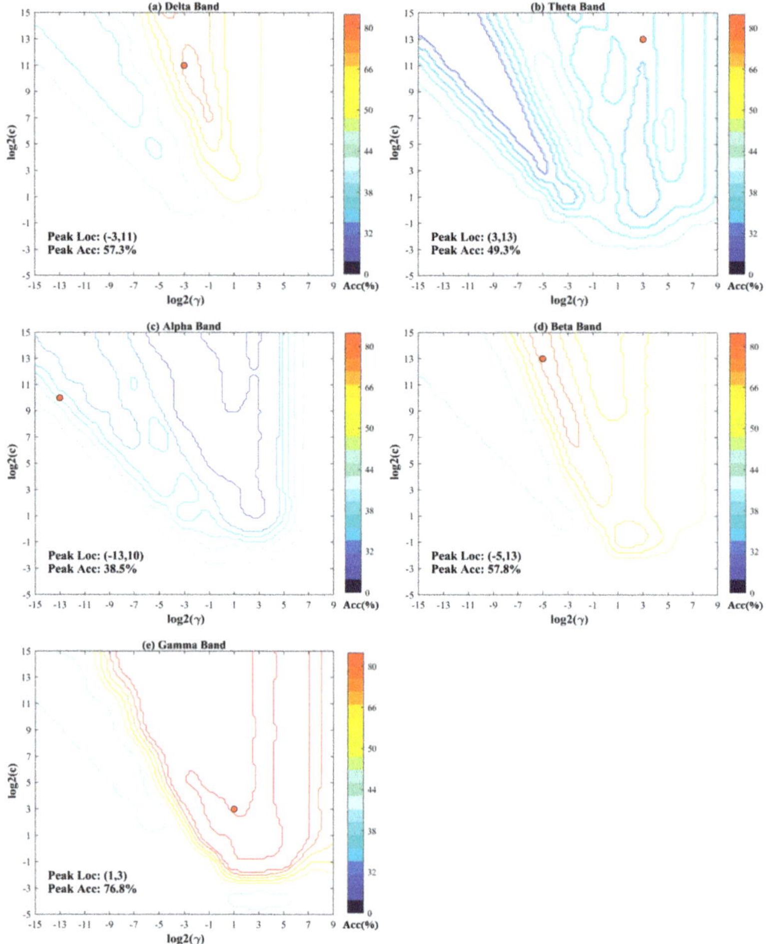

Figure 4. Grid search results of γ and C in the SVM-ML model with RBF kernel using the 21-channel EEG RSP features of the delta, theta, alpha, beta, and gamma bands. Acc: accuracy. Peak Acc: highest classification accuracy of the SVM-ML model in the predefined range of (γ, C); Peak Loc: location (γ, C) corresponding to the highest classification accuracy of the SVM-ML model.

2.4. Generalization of the SVM-ML Model

Using the RSP features during stimulation with cotton fabric as the baseline inputs, the SVM-ML model was established. Then, we first investigated the generalization performance of the model for classifying the upper-limb groups with the inputs of different fabrics, i.e., nylon, wool, and cotton. The measured RSP features in the respective stimulations were then input to the developed model, and the achieved accuracies are summarized in Table 4. The classification accuracies of the different fabric stimulations were not normally distributed ($p < 0.5$, Shapiro–Wilk test) in each frequency band. Significant intergroup differences in the accuracies ($p < 0.001$, Kruskal–Wallis test) with respect to fabric stimulation were observed in the delta, theta, alpha, beta, and gamma bands. The model achieved the highest classification accuracies of 75.4%, 83.5%, and 84.3% for the cotton, nylon, and wool stimulations, respectively, in the gamma band.

Table 2. Accuracies of the SVM-ML model for classifying the three upper-limb groups with the RBF kernel hyperparameter pairs in different frequency bands.

Number of EEG Channels		Delta	Theta	Alpha	Beta	Gamma
62	Average Acc	33.5% ± 0.05	37.8% ± 0.05	35.8% ± 0.03	38.2% ± 0.04	44.7% ± 0.11
	Peak Acc	38.5%	51.8%	40.0%	47.6%	67.4%
	Peak Loc (γ, C)	$(2^{-3}, 2^5)$	$(2^3, 2^1)$	$(2^{-5}, 2^1)$	$(2^{-7}, 2^7)$	$(2^3, 2^9)$
21	Average Acc	39.1% ± 0.06	35.6% ± 0.04	33.2% ± 0.06	41.3% ± 0.07	49.2% ± 0.16
	Peak Acc	57.3%	49.3%	38.5%	57.8%	76.8%
	Peak Loc (γ, C)	$(2^{-3}, 2^{11})$	$(2^3, 2^{13})$	$(2^{-13}, 2^{10})$	$(2^{-5}, 2^{13})$	$(2^1, 2^3)$

Note: Average Acc: average classification accuracy of the SVM-ML model with all the RBF kernel hyperparameter pairs; Peak Acc: highest classification accuracy of the SVM-ML model in the predefined range of (γ, C); Peak Loc: location (γ, C) corresponding to the highest classification accuracy of the SVM-ML model.

Table 3. Accuracies of the SVM-ML model for classifying the three upper-limb groups with different k-fold CV strategies in the different frequency bands.

CV	Accuracy				
	Delta	Theta	Alpha	Beta	Gamma
2-fold	49.6% ± 0.07	38.1% ± 0.06	27.6% ± 0.06	50.7% ± 0.07	73.8% ± 0.05
3-fold	49.1% ± 0.06	33.3% ± 0.07	26.0% ± 0.05	51.0% ± 0.05	74.5% ± 0.04
4-fold	49.7% ± 0.06	34.7% ± 0.06	23.9% ± 0.05	50.1% ± 0.05	74.8% ± 0.04
5-fold	49.7% ± 0.06	32.4% ± 0.05	26.1% ± 0.05	51.2% ± 0.04	75.0% ± 0.03
6-fold	53.2% ± 0.05	35.0% ± 0.06	22.2% ± 0.05	50.8% ± 0.05	75.4% ± 0.04
7-fold	46.7% ± 0.06	32.6% ± 0.05	32.9% ± 0.05	48.2% ± 0.06	72.6% ± 0.04
8-fold	47.0% ± 0.07	31.3% ± 0.06	28.9% ± 0.06	47.2% ± 0.06	74.8% ± 0.04
9-fold	49.5% ± 0.07	31.4% ± 0.06	27.8% ± 0.06	49.1% ± 0.06	74.7% ± 0.05
10-fold	51.7% ± 0.07	33.0% ± 0.06	25.9% ± 0.06	50.0% ± 0.05	74.8% ± 0.05
LOO	51.3%	28.2%	12.8%	53.8%	74.4%

Note: Data are given as mean ± SD. CV: cross validation; LOO: leave-one-out.

Table 4. Overall accuracies of the SVM-ML model for classifying different fabric stimulations.

Fabric Stimulation	Accuracy				
	Delta	Theta	Alpha	Beta	Gamma
Cotton	53.2% ± 0.05	35.0% ± 0.06	22.3% ± 0.05	50.8% ± 0.05	75.4% ± 0.04
Nylon	21.0% ± 0.04	40.6% ± 0.05	51.4% ± 0.04	63.2% ± 0.03	83.5% ± 0.02
Wool	30.3% ± 0.04	25.6% ± 0.06	43.0% ± 0.05	69.2% ± 0.04	84.3% ± 0.03
Significance (p-value)	<0.001 ***	<0.001 ***	<0.001 ***	<0.001 ***	<0.001 ***

Note: Data are given as mean ± SD. The significant differences are indicated by '***' ($p < 0.001$, Kruskal–Wallis test).

The comparison of the overall accuracies of the SVM-ML model with respect to fabric stimulation in each band are shown in Figure 5. Significant differences in the accuracies were observed in the delta, theta, alpha, beta, and gamma bands for pairwise comparisons among the three different fabric stimulations ($p < 0.001$, Kruskal–Wallis with Bonferroni post-hoc test), except for the difference between nylon and wool in the gamma band ($p > 0.05$, Kruskal–Wallis with Bonferroni post-hoc test). The models with nylon and wool achieved significantly higher accuracies in the beta and gamma bands than those with cotton ($p < 0.001$, Kruskal–Wallis with Bonferroni post-hoc test).

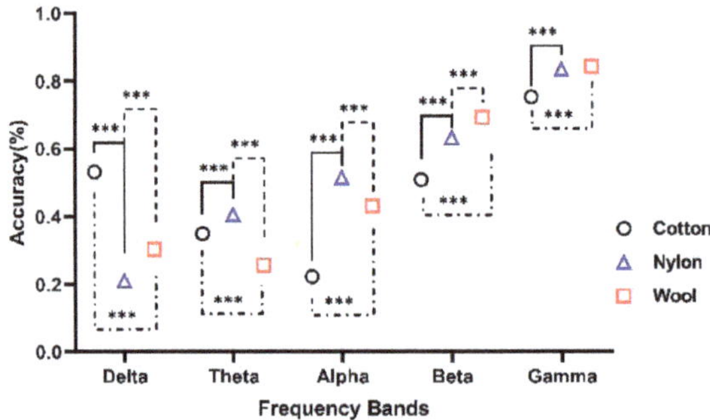

Figure 5. Comparisons of the overall classification accuracies of the SVM-ML model with respect to fabric stimulations in the delta, theta, alpha, beta, and gamma bands. The significant intergroup differences are indicated by '***' ($p < 0.001$, Kruskal–Wallis with Bonferroni post-hoc test).

The generalized performance of the model was also evaluated by considering the arm differences during stimulations with different fabrics (Table 5). The classification accuracies of each upper-limb group during stimulations with different fabrics are not normally distributed ($p < 0.5$, Shapiro–Wilk test). Significant differences in the accuracies ($p < 0.001$, Kruskal–Wallis test) with respect to fabric stimulations were observed in each band, except for the SU group in the gamma band ($p > 0.05$, Kruskal–Wallis test). The highest classification accuracy for each upper-limb group was achieved in the gamma band.

Table 5. Accuracies of the SVM-ML model for classifying the three upper-limb groups with different fabric stimulations.

Fabric Stimulation		Accuracy				
		Delta	Theta	Alpha	Beta	Gamma
SA	Cotton	47.9% ± 0.09	28.8% ± 0.11	28.5% ± 0.10	48.9% ± 0.10	59.7% ± 0.08
	Nylon	26.8% ± 0.09	31.9% ± 0.08	54.1% ± 0.11	64.3% ± 0.10	76.2% ± 0.06
	Wool	22.9% ± 0.09	21.7% ± 0.08	51.1% ± 0.11	53.6% ± 0.08	78.9% ± 0.04
	p-value	<0.001 ***	<0.001 ***	<0.001 ***	<0.001 ***	<0.001 ***
SU	Cotton	48.3% ± 0.08	40.9% ± 0.13	27.0% ± 0.09	69.3% ± 0.05	91.0% ± 0.04
	Nylon	15.5% ± 0.07	40.5% ± 0.13	61.7% ± 0.04	74.9% ± 0.01	91.2% ± 0.03
	Wool	50.3% ± 0.03	27.8% ± 0.08	40.1% ± 0.04	84.9% ± 0.04	91.6% ± 0.01
	p-value	<0.001 ***	<0.001 ***	<0.001 ***	<0.001 ***	>0.05
UD	Cotton	51.4% ± 0.14	31.6% ± 0.10	22.0% ± 0.12	35.3% ± 0.12	78.0% ± 0.10
	Nylon	24.0% ± 0.09	51.9% ± 0.07	39.2% ± 0.10	53.7% ± 0.05	83.4% ± 0.01
	Wool	19.8% ± 0.09	30.0% ± 0.11	36.9% ± 0.10	71.9% ± 0.08	84.1% ± 0.06
	p-value	<0.001 ***	<0.001 ***	<0.001 ***	<0.001 ***	<0.001 ***

Note: Data are given as mean ± SD. The significant intergroup differences are indicated by '***' ($p < 0.001$, Kruskal–Wallis test).

Based on the results in Table 4 and Figure 5, the comparisons of the accuracies of the SVM-ML model with respect to fabric stimulations when considering arm differences are presented in Figure 6. In the SA group (Figure 6a), significant differences in accuracies with respect to the fabric stimulations were obtained in the higher frequency bands, i.e., beta ($p < 0.001$, Kruskal–Wallis with Bonferroni post-hoc test) and gamma ($p < 0.05$, Kruskal–Wallis with Bonferroni post-hoc test) bands. No significant differences were found between nylon and wool in the delta, theta, and alpha bands ($p > 0.05$, Kruskal–Wallis with Bonferroni post-hoc test). In the SU group (Figure 6b), significant differences in accuracies with respect to fabric stimulations were found in the delta ($p < 0.001$, Kruskal–Wallis with Bonferroni post-hoc test), alpha ($p < 0.001$, Kruskal–Wallis with Bonferroni post-hoc test), and beta ($p < 0.05$, Kruskal–Wallis with Bonferroni post-hoc test) bands. No significant difference was observed between cotton and nylon in the theta band ($p > 0.05$, Kruskal–Wallis with Bonferroni post-hoc test). In the UD group (Figure 6c), significant differences in accuracies with respect to fabric stimulations were found in almost all frequency bands ($p < 0.001$, Kruskal–Wallis with Bonferroni post-hoc test), except for the difference between nylon and wool in the gamma band ($p > 0.05$, Kruskal–Wallis with Bonferroni post-hoc test).

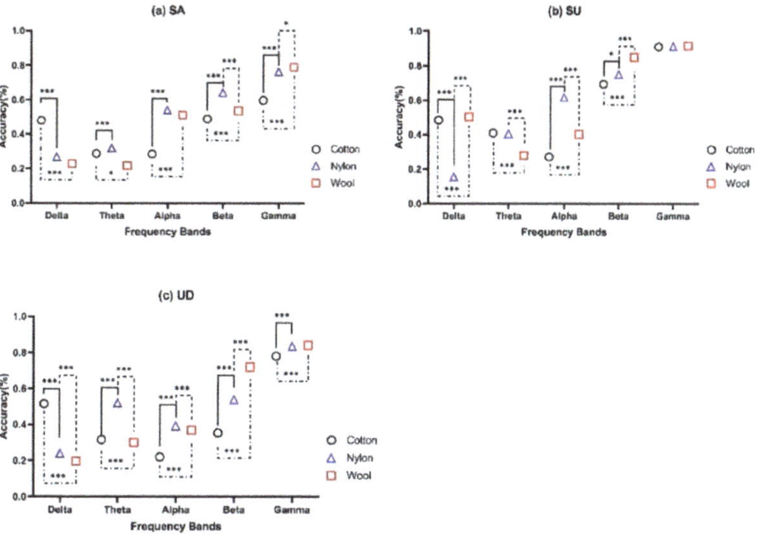

Figure 6. Comparisons of the classification accuracies of the SVM-ML model with respect to fabric stimulations when considering arm differences, i.e., (**a**) SA, (**b**) SU, and (**c**) UD. The significant differences are indicated by '*' for $p < 0.05$ and '***' for $p < 0.001$ (Kruskal–Wallis with Bonferroni post-hoc test).

3. Discussion

In this study, the EEG-based SVM-ML model was built using the RSP features during stimulation with cotton fabric as the baseline inputs. The model's generalization performance was investigated by comparing the classification accuracies during stimulations with different fabrics.

3.1. SVM-ML Configuration

3.1.1. RBF Kernel Determination

The grid search results of the hyperparameter pair (γ, C) for the RBF kernel (Figures 3 and 4) showed that the highest classification accuracies were obtained in the predefined range of the hyperparameter pair. Similar boundaries of the hyperparameter pair (γ, C) were also applied in other SVM-based studies, e.g., Chang et al. used the boundaries of ($e^{-8} \leq \gamma \leq e^{8}, e^{-8} \leq C \leq e^{8}$) [66], and Hsu et al. selected the boundaries of ($2^{-15} \leq \gamma \leq 2^{3}$, $2^{-5} \leq C \leq 2^{15}$) [42]. This showed that the selected optimal hyperparameter pair (γ, C) was in the traditional search space, and the SVM-ML model with the general search space was feasible for classifying the RSP features extracted from EEG during sensory assessments. The kernel scaling parameter γ determines the complexity of the classification decision function of the model [67]. For smaller values of γ, the decision function is nearly linear, and for larger values of γ, the function becomes more curved [67]. The optimal value of γ (2^{1}) chosen by the model was close to the upper boundary of the preset γ range, which suggested that there was a relatively strong nonlinearity among the EEGs of multiple upper-limb groups in the original feature space, and the model obtained the "curved" decision function by mapping the raw EEG to a higher dimensional space. The regularization parameter C defines the penalty degree of the model for the percentage of deviation from the misclassified trained data [67]. As the value of C increases, its penalty degree for the model becomes larger, and the percentage of deviation of the misclassified data is smaller during the training phase. The optimal value of C (2^3) selected by the model was relatively lower compared to the predefined range of C. This suggested that the model tolerated a greater percentage of misclassified training data when searching for the optimal decision function, indicating that there was an overlap among the different groups of EEG data points near the decision function. Meanwhile, the model with the optimal hyperparameter pair (γ, C) achieved an accuracy of 76.8%, which was comparable to those reported in studies on multiclass classification of EEG using SVM-ML models with accuracies exceeding 71.0% [68,69].

3.1.2. EEG Channel Selection

It was observed that the overall accuracies of the model with the 21-channel EEG were better than those of the 62-channel EEG when not considering arm differences (Table 2). The 21-channel EEG covers the sensorimotor cortex, which is the main response area to sensory stimulations [70,71]. The accuracies based on the 21-channel EEG suggested that direct cortical processing from the sensorimotor cortex was sufficient to capture the sensory differences generated by different fabric samples through the SVM-ML model. Previous studies have demonstrated that significant RSP variations for different EEG bands during sensory stimulations were mainly captured in the sensorimotor cortex for both unimpaired and stroke populations [28,72,73]. On the contrary, involuntary attention activities beyond the sensorimotor cortex were involved in passive fabric stimulation experiments [28]. This could be a hurdle to the recognition of cortical responses to fabric stimulations using the SVM-ML model. Meanwhile, voluntary cognitive activities were also found to disturb measurement of cortical responses to sensory stimulations [73,74]. For example, in post-stroke sensory evaluation by a subjective questionnaire, stroke persons with sensory impairments could distinguish different fabric stimulations because of the compensation of cognitive processing, e.g., individual experiences, to the residual sensory pathways [28]. In this work, the voluntary cognitive activities were minimized by asking the participants to stay awake but mentally inactive during the fabric stimulations. Therefore, the EEG RSP features of the sensorimotor cortex detected by the 21-channel EEG were sufficient for representing the differences in direct cortical responses to fine tactile sensations.

Table 2 also shows that when identifying different fabric stimulations without considering arm differences, the model achieved better overall accuracies in higher bands, i.e., the beta and gamma bands. This was consistent with the results of previous neurophysiologic research into how the human brain reacts to tactile sensations induced by fabrics [74,75]. The cortical responses of the brain to tactile sensations are elicited by skin–fabric interactions, which are characterized by the EEG beta and gamma band activations [76]. Beta oscillations have been shown to be involved in the phasic locking process between the primary and secondary somatosensory cortex in response to tactile sensation [77]. Meanwhile, it was observed that the neuronal assemblies of the sensorimotor cortex were joined in large-scale networks oscillating in the beta band during maintenance of a sustained hand lever press activity [78]. This phenomenon suggested that the primary somatosensory and primary motor cortex were bound together in a beta-synchronized cortical network [78]. Furthermore, Greco et al. found that beta oscillations in the sensorimotor cortex served as an informative feature characterizing affective tactile stimulation by interactions with different fabrics [79]. Singh et al. demonstrated that pleasant and unpleasant tactile sensations present different beta-oscillation patterns [74]. The gamma oscillations of the sensorimotor cortex were also observed in response to tactile sensation. This may reflect the timing code and temporal organization for higher-order somatosensory processing, which is important for sensory binding [80,81]. In addition, the study by Aya et al. showed that gamma oscillations were simultaneously evoked in the primary and secondary somatosensory cortex during sensory stimulations, thus suggesting that it is critical for forming functional cortico-cortical connections and for conveying somatosensory information from the primary and secondary somatosensory cortex [82]. Bauer et al. found that tactile-stimuli-induced gamma oscillations in the somatosensory cortex were enhanced and prolonged by spatial tactile attention [83]. This indicated that gamma-band synchronization was instrumental in the somatosensory system for processing behaviorally relevant stimulations [83]. Therefore, the RSP variations in the beta and gamma bands were sensitive input features for the SVM-ML model for both unimpaired and stroke persons.

3.2. K-Fold Cross Validation

During determination of k in CV, the model attained the highest accuracy with six-fold CV in the gamma band (Table 3). This was also observed in terms of approximately identical accuracies of the different numbers of k in the gamma band, which indicated that the SVM-ML model achieved stable classification performance with good generalization capacity for different combinations of training and testing datasets owing to the different k-fold CV strategies [84]. Furthermore, the model with the leave-one-out CV achieved an accuracy similar to the k-fold CV technique in the gamma band. This demonstrated the model's unbiased evaluation capability as a special case of k-fold CV, in which each sample has a chance to represent the entire test dataset [85]. However, the computational cost of the leave-one-out CV was greater than those of other configurations of k in the k-fold CV when evaluating the SVM-ML model performance. This was in line with previous studies [85,86] that investigated the computational efficiencies of k-fold and leave-one-out CV. Thus, it was preferable to use the optimal k-fold CV, i.e., six-fold CV during model evaluations.

4. Generalization of the SVM-ML Model

4.1. Different Fabric Stimulations

In the evaluation of model generalization, the wool and nylon fabrics in the gamma band achieved significantly higher accuracies than that of cotton fabric (Table 4 and Figure 5). This was attributable to the differences in their stimulation intensities on the skin. According to the study by Chen et al., neural oscillations in higher frequency bands, e.g., gamma band, were lower when executing an easy task; however, they increased to higher levels to obtain more information from the sensory environment when the task was difficult [87]. Cotton is the most familiar fabric that is in direct contact with skin in daily life, and it provides the lowest stimulation intensity during passive involuntary touch [28]. However, wool and nylon offer more stimulating sensory experiences because of their textile physical properties, which may require additional neural effort and cortical resources to evoke cortical responses to stimulations [28]. This was further supported by the study by Jiao et al., who found that wool elicited a relatively intense tactile stimulation in the form of scratching, resulting in the sensation of discomfort [88]. They also discovered that touching wool fabric elicited higher EEG RSP responses than cotton and nylon fabrics [88]. Hoefer et al. also observed that nylon induced significant higher event-related potential (ERP) signals than cotton, implying that there was less distraction and better cortical resources during tactile sensation [76]. As a result, the model achieved relatively higher accuracies with the RSP features of nylon and wool compared to cotton.

4.2. Different Upper-Limb Groups

When considering arm differences, the model yields various patterns for the comparison of accuracies between stimulations with different fabrics (Table 5 and Figure 6). It was observed that the significant differences in the accuracies for classifying post-stroke stimulations with different fabrics shifted to higher frequency bands, i.e., the beta and gamma bands, compared to the UD group. This pattern difference in the classification of stimulations with different fabrics among the different upper-limb groups was similar to that for manual investigations comparing the EEG RSP differences between stroke and unimpaired persons. In the manual evaluations, the post-stroke representative power spectra to fine touch stimulation shifted to higher frequency bands, i.e., the beta and gamma bands [28]. The sensitivity of the EEG-based RSP features and their capability for feature mapping by the RBF kernel allowed the SVM-ML model to detect similar pattern as manual evaluations. As the input features of the model, the average and maximal values of the RSPs represent the significant differences in RSPs among multiple upper-limb groups. The differences in the EEG RSP patterns in response to fabric stimulations have been discovered in previous manual investigations and were mainly related to neuroplastic changes after stroke [89]. For example, when the damage to the brain neurons result in post-stroke sensorimotor function deficiencies, the cortex composed of various neural subsets could be rewired [90,91]. Neural compensation to lesional functions can lead to redistributed patterns of the cortical responses to external stimulations [92]. Meanwhile, the SVM with RBF kernel can find the optimal decision boundary among multiple upper-limb groups owing to its sufficient feature mapping capability. It implicitly transforms the original RSP features to a high-dimensional feature space with fewer hyperparameters to be determined, which guarantees the generalization capability of the model when supplying new input data [37]. Previous studies have also demonstrated that the SVM with RBF kernel achieves minimal classification error rates in different clinical scenarios while controlling the complexity of the model [36,93,94]. Therefore, based on the sensitivity of the RSP features and the feature mapping capability of the RBF kernel, the proposed model was expected to achieve similar performance as in manual inspection for distinguishing between unimpaired and post-stroke persons.

5. Conclusions

In this study, an EEG-based SVM-ML model was established using the RSP features of the EEG signals, i.e., RSP_{mean} and RSP_{max}, during stimulation with cotton fabric as the baseline input. The observations demonstrated that the RSP_{mean} and RSP_{max} were sensitive to fabric stimulations and could be used as representative input features to the model. The generalization performance of the model was investigated by comparing the classification accuracies during stimulations with different fabrics while considering arm differences. The model determined that the significant differences in the accuracies of fabric stimulations after stroke were shifted toward higher bands, i.e., beta and gamma bands, similar to the differences in RSP patterns between post-stroke persons and unimpaired participants as in manual investigations, thereby implying that the model could imitate manual evaluations of cortical responses to fabric stimulations; this ability is expected to aid in automatic assessments of post-stroke fine tactile sensation.

Author Contributions: Conceptualization, J.Z. and X.H.; methodology, J.Z. and X.H.; software, J.Z.; validation, Z.L. and J.Z.; data curation, J.Z. and Y.H.; writing—original draft preparation, J.Z.; writing—review and editing, J.Z., Y.H., F.Y., B.Y. and X.H.; supervision, X.H.; funding acquisition, X.H. All authors have read and agreed to the published version of the manuscript.

Funding: This research was funded by the National Natural Science Foundation of China, China (NSFC 81771959), the University Grants Committee Research Grants Council, Hong Kong (GRF 15207120), and the Science and Technology Innovation Committee of Shenzhen, China (2021Szvup142).

Institutional Review Board Statement: The human experiments were conducted after we obtained the ethical approval from the Human Subjects Ethics Sub-Committee of the Hong Kong Polytechnic University.

Informed Consent Statement: Informed consent was obtained from all subjects involved in the study.

Data Availability Statement: Data are available upon request.

Acknowledgments: The authors would like to thank the participants who participated in this study.

Conflicts of Interest: The authors declare no conflict of interest.

References

1. Kessner, S.S.; Bingel, U.; Thomalla, G. Somatosensory deficits after stroke: A scoping review. *Top. Stroke Rehabil.* **2016**, *23*, 136–146. [CrossRef] [PubMed]
2. Sherwood, L. *Human Physiology: From Cells to Systems*; Cengage Learning: Boston, MA, USA, 2015.
3. Carey, L.M.; Matyas, T.A.; Baum, C. Effects of somatosensory impairment on participation after stroke. *Am. J. Occup. Ther.* **2018**, *72*, 7203205100p1–7203205100p10. [CrossRef] [PubMed]
4. Tyson, S.F.; Crow, J.L.; Connell, L.; Winward, C.; Hillier, S. Sensory impairments of the lower limb after stroke: A pooled analysis of individual patient data. *Top. Stroke Rehabil.* **2013**, *20*, 441–449. [CrossRef] [PubMed]
5. Carey, L.M.; Matyas, T.A.; Oke, L.E. Sensory loss in stroke patients: Effective training of tactile and proprioceptive discrimination. *Arch. Phys. Med. Rehabil.* **1993**, *74*, 602–611. [CrossRef]
6. Matsuda, K.; Satoh, M.; Tabei, K.-i.; Ueda, Y.; Taniguchi, A.; Matsuura, K.; Asahi, M.; Ii, Y.; Niwa, A.; Tomimoto, H. Impairment of intermediate somatosensory function in corticobasal syndrome. *Sci. Rep.* **2020**, *10*, 11155. [CrossRef]
7. Cunha, B.P.; Alouche, S.R.; Araujo, I.M.G.; Freitas, S.M.S.F. Individuals with post-stroke hemiparesis are able to use additional sensory information to reduce postural sway. *Neurosci. Lett.* **2012**, *513*, 6–11. [CrossRef]
8. Boonsinsukh, R.; Panichareon, L.; Phansuwan-Pujito, P. Light touch cue through a cane improves pelvic stability during walking in stroke. *Arch. Phys. Med. Rehabil.* **2009**, *90*, 919–926. [CrossRef]
9. Johannsen, L.; Wing, A.M.; Hatzitaki, V. Effects of maintaining touch contact on predictive and reactive balance. *J. Neurophysiol.* **2007**, *97*, 2686–2695. [CrossRef]
10. Campfens, S.F.; Zandvliet, S.B.; Meskers, C.G.; Schouten, A.C.; van Putten, M.J.; van der Kooij, H. Poor motor function is associated with reduced sensory processing after stroke. *Exp. Brain Res.* **2015**, *233*, 1339–1349. [CrossRef]
11. Doyle, S.D.; Bennett, S.; Dudgeon, B. Upper limb post-stroke sensory impairments: The survivor's experience. *Disabil. Rehabil.* **2014**, *36*, 993–1000. [CrossRef]
12. Pandyan, A.D.; Johnson, G.R.; Price, C.I.M.; Curless, R.H.; Barnes, M.P.; Rodgers, H. A review of the properties and limitations of the Ashworth and modified Ashworth Scales as measures of spasticity. *Clin. Rehabil.* **1999**, *13*, 373–383. [CrossRef] [PubMed]
13. Fugl-Meyer, A.R.; Jääskö, L.; Leyman, I.; Olsson, S.; Steglind, S. The post-stroke hemiplegic patient. 1. a method for evaluation of physical performance. *Scand. J. Rehabil. Med.* **1975**, *7*, 13–31. [PubMed]
14. Winward, C.E.; Halligan, P.W.; Wade, D.T. The Rivermead Assessment of Somatosensory Performance (RASP): Standardization and reliability data. *Clin. Rehabil.* **2002**, *16*, 523–533. [CrossRef] [PubMed]
15. Pan, N. Quantification and evaluation of human tactile sense towards fabrics. *Int. J. Des. Nat. Ecodyn.* **2006**, *1*, 48–60.
16. Auriat, A.M.; Neva, J.L.; Peters, S.; Ferris, J.K.; Boyd, L.A. A Review of Transcranial Magnetic Stimulation and Multimodal Neuroimaging to Characterize Post-Stroke Neuroplasticity. *Front. Neurol.* **2015**, *6*, 226. [CrossRef]
17. Lundgren, J.; Flodström, K.; Sjögren, K.; Liljequist, B.; Fugl-Meyer, A.R. Site of brain lesion and functional capacity in rehabilitated hemiplegics. *Scand. J. Rehabil. Med.* **1982**, *14*, 141–143.
18. Chollet, F.; DiPiero, V.; Wise, R.J.; Brooks, D.J.; Dolan, R.J.; Frackowiak, R.S. The functional anatomy of motor recovery after stroke in humans: A study with positron emission tomography. *Ann. Neurol.* **1991**, *29*, 63–71. [CrossRef]
19. Giaquinto, S.; Cobianchi, A.; Macera, F.; Nolfe, G. EEG recordings in the course of recovery from stroke. *Stroke* **1994**, *25*, 2204–2209. [CrossRef]
20. Broderick, P.A.; Kolodny, E.H. Biosensors for brain trauma and dual laser doppler flowmetry: Enoxaparin simultaneously reduces stroke-induced dopamine and blood flow while enhancing serotonin and blood flow in motor neurons of brain, in vivo. *Sensors* **2010**, *11*, 138–161. [CrossRef]
21. Radaelli, A.; Mancia, G.; Ferrarese, C.; Beretta, S. *New Concepts in Stroke Diagnosis and Therapy*; Bentham Science Publishers: Milan, Italy, 2017; Volume 1.
22. Chen, Q.; Xia, T.; Zhang, M.; Xia, N.; Liu, J.; Yang, Y. Radiomics in stroke neuroimaging: Techniques, applications, and challenges. *Aging Dis.* **2021**, *12*, 143. [CrossRef]
23. Militello, C.; Rundo, L.; Dimarco, M.; Orlando, A.; Woitek, R.; D'Angelo, I.; Russo, G.; Bartolotta, T.V. 3D DCE-MRI Radiomic Analysis for Malignant Lesion Prediction in Breast Cancer Patients. *Acad. Radiol.* **2021**, *29*, 830–840. [CrossRef] [PubMed]
24. Sharaev, M.; Andreev, A.; Artemov, A.; Burnaev, E.; Kondratyeva, E.; Sushchinskaya, S.; Samotaeva, I.; Gaskin, V.; Bernstein, A. *Pattern Recognition Pipeline for Neuroimaging Data*; Springer: Berlin/Heidelberg, Germany, 2018.
25. Caliandro, P.; Vecchio, F.; Miraglia, F.; Reale, G.; Della Marca, G.; La Torre, G.; Lacidogna, G.; Iacovelli, C.; Padua, L.; Bramanti, P. Small-world characteristics of cortical connectivity changes in acute stroke. *Neurorehabil. Neural Repair* **2017**, *31*, 81–94. [CrossRef] [PubMed]
26. Bentes, C.; Peralta, A.R.; Viana, P.; Martins, H.; Morgado, C.; Casimiro, C.; Franco, A.C.; Fonseca, A.C.; Geraldes, R.; Canhão, P. Quantitative EEG and functional outcome following acute ischemic stroke. *Clin. Neurophysiol.* **2018**, *129*, 1680–1687. [CrossRef] [PubMed]
27. Ahn, S.-N.; Lee, J.-W.; Hwang, S. Tactile Perception for Stroke Induce Changes in Electroencephalography. *Hong Kong J. Occup. Ther.* **2016**, *28*, 1–6. [CrossRef] [PubMed]
28. Huang, Y.; Jiao, J.; Hu, J.; Hsing, C.; Lai, Z.; Yang, Y.; Hu, X.J. Measurement of sensory deficiency in fine touch after stroke during textile fabric stimulation by electroencephalography (EEG). *J. Neural Eng.* **2020**, *17*, 045007. [CrossRef]

29. Roy, S.; Kiral-Kornek, I.; Harrer, S. ChronoNet: A deep recurrent neural network for abnormal EEG identification. In *Conference on Artificial Intelligence in Medicine in Europe*; Springer: Berlin/Heidelberg, Germany, 2019.
30. Golmohammadi, M.; Ziyabari, S.; Shah, V.; de Diego, S.L.; Obeid, I.; Picone, J. Deep architectures for automated seizure detection in scalp EEGs. *arXiv* **2017**, arXiv:1712,09776.
31. Sirsat, M.S.; Fermé, E.; Câmara, J. Machine Learning for Brain Stroke: A Review. *J. Stroke Cerebrovasc. Dis.* **2020**, *29*, 105162. [CrossRef]
32. Hosseini, M.-P.; Hemingway, C.; Madamba, J.; McKee, A.; Ploof, N.; Schuman, J.; Voss, E. Review of Machine Learning Algorithms for Brain Stroke Diagnosis and Prognosis by EEG Analysis. *arXiv* **2020**, arXiv:2008.08118.
33. Jochumsen, M.; Rovsing, C.; Rovsing, H.; Niazi, I.K.; Dremstrup, K.; Kamavuako, E.N. Classification of Hand Grasp Kinetics and Types Using Movement-Related Cortical Potentials and EEG Rhythms. *Comput. Intell. Neurosci.* **2017**, *2017*, 7470864. [CrossRef]
34. Usama, N.; Niazi, I.K.; Dremstrup, K.; Jochumsen, M. Detection of Error-Related Potentials in Stroke Patients from EEG Using an Artificial Neural Network. *Sensors* **2021**, *21*, 6274. [CrossRef]
35. Iáñez, E.; Azorín, J.M.; Úbeda, A.; Fernández, E.; Sirvent, J.L. LDA-based classifiers for a mental tasks-based brain-computer interface. In Proceedings of the 2010 IEEE International Conference on Systems, Man and Cybernetics, Istanbul, Turkey, 10–13 October 2010.
36. Garrett, D.; Peterson, D.A.; Anderson, C.W.; Thaut, M.H. Comparison of linear, nonlinear, and feature selection methods for EEG signal classification. *IEEE Trans. Neural Syst. Rehabil. Eng.* **2003**, *11*, 141–144. [CrossRef]
37. Cao, J.; Fang, Z.; Qu, G.; Sun, H.; Zhang, D. An accurate traffic classification model based on support vector machines. *Int. J. Netw. Manag.* **2017**, *27*, e1962. [CrossRef]
38. Vapnik, V. *The Nature of Statistical Learning Theory*; Springer Science & Business Media: Berlin/Heidelberg, Germany, 2013.
39. Liu, Y.; Zhang, H.; Chen, M.; Zhang, L. A boosting-based spatial-spectral model for stroke patients' EEG analysis in rehabilitation training. *IEEE Trans. Neural Syst. Rehabil. Eng.* **2015**, *24*, 169–179. [CrossRef]
40. Ghumman, M.K.; Singh, S.; Singh, N.; Jindal, B. Optimization of parameters for improving the performance of EEG-based BCI system. *J. Reliab. Intell. Environ.* **2021**, *7*, 145–156. [CrossRef]
41. Bousseta, R.; Tayeb, S.; El Ouakouak, I.; Gharbi, M.; Regragui, F.; Himmi, M.M. EEG efficient classification of imagined hand movement using RBF kernel SVM. In Proceedings of the 2016 11th International Conference on Intelligent Systems: Theories and Applications (SITA), Mohammedia, Morocco, 19–20 October 2016.
42. Hsu, C.-W.; Chang, C.-C.; Lin, C.-J. *A Practical Guide to Support Vector Classification*; National Taiwan University: Taipei, Taiwan, 2003.
43. Farid, N.; Elbagoury, B.; Roushdy, M.; Salem, A.-B.M. A comparative analysis for support vector machines for stroke patients. *Recent Adv. Inf. Sci.* **2013**, *41*, 71–76.
44. Kim, M.-K.; Cho, J.-H.; Jeong, J.-H. Classification of Tactile Perception and Attention on Natural Textures from EEG Signals. In Proceedings of the 2021 9th International Winter Conference on Brain-Computer Interface (BCI), Gangwon, Korea, 22–24 February 2021.
45. Lilliefors, H.W. On the Kolmogorov-Smirnov test for the exponential distribution with mean unknown. *J. Am. Stat. Assoc.* **1969**, *64*, 387–389. [CrossRef]
46. Homan, R.W.; Herman, J.; Purdy, P. Cerebral location of international 10–20 system electrode placement. *Electroencephalogr. Clin. Neurophysiol.* **1987**, *66*, 376–382. [CrossRef]
47. Welch, P. The use of fast Fourier transform for the estimation of power spectra: A method based on time averaging over short, modified periodograms. *IEEE Trans. Audio Electroacoust.* **1967**, *15*, 70–73. [CrossRef]
48. Teplan, M. Fundamentals of EEG measurement. *Meas. Sci. Rev.* **2002**, *2*, 1–11.
49. Bronzino, J.D. *Biomedical Engineering Handbook 2*; Springer Science & Business Media: Berlin/Heidelberg, Germany, 2000; Volume 2.
50. Patro, S.; Sahu, K.K. Normalization: A preprocessing stage. *arXiv* **2015**, arXiv:1503,06462. [CrossRef]
51. Liao, X.; Li, Y.; Hu, J.; Li, Q.; Wu, X. Psychophysical Relations between Interacted Fabric Thermal-Tactile Properties and Psychological Touch Perceptions. *J. Sens. Stud.* **2016**, *31*, 181–192. [CrossRef]
52. Hu, J.Y.; Hes, L.; Li, Y.; Yeung, K.W.; Yao, B.G. Fabric Touch Tester: Integrated evaluation of thermal–mechanical sensory properties of polymeric materials. *Polym. Test.* **2006**, *25*, 1081–1090. [CrossRef]
53. Bishop, C.M. *Pattern Recognition and Machine Learning*; Springer: Berlin/Heidelberg, Germany, 2006.
54. Syarif, I.; Prugel-Bennett, A.; Wills, G. SVM parameter optimization using grid search and genetic algorithm to improve classification performance. *Telkomnika* **2016**, *14*, 1502. [CrossRef]
55. Zhang, Y.; Ji, X.; Liu, B.; Huang, D.; Xie, F.; Zhang, Y. Combined feature extraction method for classification of EEG signals. *Neural Comput. Appl.* **2017**, *28*, 3153–3161. [CrossRef]
56. Guler, I.; Ubeyli, E.D. Multiclass support vector machines for EEG-signals classification. *IEEE Trans. Inf. Technol. Biomed.* **2007**, *11*, 117–126. [CrossRef]
57. Avelino, J.; Paulino, T.; Cardoso, C.; Moreno, P.; Bernardino, A. Human-aware natural handshaking using tactile sensors for Vizzy, a social robot. In Proceedings of the Workshop on Behavior Adaptation, Interaction and Learning for Assistive Robotics at RO-MAN, Lisbon, Portugal, 28 August–1 September 2017.
58. Pedregosa, F.; Varoquaux, G.; Gramfort, A.; Michel, V.; Thirion, B.; Grisel, O.; Blondel, M.; Prettenhofer, P.; Weiss, R.; Dubourg, V. Scikit-learn: Machine learning in Python. *J. Mach. Learn. Res.* **2011**, *12*, 2825–2830.

59. Kattenstroth, J.-C.; Kalisch, T.; Peters, S.; Tegenthoff, M.; Dinse, H. Long-term sensory stimulation therapy improves hand function and restores cortical responsiveness in patients with chronic cerebral lesions. Three single case studies. *Front. Hum. Neurosci.* **2012**, *6*, 244. [CrossRef]
60. Pfurtscheller, G.; Neuper, C. Motor imagery activates primary sensorimotor area in humans. *Neurosci. Lett.* **1997**, *239*, 65–68. [CrossRef]
61. James, G.; Witten, D.; Hastie, T.; Tibshirani, R. *An Introduction to Statistical Learning*; Springer: Berlin/Heidelberg, Germany, 2013; Volume 112.
62. Breiman, L. Heuristics of instability and stabilization in model selection. *Ann. Stat.* **1996**, *24*, 2350–2383. [CrossRef]
63. Witten, I.H.; Frank, E. Data mining: Practical machine learning tools and techniques with Java implementations. *Acm Sigmod Rec.* **2002**, *31*, 76–77. [CrossRef]
64. Bouckaert, R.R. Choosing between two learning algorithms based on calibrated tests. *ICML* **2003**, *3*, 51–58.
65. Kuhn, M.; Johnson, K. *Applied Predictive Modeling*; Springer: Berlin/Heidelberg, Germany, 2013; Volume 26.
66. Chang, M.-W.; Lin, C.-J. Leave-one-out bounds for support vector regression model selection. *Neural Comput.* **2005**, *17*, 1188–1222. [CrossRef]
67. Burges, C.J.C. A tutorial on support vector machines for pattern recognition. *Data Min. Knowl. Discov.* **1998**, *2*, 121–167. [CrossRef]
68. Vivaldi, N.; Caiola, M.; Solarana, K.; Ye, M. Evaluating performance of eeg data-driven machine learning for traumatic brain injury classification. *IEEE Trans. Biomed. Eng.* **2021**, *68*, 3205–3216. [CrossRef]
69. Gao, L.; Cheng, W.; Zhang, J.; Wang, J. EEG classification for motor imagery and resting state in BCI applications using multi-class Adaboost extreme learning machine. *Rev. Sci. Instrum.* **2016**, *87*, 085110. [CrossRef] [PubMed]
70. Kira, K.; Rendell, L.A. A practical approach to feature selection. In *Machine Learning Proceedings*; Elsevier: Amsterdam, The Netherlands, 1992; pp. 249–256.
71. Kitada, R.; Hashimoto, T.; Kochiyama, T.; Kito, T.; Okada, T.; Matsumura, M.; Lederman, S.J.; Sadato, N. Tactile estimation of the roughness of gratings yields a graded response in the human brain: An fMRI study. *Neuroimage* **2005**, *25*, 90–100. [CrossRef]
72. Wu, J.; Srinivasan, R.; Quinlan, E.B.; Solodkin, A.; Small, S.L.; Cramer, T.C. Utility of EEG measures of brain function in patients with acute stroke. *J. Neurophysiol.* **2016**, *115*, 2399–2405. [CrossRef]
73. Singh, H.; Bauer, M.; Chowanski, W.; Sui, Y.; Atkinson, D.; Baurley, S.; Fry, M.; Evans, J.; Bianchi-Berthouze, N. The brain's response to pleasant touch: An EEG investigation of tactile caressing. *Front. Hum. Neurosci.* **2014**, *8*, 893. [CrossRef]
74. Merabet, L.B.; Pascual-Leone, A. Neural reorganization following sensory loss: The opportunity of change. *Nat. Rev. Neurosci.* **2010**, *11*, 44–52. [CrossRef]
75. Ackerley, R.; Carlsson, I.; Wester, H.; Olausson, H.; Backlund Wasling, H. Touch perceptions across skin sites: Differences between sensitivity, direction discrimination and pleasantness. *Front. Behav. Neurosci.* **2014**, *8*, 54. [CrossRef]
76. Hoefer, D.; Handel, M.; Müller, K.M.; Hammer, T.R. Electroencephalographic study showing that tactile stimulation by fabrics of different qualities elicit graded event-related potentials. *Ski. Res. Technol.* **2016**, *22*, 470–478. [CrossRef] [PubMed]
77. Simões, C.; Jensen, O.; Parkkonen, L.; Hari, R. Phase locking between human primary and secondary somatosensory cortices. *Proc. Natl. Acad. Sci. USA* **2003**, *100*, 2691–2694. [CrossRef] [PubMed]
78. Brovelli, A.; Ding, M.; Ledberg, A.; Chen, Y.; Nakamura, R.; Bressler, S.L. Beta oscillations in a large-scale sensorimotor cortical network: Directional influences revealed by Granger causality. *Proc. Natl. Acad. Sci. USA* **2004**, *101*, 9849–9854. [CrossRef] [PubMed]
79. Greco, A.; Guidi, A.; Bianchi, M.; Lanata, A.; Valenza, G.; Scilingo, E.P. Brain dynamics induced by pleasant/unpleasant tactile stimuli conveyed by different fabrics. *IEEE J. Biomed. Health Inform.* **2019**, *23*, 2417–2427. [CrossRef] [PubMed]
80. Hasenstaub, A.; Shu, Y.; Haider, B.; Kraushaar, U.; Duque, A.; McCormick, D.A. Inhibitory postsynaptic potentials carry synchronized frequency information in active cortical networks. *Neuron* **2005**, *47*, 423–435. [CrossRef]
81. Engel, A.K.; Singer, W. Temporal binding and the neural correlates of sensory awareness. *Trends Cogn. Sci.* **2001**, *5*, 16–25. [CrossRef]
82. Ihara, A.; Hirata, M.; Yanagihara, K.; Ninomiya, H.; Imai, K.; Ishii, R.; Osaki, Y.; Sakihara, K.; Izumi, H.; Imaoka, H.; et al. Neuromagnetic gamma-band activity in the primary and secondary somatosensory areas. *NeuroReport* **2003**, *14*, 273–277. [CrossRef] [PubMed]
83. Bauer, M.; Oostenveld, R.; Peeters, M.; Fries, P. Tactile spatial attention enhances gamma-band activity in somatosensory cortex and reduces low-frequency activity in parieto-occipital areas. *J. Neurosci.* **2006**, *26*, 490–501. [CrossRef]
84. Xiong, Z.; Cui, Y.; Liu, Z.; Zhao, Y.; Hu, M.; Hu, J. Evaluating explorative prediction power of machine learning algorithms for materials discovery using k-fold forward cross-validation. *Comput. Mater. Sci.* **2020**, *171*, 109203. [CrossRef]
85. Refaeilzadeh, P.; Tang, L.; Liu, H. Cross-validation. *Encycl. Database Syst.* **2009**, *5*, 532–538.
86. Jung, Y. Multiple predicting K-fold cross-validation for model selection. *J. Nonparametr. Stat.* **2018**, *30*, 197–215. [CrossRef]
87. Chen, A.; Wang, A.; Wang, T.; Tang, X.; Zhang, M. Behavioral oscillations in visual attention modulated by task difficulty. *Front. Psychol.* **2017**, *8*, 1630. [CrossRef] [PubMed]
88. Jiao, J.; Hu, X.; Huang, Y.; Hu, J.; Hsing, C.; Lai, Z.; Wong, C.; Xin, J.H. Neuro-perceptive discrimination on fabric tactile stimulation by Electroencephalographic (EEG) spectra. *PLoS ONE* **2020**, *15*, e0241378. [CrossRef] [PubMed]
89. Snyder, D.B.; Schmit, B.D.; Hyngstrom, A.S.; Beardsley, S.A. Electroencephalography resting-state networks in people with Stroke. *Brain Behav.* **2021**, *11*, e02097. [CrossRef]

90. Dąbrowski, J.; Czajka, A.; Zielińska-Turek, J.; Jaroszyński, J.; Furtak-Niczyporuk, M.; Mela, A.; Poniatowski, Ł.A.; Drop, B.; Dorobek, M.; Barcikowska-Kotowicz, M. Brain functional reserve in the context of neuroplasticity after stroke. *Neural Plast.* **2019**, *2019*, 9708905. [CrossRef]
91. Voss, P.; Thomas, M.E.; Cisneros-Franco, J.M.; de Villers-Sidani, É. Dynamic brains and the changing rules of neuroplasticity: Implications for learning and recovery. *Front. Psychol.* **2017**, *8*, 1657. [CrossRef]
92. Lin, M.P.; Liebeskind, D.S. Imaging of ischemic stroke. *Contin. Lifelong Learn. Neurol.* **2016**, *22*, 1399. [CrossRef]
93. Sun, S.; Zhang, C. Adaptive feature extraction for EEG signal classification. *Med. Biol. Eng. Comput.* **2006**, *44*, 931–935. [CrossRef]
94. Al-Qazzaz, N.K.; Ali, S.H.B.M.; Ahmad, S.A.; Islam, M.S.; Escudero, J. Discrimination of stroke-related mild cognitive impairment and vascular dementia using EEG signal analysis. *Med. Biol. Eng. Comput.* **2018**, *56*, 137–157. [CrossRef]

applied sciences

MDPI

Article

Unsupervised Segmentation in NSCLC: How to Map the Output of Unsupervised Segmentation to Meaningful Histological Labels by Linear Combination?

Cleo-Aron Weis [1,*,†], Kian R. Weihrauch [1,†], Katharina Kriegsmann [1,2,‡] and Mark Kriegsmann [3,‡]

1 Institute of Pathology, University Medical Centre Mannheim, Medical Faculty Mannheim, Heidelberg University, 68167 Mannheim, Germany; kianw@umich.edu (K.R.W.); katharina.kriegsmann@med.uni-heidelberg.de (K.K.)
2 Department of Hematology, Oncology and Rheumatology, University Hospital Heidelberg, 69120 Heidelberg, Germany
3 Institute of Pathology, University Medical Hospital Heidelberg, Heidelberg University, 69120 Heidelberg, Germany; mark.kriegsmann@med.uni-heidelberg.de
* Correspondence: cleo-aron.weis@medma.uni-heidelberg.de; Tel.: +49-621-383-4072
† Current address: Institute of Pathology, Medical Faculty Mannheim, Heidelberg University, 68167 Mannheim, Germany.
‡ These authors contributed equally to this work.

Abstract: Background: Segmentation is, in many Pathomics projects, an initial step. Usually, in supervised settings, well-annotated and large datasets are required. Regarding the rarity of such datasets, unsupervised learning concepts appear to be a potential solution. Against this background, we tested for a small dataset on lung cancer tissue microarrays (TMA) if a model (i) first can be in a previously published unsupervised setting and (ii) secondly can be modified and retrained to produce meaningful labels, and (iii) we finally compared this approach to standard segmentation models. Methods: (ad i) First, a convolutional neuronal network (CNN) segmentation model is trained in an unsupervised fashion, as recently described by Kanezaki et al. (ad ii) Second, the model is modified by adding a remapping block and is retrained on an annotated dataset in a supervised setting. (ad iii) Third, the segmentation results are compared to standard segmentation models trained on the same dataset. Results: (ad i–ii) By adding an additional mapping-block layer and by retraining, models previously trained in an unsupervised manner can produce meaningful labels. (ad iii) The segmentation quality is inferior to standard segmentation models trained on the same dataset. Conclusions: Unsupervised training in combination with subsequent supervised training offers for histological images here no benefit.

Keywords: histopathology; lung cancer; supervised segmentation; unsupervised segmentation

Citation: Weis, C.-A.; Weihrauch, K.R.; Kriegsmann, K.; Kriegsmann, M. Unsupervised Segmentation in NSCLC: How to Map the Output of Unsupervised Segmentation to Meaningful Histological Labels by Linear Combination? *Appl. Sci.* **2022**, *12*, 3718. https://doi.org/10.3390/app12083718

Academic Editors: Leonardo Rundo and Andrea Prati

Received: 9 February 2022
Accepted: 29 March 2022
Published: 7 April 2022

1. Introduction

After the emergence of immunohistochemistry in the 1980s, molecular pathology in the 2000s, and next-generation sequencing in the 2010s, the implementation of image analysis tools into the methodical arsenal of pathology appears to be the next level of development. Digital Pathology, Computational Pathology, and Pathomics are several names for this new branch of expertise, and each term represents a slightly different focus [1–3]. Pathomics, for example, focuses on the extraction of image features that can act as biomarkers in the context of, e.g., neoplastic diseases. In this context, image segmentation is one of the early but essential steps. On the basis of segmented images, image features are extracted and used in further analysis [1]. With machine learning-based image segmentation techniques such as convolutional neuronal networks (CNNs), high-quality and reliable image segmentation is possible. These CNN-based segmentation approaches typically comprise four development phases: Phase 1 is the creation of a well-labelled dataset; phase 2 is the choice of the model architecture; phase 3 is the design or choice of an appropriate loss function; and phase 4 is choosing or defining an appropriate optimiser [4]. For phase 1, in a usual supervised setting, to avoid overfitting, typically, large annotated datasets are necessary. Creating a representative, large training database tends to be tedious, especially the segmentation tasks; therefore, good datasets are scarce [3,5–7]. To overcome this limitation, several publicly available databases are available online—for example, the Atlas of Digital Pathology [8]. Unfortunately, such databases do not help with more specific questions than segmenting different, non-neoplastic tissues. In addition, rare entities cannot be covered. Many technically different methods have been implemented to overcome the dependency on laboriously generated huge databases. These methods either reduce the number of annotated data needed or are completely independent of labelled data. In addition to approaches based on generated features, of particular interest in this study are machine learning methods that learn the features independently [5,9,10]. On one hand, some methods apply machine learning on small datasets, such as few-shot learning [11,12] or zero-shot learning [13,14]. On the other hand, there are completely unsupervised learning methods for classification or segmentation tasks [5].

Against this background, recent publications by Kanezaki et al. on unsupervised image segmentation are of substantial interest. They describe a framework to train CNN segmentation models that is completely unsupervised [15,16].

In this work, (i) we tested this approach for the segmentation of non-small cell lung carcinoma in tissue microarrays as an example. Furthermore, (ii) we addressed the problem that unsupervised segmentation leads to undefined labels. To map the labels by the unsupervised training to known, meaningful labels (e.g., adenocarcinoma), we tested a second training step with a small human-labelled dataset.

2. Materials and Methods

2.1. Data Collection and Management

Whole-slide tissue specimens of formalin-fixed paraffin-embedded tumour tissue and tissue microarrays (TMAs) were retrieved (Institute of Pathology, Medical Faculty Heidelberg, Heidelberg University) and used in a completely anonymous manner. No patient information—for example, age or sex—was included. Only the histological diagnoses (e.g., normal lung tissue, adenocarcinoma of the lung, squamous cell carcinoma of the lung) were used. This study was approved by the local ethics committee (#S-207/2005 and #S315/2020).

2.2. Whole-Slide Image and TMA Image Preparation

The whole tissue sections (haematoxylin–eosin(HE)-stained) and TMAs (HE-stained and stained by immunohistochemistry (IHC) for panCK) were scanned by a Leica whole-slide scanner or by a PreciPoint M8-scanner. The resulting whole-slide images were saved in the .svs format. For model training and validation, the TMA cores were automatically cropped and saved to 2600 × 2600 pixel-sized images by using QuPath implemented functions [17]. The whole-slide images were automatically cropped into tiles of the same size by a QuPath script published by Peter Bankhead.

2.3. Training and Validation Dataset (Dataset #1)

Dataset #1 was used for model training and validation. It is based on IHC-HE-TMA core pairs. In this case, every core has a clinical label (normal tissue, adenocarcinoma, squamous cell carcinoma). This dataset was created in a multi-step approach: Step #1: From the included TMA paraffin block, two subsequent sections were produced: the first was HE-stained and the second was IHC-stained (panCK). Both slides were scanned, and the TMA cores were extracted as described above. Step #2: On the basis of their location on the TMA grid, the HE- and the IHC-stained cores can be assigned to each other (e.g., TMA grid position A-1 in HE stain corresponds to TMA grid position A-1 in panCK stain). Next, these images, containing a single TMA core each, were registered, resulting in the IHC-HE-TMA core pair. For registration, the airlab tool published by Sandkuehler et al. was used (https://github.com/airlab-unibas/airlab accessed on 1 February 2022). Step #3: The IHC-positive area of each image (containing epithelium) was extracted by using a combination of colour deconvolution [18] and thresholding, resulting in a map for background tissue and IHC-positive tissue. Based on the clinical annotation (every TMA grid position is assigned to one case), the IHC-positive areas are assigned to the defined labels: 1 non-tumourous tissue (NT), 2 adenocarcinoma (ADC), and 3 squamous cell carcinoma (SqCC). IHC-negative areas are assigned based on thresholding to the labels background (0) and non-tumourous tissue (1). Notably, based on this assignment approach, IHC-positive epithelium in normal tissue is labelled 1 together with the IHC-negative tissue in the same cases. The label ratio between the area per label is approximately 19.2 (background (BG)) to 4.7 (normal tissue or non-tumourous tissue (NT)) to 1.5 (adenocarcinoma (ADC)) to 1.0 (sqqmous cell carcinoma (SqCC)). Because of the image pair production in steps #1 and #2, the labels produced in step #3 can be used for the HE-stained images. By doing so, the advantage here is that segmentation data are produced without the need for human experts to have laboriously drawn each class per TMA core. However, this advantage is at the cost of errors due to, for example, poor registration or false thresholding between IHC-negative and -positive areas. The multi-stage process only produces a rough visual inspection of the results. In summary, dataset #1 contained n = 247 images (n = 108 for NT, n = 84 for ADC, n = 55 for SqCC). Nine examples are shown in Figure 1. This dataset was used for the training and validation of the modified and retrained unsupervised models (hereafter Kanezaki models) and the supervised model (a UNet-Variant). Dataset #1 is available at HeiData: https://heidata.uni-heidelberg.de/privateurl.xhtml?token=0129f05c-b1a7-4927-a841-2440eb0b3cc4.

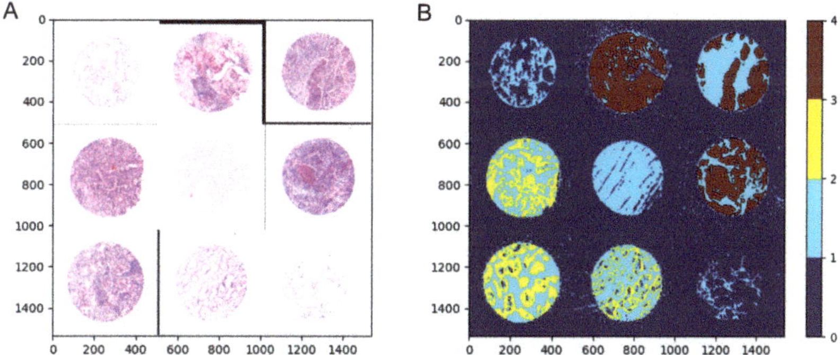

Figure 1. Example for training and validation dataset (dataset #1). Based on registered pairs of HE- and IHC-stained sections from TMA cores, tumour segmentation or rather tumour mask generation is performed by a combination of colour deconvolution and thresholding. (**A**) Composite image of nine HE-stained TMA cores. For each of the three classes (NT, ADC, SqCC), there are three images. (**B**) Corresponding IHC(panCK)-stained images were registered on the HE-stained cores. Based on the IHC-positive area and the diagnosis per core, the according image and image regions were labelled: 0 background, 1 normal tissue or non-tumourous (NT), 2 adenocarcinoma (ADC), and 3 squamous cell carcinoma (SqCC)

2.4. Testing Dataset (Dataset #2)

Dataset #2 was used for model testing only and was based on manual segmentation (examples shown in Figure 2). Therefore, TMA images were manually annotated, segmented, and further prepared in QuPath [17]. For this manual segmentation, the following labels were defined (in accordance with the definitions for dataset #1): 0 background, 1 non-tumourous tissue (NT), 2 adenocarcinoma (ADC), and 3 squamous cell carcinoma (SqCC). The ratio between the area per label was approximately 3.7 (background) to 1.2 (NT) to 1.1 (ADC) to 1.0 (SqCC). Dataset #2 contained n = 40 images (n = 3 for NT, n = 18 for ADC, n = 19 for SqCC) and is available at HeiData: https://heidata.uni-heidelberg.de/privateurl.xhtml?token=0129f05c-b1a7-4927-a841-2440eb0b3cc4.

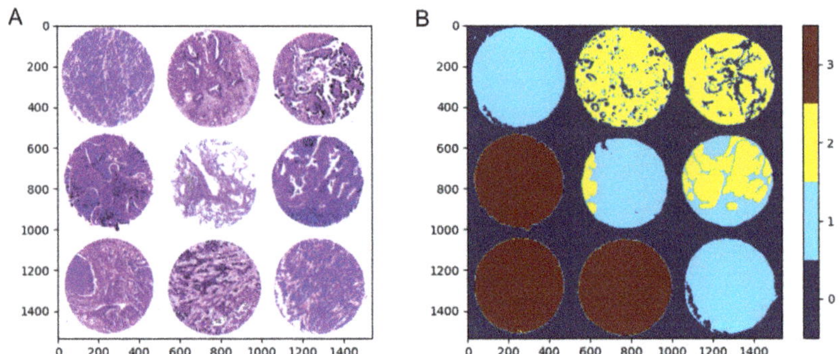

Figure 2. Example for testing dataset (dataset #2). Example of human-labelled ground truth images used for testing the models. (**A**) Composite image of nine TMA cores. For each of the three classes (NT, ADC, Sqcc), there were three images. (**B**) Composite image of corresponding labelled images with 0 for background, 1 for normal tissue or non-tumourous tissue (NT), 2 for adenocarcinoma (ADC), and 3 for squamous cell carcinoma (SqCC).

2.5. Model Training

Machine learning was performed in Python with PyTorch [19]. For supervised learning, the Segmentation Models toolbox from Yakubovskiy et al. [20] was used (https://github.com/qubvel/segmentation_models.pytorch, accessed on 1 February 2022). For unsupervised training, the models and scripts from Kanezaki et al. [15] were adapted (https://github.com/kanezaki/pytorch-unsupervised-segmentation, accessed on 1 February 2022) and used.

2.6. Loss Functions

Against the background of unbalanced labels and heterogeneously shaped objects, different loss functions selected from the plethora of published functions were used. These loss functions are differently well suited for imbalanced datasets. Furthermore, the loss functions are differently well suited for different models [21,22] (Table 1).

Table 1. List of different loss functions here tested. A set of loss functions is tested against the background of imbalanced labels and heterogeneous objects.

Loss Function	Source
Cross-Entropy Loss	Pytorch implementation [19]
Dice Loss	Pytorch implementation [19]
Focal Loss	pytorch-toolbelt implementation [23]
Tversky Loss	pywick implementation [24]
Boundary loss	proposed by Bokhovkin et al. [25]
Surface Loss with Cross-Entropy Loss	proposed by Kervadec et al. [26]
Surface Loss with Focal Loss	proposed by Kervadec et al. [26]

2.7. Segmentation Quality Assessment

The segmentation quality per image was evaluated by calculating the accuracy and the F1 score, each in its scikit-learn implementation [27]. As ground truth for the calculations, the validation set (see Section 2.3) and the test set (see Section 2.4) were used.

3. Results

3.1. How Can Labels from a Model Trained in an Unsupervised Manner Be Converted to Meaningful Labels?

For unsupervised image segmentation, Kanezaki et al. (Figure 3) published a training approach in 2018 based on similarity [15] and another in 2020 based on differentiable feature clustering [28]. Here, we use the first approach and the framework described for it, which we refer to as the Kanezaki framework. It starts with a high number of classes and minimises the label classes in every training epoch, until a predefined number of classes is reached. The label classes are merged based on the similarity in the segmented image region. Finally, there are a given number of label classes segmented per image. However, these labels do not correspond to meaningful labels. Which label class (e.g., 1) belongs to which histological structure (e.g., alveolar epithelium) is unclear. We mapped such a model to defined classes by a two-step process.

First, the Kanezaki training approach was reproduced using the histological images available. The excised HE-stained cores were used in random order for training. As we have described, the approach of Kanezaki et al. starts with a predefined number of labels—in our case, 100. Next, at each epoch, the number of labels was reduced or the labels were merged. The training process ended when the previously specified expected number of labels was reached or undercut. Because, in our setting, the classes 'background', 'tumour stroma', 'squamous cell carcinoma', and 'adenocarcinoma' were expected, the number of expected labels was set to 10.

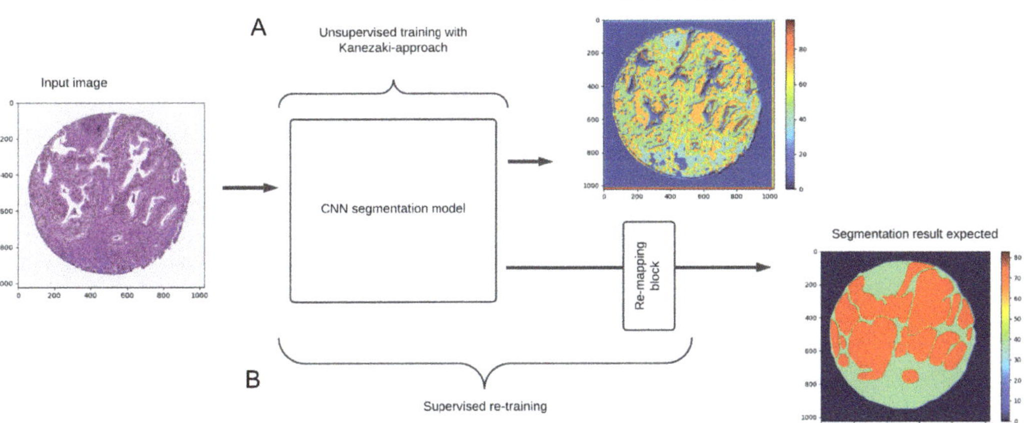

Figure 3. Schematic of the remapping approach. Kanezaki et al. described an approach for unsupervised segmentation [15]. This leads to meaningless morphological labelling. For example, the epithelial and stromal structures are segmented. In this manner, homogeneous labels (in terms of texture, for example) are created. Depending on the resolution, this can lead to tumour formation and splitting of the tumour stroma into different partial labels. The main hypothesis of the underlying work is that labels such as 'adenocarcinoma' are composed of a distinct set of morphological labels produced by unsupervised training. The approach described herein is divided into two main parts. (**A**) First, a CNN model (e.g., consisting of several convolutional and batch normalization blocks) is trained in an unsupervised manner, as described by Kanezaki et al. This training was performed on the image batches to ensure that all classes were represented. (**B**) An additional block was added to the model to map the classes of the model to patho-histological labels. The mapping was trained in a supervised manner. There were four components for the label frequency vectors per TMA core.

For the CNN model, a simple model composed of a linear combination of convolution and batch normalisation blocks was used, as described by Kanezaki et al. [15]. This simple

CNN model is henceforth referred to as concise_CNN (visualised in Figure A1A). Other more complex segmentation models, such as UNets or FCNs, have also been trained in this framework. However, only concise models with a few layers (called concise_UNET and concise_FCN; visualised in Figure A1B,C) converged. Standard UNet-variants such as those published in Yakubovskiy et al. (henceforth called Standard_UNET) cannot be trained [20].

Second, to map the labels to meaningful labels, a fully connected layer is added to the model previously trained in an unsupervised setting. This layer is supposed to map the learned labels to given, defined labels, such as 'stroma'. The extended model is then retrained again in a supervised setting on a small, labelled dataset. This two-step approach is performed for three image sizes (256 × 256, 512 × 512, and 1024 × 1024 pixels), to test whether the image or object size affects the segmentation performance.

3.2. *Do Different Loss Functions Affect Retraining?*

The used lung cancer datasets (dataset #1 based on IHC annotations (Section 2.3) and dataset #2 based on manual segmentation (Section 2.4)) are highly heterogeneous. For example, the background area (label 0) was three-times more frequent than the other three labels (1–3) in the overall dataset. In a single image, the ratio of, e.g., adenocarcinoma to stroma can easily exceed 1 to 10. Furthermore, the shape and histological characteristics of tumour formations of one entity (e.g., SqCC) can be diverse. To compensate for the imbalanced dataset with regard to the area per label, we compared different error functions and metrics.

The loss functions are as follows: (1) the PyTorch-implemented weighted cross-entropy loss function [19]; (2) the dice loss and (3) the focal loss function (with the pytorch-toolbelt implementation [23]); (4) the Tversky loss function (with its pywick implementation [24]); (5) the boundary loss function proposed by Bokhovkin et al. [25]; and (6) the surface loss function proposed by Kervadec et al. [26].

As a readout, the segmentation quality was measured by calculating the accuracy and the F1 score. These parameters were assessed for the validation dataset (being 0.25 for dataset #1) and the test dataset (dataset #2). Notably, there is a morphological or quality difference between IHC-based and manual segmentation. Thus, for the models, it is a certain transfer task, because the training, validation, or test data differ.

Independent of the image size, the segmentation quality reaches its highest value when the cross-entropy loss function is used alone or in combination (for both validation and testing). Notably, the segmentation quality was only moderate even for the best models. For example, for the validation dataset (see Table 2), we observed unbalanced cross-entropy with an accuracy of 0.88 $\pm$ 0.11 and an F1 score of 0.63 $\pm$ 0.19; for balanced cross-entropy, we observed an accuracy of 0.82 $\pm$ 0.13 and an F1 score of 0.47 $\pm$ 0.16; and for surface loss (in combination with balanced cross-entropy), as described by Kervadec et al. [26], we observed an accuracy of 0.86 $\pm$ 0.22 and an F1 score of 0.39 $\pm$ 0.12).

3.3. *In Comparison, What Are the Results of an Often-Used Segmentation Model Trained in a Supervised Manner?*

For comparing the segmentation quality, a standard UNet-variant [20] was trained. This model was trained and validated on dataset #1 (Section 2.3) and tested on dataset #2 (Section 2.4) under the same conditions as described above for the Kanezaki models. Different image sizes and error functions were used to visualise their effects on the segmentation quality (measured with accuracy and the F1 score).

Table 2. Retrained unsupervised segmentation results. An adapted Kanezaki model was retrained on different image sizes (256 × 256 and 512 × 512 pixels) with five different loss functions: (1) cross-entropy, (2) dice loss, (3) focal loss, (4) Tversky loss, and (5) boundary loss function. The validation set corresponds to 0.25 from dataset #1 (compare Section 2.3) and the test set corresponds to the entire dataset #2 (compare Section 2.4). As a metric for the segmentation quality, the accuracy and F1 score are calculated.

Image Size	Loss Function	Validation Set (Accuracy/F1)		Test Set (Accuracy/F1)	
256	Cross-Entropy (balanced)	0.82 ± 0.13	0.47 ± 0.16	0.63 ± 0.21	0.37 ± 0.09
	Cross-Entropy (unbalanced)	0.88 ± 0.11	0.63 ± 0.19	0.59 ± 0.21	0.43 ± 0.16
	Surface Loss (with Cross-Entropy)	0.86 ± 0.22	0.39 ± 0.12	0.51 ± 0.17	0.33 ± 0.10
	Surface Loss (with Dice Loss)	0.32 ± 0.12	0.28 ± 0.11	0.44 ± 0.23	0.30 ± 0.11
	Focal Loss	0.88 ± 0.12	0.63 ± 0.21	0.59 ± 0.20	0.43 ± 0.18
	Tversky Loss	0.87 ± 0.12	0.62 ± 0.21	0.57 ± 0.19	0.39 ± 0.16
	Dice Loss	0.86 ± 0.12	0.60 ± 0.21	0.61 ± 0.16	0.37 ± 0.21
	Boundary Loss	0.13 ± 0.07	0.15 ± 0.06	0.27 ± 0.17	0.17 ± 0.06
512	Cross-Entropy (balanced)	0.86 ± 0.12	0.53 ± 0.17	0.60 ± 0.11	0.40 ± 0.08
	Cross-Entropy (unbalanced)	0.87 ± 0.12	0.56 ± 0.18	0.60 ± 0.16	0.37 ± 0.10
	Surface Loss (with Cross-Entropy)	0.86 ± 0.11	0.56 ± 0.20	0.66 ± 0.16	0.41 ± 0.10
	Surface Loss (with Dice Loss)	0.28 ± 0.09	0.26 ± 0.08	0.33 ± 0.10	0.25 ± 0.07
	Focal Loss	0.88 ± 0.12	0.58 ± 0.21	0.64 ± 0.18	0.36 ± 0.11
	Tversky Loss	0.88 ± 0.11	0.56 ± 0.20	0.60 ± 0.17	0.36 ± 0.10
	Dice Loss	0.88 ± 0.11	0.54 ± 0.20	0.60 ± 0.15	0.37 ± 0.10
	Boundary Loss	0.13 ± 0.08	0.14 ± 0.06	0.20 ± 0.11	0.15 ± 0.07

3.4. How Does the Training Dataset Size Affect the Segmentation Quality of the Models Trained under Unsupervised and Supervised Conditions?

To test whether the segmentation performance of the models tested depends on the size of the training data as expected, the modified concise_CNN and the UNet-model were pretrained with eight subsets of different size from the previous datasets (see x-axis in Figure 4: The first subset, named selection, was a manual image selection with three images per diagnosis (NT, ADC, and SqCC) from dataset #2 (see Section 2.4). The other eight subsets were named with 1.0, 0.75, 0.5, 0.25, 0.1, 0.05, and 0.01, respectively, after the fractions from dataset #1 (see Section 2.3). In addition, to test the effect of the pretraininig of the UNet-model, two UNet-models differing in the means of pretraining were tested: one naive for histological images and henceforth referred to as UNet_naive (orange bars in Figure 4), and the other with a ResNet-model [20] pretrained on a image tile classification task with the classes normal tissue, adenocarcinoma, and squamous cell carcinoma, hereafter referred to as UNet_histo (green bars in Figure 4).

All models were trained with the surface loss function (with the combination of boundary loss and cross-entropy loss) [26] for 50 epochs. All image tiles used for training and testing had a size of 256 × 256 pixels.

Segmentation quality was assessed based on the testing images (dataset #2), as in the prior section.

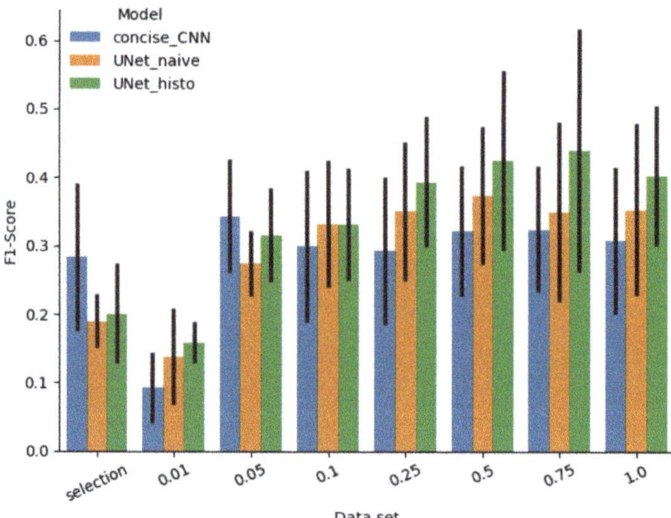

Figure 4. Effect of the number of images used for training. Three models were trained on eight different datasets. The three models are: a small CNN variant as described by Kanezaki et al. [15] (called concise_CNN), trained in an unsupervised approach and then modified and retrained as described in this work to produce histologically meaningful labels; two UNet models [20] with a ResNet backbone pretrained on a tissue classification task (called UNet_histo) and with a ResNet backbone without pretraining (called UNet_histo). The eight datasets are a manual selection (with nine images, three per diagnosis; compare Figure 1) from dataset #1 (called selection), and seven increasingly larger fractions from dataset #2 (called 0.01, 0.05, 0.1, 0.25, 0.5, 0.75, and 1.0) Subsequently, the different models were tested using dataset #2. As a segmentation metric, the F1 score is plotted.

The models trained on dataset #2 had equal segmentation quality, as in the prior section. For the Concise_CNN, the accuracy was 0.49 $\pm$ 0.22 and the F1 score was 0.31 $\pm$ 0.10. For the UNet_naive, the accuracy and the F1 score were 0.53 $\pm$ 0.20 and 0.35 $\pm$ 0.12. Finally, for the UNet_histo, the accuracy was 0.63 $\pm$ 0.18 and F1 score was 0.40 $\pm$ 0.10 (right end of the box plot in Figure 4). Notably, no statistically significant differences were observed in the accuracy and F1 score for the models in the range 0.1 to 1.0: for 0.1 of dataset #1, the F1 score was 0.30 $\pm$ 0.11; for the UNet_naive, it was 0.33 $\pm$ 0.09; and for the UNet_histo, it was 0.31 $\pm$ 0.07. Only for the small fractions (0.01 and the nine images from the selection) were the models' performance reduced (left end of the box plot in Figure 4). For the selection with nine images, there was a trend for better performance of the Kanezaki model, since its accuracy (0.43 $\pm$ 0.23) and F1 score (0.29$\pm$0.10) were not reduced as much as those for the UNet models (for UNet_naive, 0.26 $\pm$ 0.06 and 0.20 $\pm$ 0.04, respectively, and for UNet_histo, 0.13 $\pm$ 0.19 and 0.20 $\pm$ 0.07, respectively)

3.5. Does the Model Architecture Trained in an Unsupervised Manner Influence the Segmentation Quality?

Kanezaki et al. [15] demonstrated that with their approach, different CNN models can be trained. As we have described, complex models such as the UNet variant by Yakubovskiy et al. [20] do not converge. However, simple model variants for FCN and UNet can be trained and do converge. To compare three different architectures, we first used the unsupervised Kanezaki approach to train the aforementioned models, which consist of a linear combination of convolution blocks (called concise_CNN; sketched in Appendix A Figure A1A), a relatively simple FCN variant (called concise_FCN; Appendix A Figure A1B), and a relatively simple UNet variant (called concise_UNet; Appendix A Figure A1C).

Next, these three models were retrained in a supervised setting (as described in Section 4.1). To test if there is an advantage of such pretrained models for smaller datasets, we performed retraining by using two training datasets: (i) a manual selection of nine images (three per NT, ADC, and SqCC; called selection) and (ii) the entire dataset #1.

Regarding segmentation quality measurement, these models were again tested on dataset #2 (the testing dataset).

For very small retraining datasets (selection; n = 9 images), the Concise_CNN shows better results if only the last layers are retrained (accuracy 0.53 ± 0.21 and F1 score 0.34 ± 0.11). For the more complex concise_FCN (accuracy 0.50 ± 0.24 and F1 score 0.30 ± 0.10) and concise_UNet (accuracy 0.43 ± 0.11 and F1 score 0.24 ± 0.09), the models only show moderate segmentation quality, if the entire models are retrained (Figure A2A).

For larger retraining datasets (dataset #1; n = 247 images (with 0.8 for training and 0.2 for validation)), there was no significant difference for all three models (Figure A2B). The best model was the simple_FCN. The accuracy was 0.63 ± 0.21 and the F1 score was 0.38 ± 0.10 when only the last layers were retrained, and the accuracy was 0.61 ± 0.15 and the F1 score was 0.387 ± 0.07 when all layers were retrained. The worst model was the simple_CNN. For retraining only the last layers, the accuracy was 0.57 ± 0.17 and the F1 score was 0.36 ± 0.10. For retraining all layers, the accuracy was 0.57 ± 0.19 and the F1 score was 0.37 ± 0.10.

Notably, segmentation results were more than 0.1 worse than the results for the complex UNet model variants, such as the Standard_UNET (compare Table 3) by Yakubovskiy et al. [20].

Table 3. Supervised segmentation results. A UNet model was trained on different image sizes (256x256 and 512x512 pixels) with five different loss functions: (1) cross-entropy, (2) dice loss, (3) focal loss, (4) Tversky loss, and (5) boundary loss function. The validation set corresponds to 0.25 from dataset #1 (compare Section 2.3) and the test set corresponds to the entire dataset #2 (compare Section 2.4). As a metric for the segmentation quality, the accuracy and the F1 score are calculated.

Image Size	Loss Function	Validation Set (Accuracy/F1)		Test Set (Accuracy/F1)	
256	Cross-Entropy (balanced)	0.86 ± 0.13	0.52 ± 0.20	0.71 ± 0.17	0.43 ± 0.08
	Cross-Entropy (unbalanced)	0.89 ± 0.11	0.64 ± 0.23	0.64 ± 0.19	0.45 ± 0.16
	Surface Loss (with Cross-Entropy)	0.87 ± 0.12	0.57 ± 0.17	0.86 ± 0.17	0.41 ± 0.08
	Surface Loss (with Dice Loss)	0.78 ± 0.18	0.41 ± 0.08	0.57 ± 0.16	0.34 ± 0.07
	Focal Loss	0.88 ± 0.11	0.63 ± 0.23	0.70 ± 0.17	0.42 ± 0.08
	Tversky Loss	0.88 ± 0.12	0.64 ± 0.23	0.70 ± 0.19	0.42 ± 0.09
	Dice Loss	0.88 ± 0.12	0.54 ± 0.65	0.71 ± 0.22	0.51 ± 0.21
	Boundary Loss	0.13 ± 0.14	0.06 ± 0.16	0.24 ± 0.04	0.19 ± 0.03
512	Cross-Entropy (balanced)	0.86 ± 0.12	0.47 ± 0.09	0.70 ± 0.14	0.44 ± 0.08
	Cross-Entropy (unbalanced)	0.90 ± 0.10	0.66 ± 0.22	0.66 ± 0.15	0.42 ± 0.10
	Surface Loss (with Cross-Entropy)	0.90 ± 0.12	0.59 ± 0.19	0.68 ± 0.12	0.42 ± 0.08
	Surface Loss (with Dice Loss)	0.90 ± 0.09	0.54 ± 0.14	0.66 ± 0.14	0.40 ± 0.17
	Focal Loss	0.91 ± 0.10	0.61 ± 0.19	0.66 ± 0.15	0.40 ± 0.08
	Tversky Loss	0.90 ± 0.11	0.65 ± 0.21	0.66 ± 0.19	0.43 ± 0.13
	Dice Loss	0.90 ± 0.10	0.64 ± 0.21	0.64 ± 0.16	0.39 ± 0.08
	Boundary Loss	0.11 ± 0.06	0.12 ± 0.05	0.23 ± 0.11	0.15 ± 0.05

3.6. Are the Labels Learned in an Unsupervised Fashion Already Meaningful?

As we have described, the models trained without supervision produce a predefined number of labels that are not directly connected to labels defined by humans. These labels are based on texture or morphological similarity. To test whether these labels alone correlate to the known structures (e.g., tumour glands) or more to the diagnoses of non-tumourous tissue (NT), adenocarcinoma (ADC), or squamous cell carcinoma (SqCC), in this context, the frequency of these labels per diagnosis was examined.

The basic idea was that each TMA core has its own label composition or frequency that correlates with the diagnosis. For example, a TMA core from a case with ADC should contain (only) the labels NT and/or ADC. To test this assumption in principle, for human-generated labels, we plotted the label composition (frequency of labels per TMA core) against the known diagnosis per TMA kernel (see A1-2 in Figure 5). Plotting the label frequency per diagnosis (A1 in Figure 5) or running a principal component analysis (PCA) to compare the frequency vectors per TMA core (A2 in Figure 5) verifies this assumption. As expected, cases with, for example, a diagnosis of ADC differ in that only in these cases does the label ADC occur alongside the image background (BG) and non-tumourous tissue (NT).

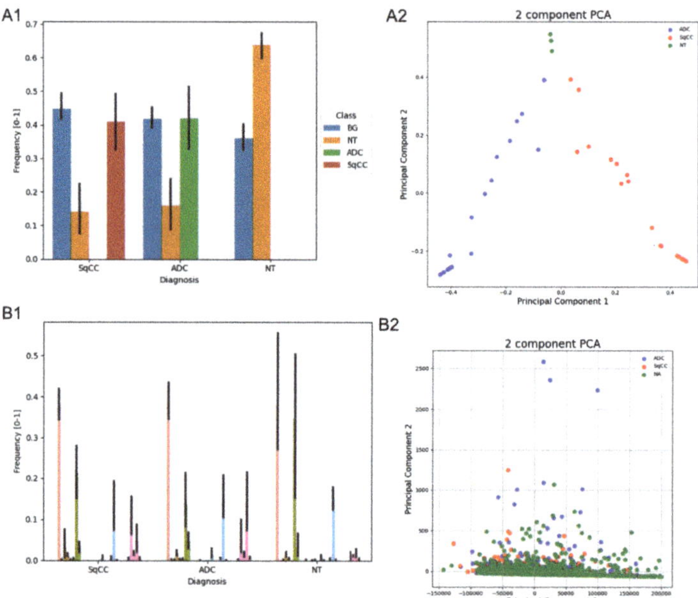

Figure 5. Segmentation results for different models trained. Kanezaki et al. [15] described a method for training CNN models in an unsupervised fashion, resulting in some labels that are not assigned to certain structures by a human. For testing whether these labels or a combination are already meaningful, the label frequency per TMA core was analysed with respect to the known diagnosis per core (normal tissue (NT), adenocarcinoma (ADC), and squamous cell carcinoma (SqCC)). (**A1**,**A2**) For testing whether the label frequency per TMA core could correlate with the diagnosis, the label frequency of the manually annotated TMA cores was examined. (**A1**) shows the label frequency (background (BG), normal tissue (NT), adenocarcinoma (ADC), and squamous cell carcinoma (SqCC)) plotted against the diagnosis. (**A2**) shows a PCA (two components) for the label frequency vector per TMA core. (**B1**,**B2**) For the labels produced by a simple CNN model (previously called concise_CNN) trained in an unsupervised manner, the label frequency was also analysed in regard to the known diagnosis per TMA core. (**B1**) shows the frequency distribution per label and diagnosis. (**B2**) shows a PCA (two components) for the label frequency vectors per TMA core.

In the next step, we examined the label frequency for the labels generated by the model after unsupervised training. Here, it can be seen that neither plotting the label frequency per diagnosis (B1 in Figure 5)) nor a PCA analysis (B2 in Figure 5) show a reliable correlation between label composition and diagnosis. A distinction between ADC and SqCC is not possible on this basis. Based on these plots, only TMA cores with and without tumour infiltration can be distinguished. A comparable analysis for the standard UNet [20] trained in a supervised setting also showed no sharp separation of diagnosis based on label frequency (compare Appendix A Figure A3). This fits well with the overall moderate segmentation quality of the models trained and validated. In the unsupervised trained or retrained models, the small network size might be causative. In the standard UNet, the training database may not be sufficiently large.

4. Discussion

Digital and complex medical data are available from various medical specialities. For patients with tumours, for example, there are molecular data, and radiological and pathological image data [10,29]. The analysis of these vast data from one field—or, better, combined—leads to an opportunity to find next-generation, data-driven biomarkers [29]. In pathology, Pathomics is the subdiscipline dedicated to mapping image data to clinical information such as nodal status. In other words, new image-based biomarkers are being sought in Pathomics. In this context, the segmentation of histological images (e.g., in tumour and stroma) is an early and major step in many projects [1,4]. For supervised segmentation approaches, the scarcity of large, properly annotated datasets is a common obstacle [3,7,30]. Not only is annotation tedious per se, but in many cases, the number of available images is limited. For example, managing thymoma, a rare disease, significantly limits the number of cases to be included [31]. Data augmentation techniques alone have usually not solved the problem of small numbers [7,10]. In addition to supervised training approaches, there are weakly supervised and unsupervised approaches that can help to overcome this constraint. However, mapping the results produced by unsupervised approaches to meaningful labels is a non-trivial task. Against this background, we tested, in the complex setting of the distinction between pulmonary (solid-growing) ADC and (non-keratinising) SqCC [32], whether a segmentation model could (i) be trained in an unsupervised approach and (ii) modified and retrained in a supervised setting to produce meaningful histological labels such as 'ADC' or 'SqCC'. In the best case, these labels should handle the aforementioned non-trivial distinction between solid-growing ADC and non-keratinising SqCC. (iii) We compared the the segmentation results to standard segmentation models for the same datasets.

(Ad i), we show that unsupervised image segmentation techniques or training frameworks as described by Kanezaki et al. can be used for the unsupervised segmentation of histological images (Figure 3A) [15,16]. Here, only aspects such as the ratio of filter size to object size in the image need to be considered.

(Ad ii), we demonstrate that these models can be extended by another block, which can, after a second supervised training, remap (by a linear combination) the produced labels to meaningful labels such as 'ADC' or 'SqCC' (Figure 3B). Our new contribution is that a simple linear combination of the different labels previously recognised based on unsupervised training is applied to predict difficult labels such as 'ADC'.

(Ad iii), finally, we compare the results to conventional training approaches and demonstrate that our approach of remapping the labels is not superior to conventional supervised learning. It is indeed inferior and there are only limited settings where it can be useful.

4.1. *Unsupervised Segmentation in Pathology and the Problem of Obtaining Meaningful Labels (Ad i)*

Image segmentation is, for many projects in the realm of Digital Pathology or Pathomics, an important early step. There is a legion of different approaches that, based on the training setting, can be broadly categorised as fully supervised, weakly supervised, and unsupervised approaches. For the supervised training approaches, the necessary annotations are time-consuming and tedious to produce. Indeed, the shortage of such annotated datasets is a well-known obstacle [3,7,30]. Needing less annotated data for weakly supervised or no annotated data at all for unsupervised methods sounds, in this context, very promising. In addition, by using unsupervised approaches, the need to tailor to every project a well-annotated training set for machine learning models will be reduced to gathering a fitting image collection [10].

Regarding weakly supervised approaches in pathology, generative adversarial networks can be used (after training in a weakly supervised setting) to generate synthetic data based on a small dataset [5,33]. However, this would mean adding a training cost-intensive step before the actual segmentation model training.

Regarding unsupervised image segmentation, in pathology, there are several published approaches [5,9]. These approaches cover a vast methodological spectrum with, for example, the combination of feature extraction and subsequent clustering [9,34] or the application of auto-encoders for classification or staining adaption [5,35–37].

In a nutshell, there are various working, published, easily adaptable, unsupervised approaches for segmenting (histological) images into different morphological regions. This would then overcome the problem of data scarcity. Unfortunately, this advantage brings a new problem. The labels generated based on morphological aspects (e.g., 'blue granular area') cannot in every case be simply mapped to (histologically) meaningful labels such as 'carcinoma'. One solution to this is to assign names to the labels by human experts. For example, we could allow the expert to define blue areas with many small cells as lymphoid infiltration. However, this expert approach only works if exactly one label is generated per annotation. A multiphase process such as a tumour consisting of tumour cells, stroma, and inflammatory infiltrate, etc., will not be nameable in this manner. For such multiphase entities, the true annotation can be considers as a combination of the morphological labels. However, this linear combination is too simplistic for many areas, as the context is then missing. For example, a homogeneous, blue area can be part of the sky or a blue car. In this regard, there are works that use graphs to include the neighbourhood relationships of the individual labels. For example, Pourian et al. used graphs of regions to combine the visual and spatial characteristics of different image parts to meaningful image-part groupings [38]. Alternatively, Wigness et al. used local graphs to combine labels in image regions [39].

Our approach, by contrast, is a simple linear combination of the different labels generated by unsupervised learning based on morphological similarity (by a adding a fully connected layer to a CNN model; see B in Figure 3). This linear combination is in analogy to the pathological thinking of tissue or organs as a combination of different structures such as epithelium, stroma, blood vessels, etc. [40,41]. However, this approach ignores neighbourhood relations or local aspects.

4.2. CNN Models Previously Trained in an Unsupervised Manner Can Be Adapted to Produce Meaningful Histological Labels (Ad ii)

As we have discussed, unsupervised training approaches can be used for histological images, but they produce distinct regions or labels based on morphology (e.g., reddish area with little texture) without histologically meaningful labels (such as, e.g., 'fibrosis'). We have successfully trained several CNN model variations (a combination of convolutional blocks (called concise_CNN), a shallow UNET variant (called concise_UNET), and a shallow FCN variant (called concise_FCN)) in an unsupervised approach, as described by Kanezaki et al. [15,16]. The produced image regions, or rather labels, however, are not mapped to the conventional histological structures. For instance, gland structures are composed of an epithelial layer (one label) and the luminal space (another label). Pathologists would usually annotate these structures together as a gland, in analogy to the typical thinking of tissues and organs as combinations of a limited number of substrata [40,41].

For remapping the labels produced by the model trained in an unsupervised manner, we added a block (a fully connected layer) and then trained it on mapping the labels to human-produced annotations (compare B in Figure 3). However, this again necessitates the presence of (a small amount of) annotated data. Notably, this approach is therefore no longer an unsupervised but a weakly supervised approach. By adding another block and retraining, we can show that a model can produce meaningful annotations. However, compared to other segmentation models (such as the UNet implementation by Yakubovskiy et al. [20]) trained on the same dataset, the approach proposed here leads to inferior results.

4.3. The Combination of Unsupervised and Subsequent Supervised Label Mapping Is Not Better than Conventional CNN-Based Segmentation (Ad iii)

The inferior segmentation results in combination with again the need of a labelled dataset argue against the herein proposed remapping of labels by adding an additional block and by retraining with a small annotated dataset (compare Figure 3).

There are several potential explanations for the rather moderate segmentation results of the proposed remapping approach:

(1) The model complexity is maybe not fitting with the task. Of note, the used framework for unsupervised training described by Kanezaki et al. [15,16] only works with shallow CNN models. Large models such as the UNet model implemented by Yakubovskiy et al. [20] do not converge.

(2) Another idea would be that the ratio of the CNN filter size to the object size in the images is either too small or too large. Therefore, we tested different images sizes (256 × 256, 512 × 512, and 1024 × 1024 pixels) and found no significant difference. Likewise, it would be possible in principle that the CNN models trained in this way were too shallow. However, this is contradicted by the fact that the tumour sub-type differentiation also did not work well in the UNet models used by other groups [20], which are frequently used and perform well.

(3) The task itself is non-trivial since neither models adapted as described nor standard segmentation models can make the distinction between different tumour types, particularly between (non-keratinising) SqCc and (solid) ADC. In the used dataset, on which a work on tumour classification has been published recently [32], the models can only distinguish background from tissue and normal tissue from tumour parts. Of course, the models trained supervised perform better (see Tables 2 and 3); however, their segmentation results are also only moderate with assigning mixed labels (such as ADC and SqCC) per tumour infiltration. This could be due to several reasons. For example, the task of distinguishing between a non-glandular growing ADC and a non-keratinising SqCC is non-trivial, even for an experienced pathologist, on the basis of HE-stained images alone. Moreover, unlike the previously published work on this dataset for classification [32], where each image must be assigned to one class, now, each pixel must be assigned the correct label.

(4) Finally, maybe the labels or morphological clusters segmented by the models after unsupervised training are found within non-neoplastic and neoplastic structures. For example, glandular structures are found in both. To test this, we looked at the distribution of morphological labels generated by such a model compared to the diagnoses per TMA core (see Figure 5). Based on plotting for every TMA core the frequency of the labels background (BG), normal tissue (NT), adenocarcinoma (ADC), and squamous cell carcinoma (SqCC) in a two-dimensional PCA, we were able to classify the TMA cores into the three classes of normal tissue, adenocarcinoma, and squamous cell carcinoma (see Figure 5A1,A2). Interestingly, for the models initially trained unsupervised and then retrained and for the UNets trained supervised, based on the label distribution, such a classification is not possible (see Figure 5B1,B2 and Appendix A Figure A3A,B). This is an argument for the assumption that solely morphological labels are not enough for the herein analysed task, in line with the discussion in the section above.

The limitations described above raise the question of whether the method can be improved. Increasing the complexity of the models pretrained in an unsupervised manner alone seems not promising, since even complex models such as UNets are not able to adequately solve the task. The most promising approaches, in view of the work of Pourian et al. [38] or Wigness et al. [39], seem to involve the neighbourhood when combining the individual labels into meaningful histological annotations. This should certainly be followed up in future work. Moreover, the approach described here should certainly be tested on a simple histological task as a proof of principle.

4.4. Are There Arguments for Using the Herein Proposed Remapping Approach?

Regarding the only moderate segmentation quality on one hand and the greater effort of retraining on the other hand, one could ask if there are arguments for using such an approach. Having a well-annotated dataset at hand, there are no arguments against using a standard supervised training setting. Of course, large and good datasets are rather scarce for histology [3,5–7]. The approach proposed herein, which, in combination, is more akin to weakly supervised approaches, might provide an initial advantage if there is only a dataset of limited size. Moreover, in a scenario in which there is only a small dataset, one could also consider using methods such as generative adversarial networks to produce synthetic data, on which then the segmentation model is trained [5,33].

Author Contributions: Conceptualisation, C.-A.W., K.K. and M.K.; methodology and formal analysis, C.-A.W.; data generation and curation, K.R.W., C.-A.W., K.K. and M.K.; manuscript writing and preparation, C.-A.W., K.K. and M.K. All authors have read and agreed to the published version of the manuscript.

Institutional Review Board Statement: The study was approved by the ethics committee of the Medical Faculty Heidelberg, Heidelberg University (#S-207/2005 and #S315/2020).

Data Availability Statement: Image data (for training, validation, and testing) along with ground truth are available at HeiData: https://heidata.uni-heidelberg.de/privateurl.xhtml?token=0129f05c-b1a7-4927-a841-2440eb0b3cc4. The code for the work is available at GitHub: GitHub-Link/The repo is created when the publication title is clear so that it can be generated under the same name.

Acknowledgments: The authors gratefully acknowledge the data storage service SDS@hd supported by the Ministry of Science, Research and the Arts Baden-Württemberg (MWK) and the German Research Foundation (DFG) through grant INST 35/1314-1 FUGG and INST 35/1503-1 FUGG. Furthermore, the authors also thank the IT department staff of the Medical Faculty Mannheim and especially Bohne-Lang for supervising the computer administration and infrastructure.

Conflicts of Interest: The authors declare no conflicts of interest.

Abbreviations

The following abbreviations are used in this manuscript:

MDPI	Multidisciplinary Digital Publishing Institute
CNN	Convolutional Neuronal Network
PCA	Principal Component Analysis
GAN	Generative Adversial Network

Appendix A

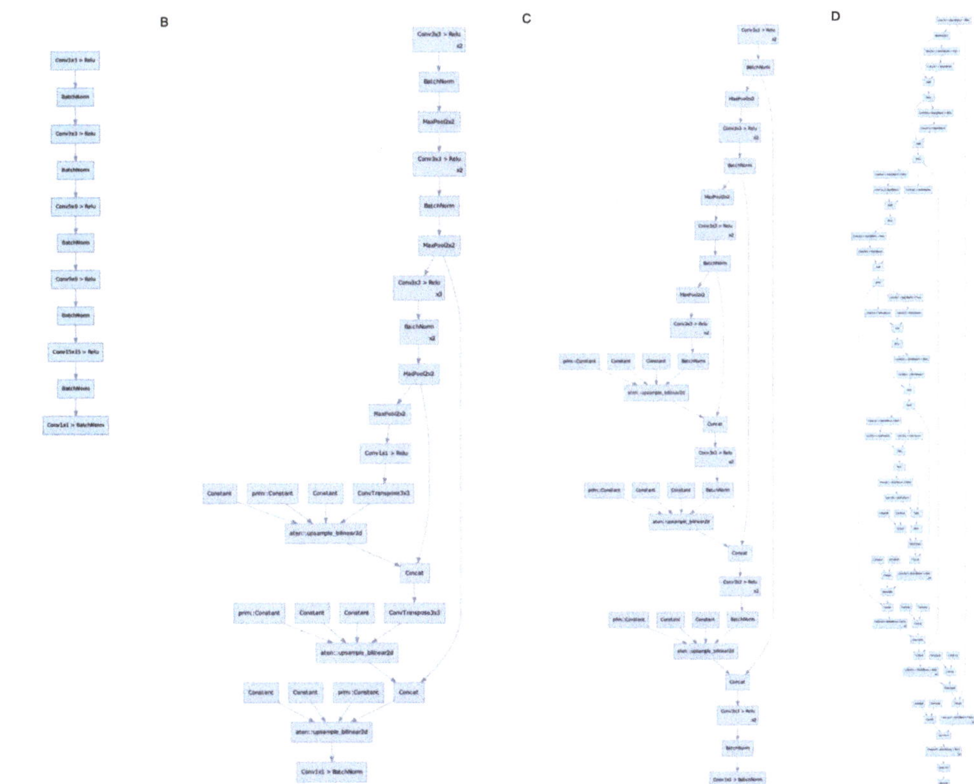

Figure A1. Schematic representation of the models used. The models (**A–C**) were first trained in an unsupervised fashion and subsequently trained on mapping the labels to meaningful labels. In contrast, the model in (**D**) was trained solely in a supervised fashion. (**A**) Concise_CNN: Schematic plot of the simple CNN model used by Kanezaki et al. [15], which is composed of a linear combination of convolutional, ReLu, and batch normalization layers. (**B**,**C**) Concise_UNET and Concise_FCN: Schematic representation of concise UNET and FCN model variants. (**D**) Standard_UNet: Schematic representation of the UNet model designed by Yakubovskiy et al. [20], here used for supervised segmentation.

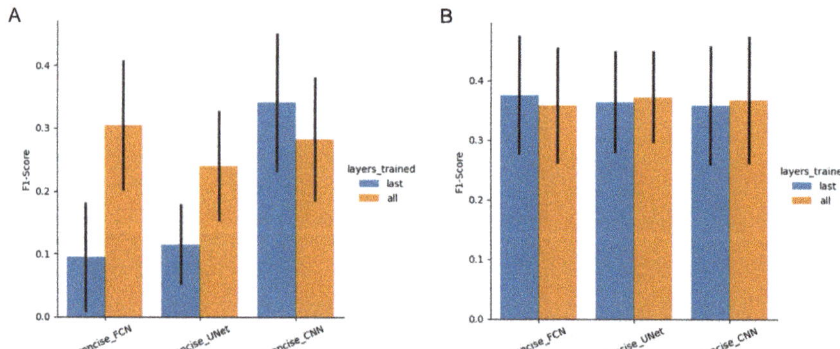

Figure A2. Segmentation results for different models trained. Kanezaki et al. [15] described a method for training CNN models in an unsupervised fashion. In this approach, different CNN models can be trained. Here, a simple CNN model composed of several convolution blocks (simple_CNN), a simple FCN variant (simple_FCN), and a simple UNet variant (simple_UNet) are trained and retrained on two datasets (**A**,**B**). Furthermore, in the retraining, only the last layers (last) or the entire model (all) are retrained. (**A**) The models are retrained on a selection of nine images (three images per diagnosis: NT, ADC, and SqCC). (**B**) The models are retrained on the entire dataset #2.

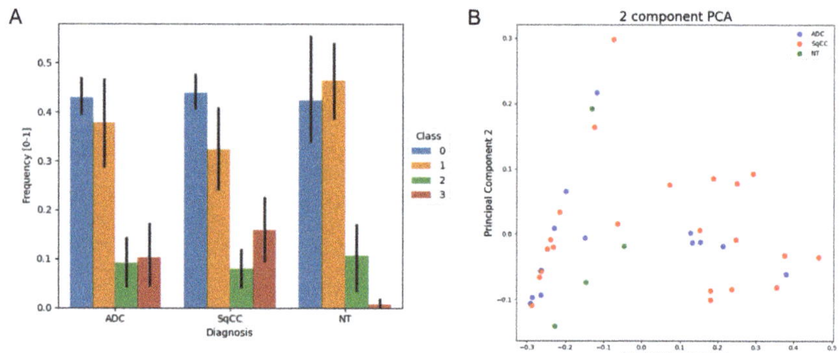

Figure A3. Segmentation results for a UNet model. The labels produced per image by a UNet model [20] trained in a supervised fashion are plotted. (**A**) shows the label frequency (background (BG), normal tissue (NT), adenocarcinoma (ADC), and squamous cell carcinoma (SqCC)) plotted against the diagnosis. (**B**) shows a PCA (two components) for the label frequency vector per TMA core.

References

1. Gupta, R.; Kurc, T.; Sharma, A.; Almeida, J.S.; Saltz, J. The Emergence of Pathomics. *Curr. Pathobiol. Rep.* **2019**, *7*, 73–84. [CrossRef]
2. Bui, M.M.; Asa, S.L.; Pantanowitz, L.; Parwani, A.; van der Laak, J.; Ung, C.; Balis, U.; Isaacs, M.; Glassy, E.; Manning, L. *Digital and Computational Pathology: Bring the Future into Focus;* Wolters Kluwer–Medknow Publications: Mumbai, India, 2019; Volume 10.
3. Abels, E.; Pantanowitz, L.; Aeffner, F.; Zarella, M.D.; van der Laak, J.; Bui, M.M.; Vemuri, V.N.; Parwani, A.V.; Gibbs, J.; Agosto-Arroyo, E. *Computational Pathology Definitions, Best Practices, and Recommendations for Regulatory Guidance: A White Paper from the Digital Pathology Association;* Wiley Online Library: Hoboken, NJ, USA, 2019; Volume 249, pp. 286–294. ISBN 0022-3417.
4. Ma, B.; Guo, Y.; Hu, W.; Yuan, F.; Zhu, Z.; Yu, Y.; Zou, H. Artificial Intelligence-Based Multiclass Classification of Benign or Malignant Mucosal Lesions of the Stomach. *Front. Pharmacol.* **2020**, *11*, 1542. [CrossRef] [PubMed]
5. McAlpine, E.D.; Michelow, P.; Celik, T. *The Utility of Unsupervised Machine Learning in Anatomic Pathology;* Oxford University Press: New York, NY, , USA, 2022; Volume 157, pp. 5–14.
6. Hou, L.; Agarwal, A.; Samaras, D.; Kurc, T.M.; Gupta, R.R.; Saltz, J.H. Robust histopathology image analysis: To label or to synthesize? In Proceedings of the IEEE/CVF Conference on Computer Vision and Pattern Recognition, Long Beach, CA, USA, 15–20 June 2019; pp. 8533–8542.

7. Ravì, D.; Wong, C.; Deligianni, F.; Berthelot, M.; Andreu-Perez, J.; Lo, B.; Yang, G.Z. Deep learning for health informatics. *IEEE J. Biomed. Health Inf.* **2016**, *21*, 4–21. [CrossRef] [PubMed]
8. Hosseini, M.S.; Chan, L.; Tse, G.; Tang, M.; Deng, J.; Norouzi, S.; Rowsell, C.; Plataniotis, K.N.; Damaskinos, S. Atlas of digital pathology: A generalized hierarchical histological tissue type-annotated database for deep learning. In Proceedings of the IEEE/CVF Conference on Computer Vision and Pattern Recognition, Long Beach, CA, USA, 16–17 June 2019; pp. 11747–11756.
9. Roohi, A.; Faust, K.; Djuric, U.; Diamandis, P. Unsupervised machine learning in pathology. *Surg. Pathol. Clin.* **2020**, *13*, 349–358. [CrossRef] [PubMed]
10. Rundo, L.; Militello, C.; Vitabile, S.; Russo, G.; Sala, E.; Gilardi, M.C. *A Survey on Nature-Inspired Medical Image Analysis: A Step Further in Biomedical Data Integration*; IOS Press: Amsterdam, The Netherlands, 2020; Volume 171, pp. 345–365.
11. Medela, A.; Picon, A.; Saratxaga, C.L.; Belar, O.; Cabezón, V.; Cicchi, R.; Bilbao, R.; Glover, B. Few shot learning in histopathological images: Reducing the need of labeled data on biological datasets. In Proceedings of the 2019 IEEE 16th International Symposium on Biomedical Imaging (ISBI 2019), Venice, Italy, 8–11 April 2019; pp. 1860–1864.
12. Deuschel, J.; Firmbach, D.; Geppert, C.I.; Eckstein, M.; Hartmann, A.; Bruns, V.; Kuritcyn, P.; Dexl, J.; Hartmann, D.; Perrin, D.; et al. Multi-Prototype Few-shot Learning in Histopathology. In Proceedings of the IEEE/CVF International Conference on Computer Vision, Montreal, BC, Canada, 11–17 October 2021; pp. 620–628.
13. Mahapatra, D.; Bozorgtabar, B.; Ge, Z. Medical Image Classification Using Generalized Zero Shot Learning. In Proceedings of the IEEE/CVF International Conference on Computer Vision, Montreal, BC, Canada, 11–17 October 2021; pp. 3344–3353.
14. Mahapatra, D.; Bozorgtabar, B.; Kuanar, S.; Ge, Z. Self-supervised multimodal generalized zero shot learning for gleason grading. In *Domain Adaptation and Representation Transfer, and Affordable Healthcare and AI for Resource Diverse Global Health*; Springer: Berlin/Heidelberg, Germany, 2021; pp. 46–56.
15. Kanezaki, A. Unsupervised Image Segmentation by Backpropagation. In Proceedings of the 2018 IEEE International Conference on Acoustics, Speech and Signal Processing (ICASSP), Calgary, AB, Canada, 15–20 April 2018; pp. 1543–1547. [CrossRef]
16. Kim, W.; Kanezaki, A.; Tanaka, M. Unsupervised Learning of Image Segmentation Based on Differentiable Feature Clustering. *IEEE Trans. Image Process.* **2020**, *29*, 8055–8068. [CrossRef]
17. Bankhead, P.; Loughrey, M.B.; Fernández, J.A.; Dombrowski, Y.; McArt, D.G.; Dunne, P.D.; McQuaid, S.; Gray, R.T.; Murray, L.J.; Coleman, H.G.; et al. QuPath: Open source software for digital pathology image analysis. *Sci. Rep.* **2017**, *7*, 16878. [CrossRef] [PubMed]
18. Ruifrok, A.C.; Johnston, D.A. Quantification of histochemical staining by color deconvolution. *Anal. Quant. Cytol. Histol.* **2001**, *23*, 291–299. [PubMed]
19. Paszke, A.; Gross, S.; Massa, F.; Lerer, A.; Bradbury, J.; Chanan, G.; Killeen, T.; Lin, Z.; Gimelshein, N.; Antiga, L.; et al. PyTorch: An Imperative Style, High-Performance Deep Learning Library. In *Advances in Neural Information Processing Systems*; Wallach, H., Larochelle, H., Beygelzimer, A., Alché-Buc, F.D., Fox, E., Garnett, R., Eds.; Curran Associates, Inc.: Red Hook, NY, USA, 2019; Volume 32.
20. Yakubovskiy, P. *Segmentation Models Pytorch*; GitHub Repository: Online, 2020.
21. Ma, J.; Chen, J.; Ng, M.; Huang, R.; Li, Y.; Li, C.; Yang, X.; Martel, A.L. Loss odyssey in medical image segmentation. *Med. Image Anal.* **2021**, *71*, 102035. [CrossRef] [PubMed]
22. Yeung, M.; Sala, E.; Schönlieb, C.B.; Rundo, L. Unified focal loss: Generalising dice and cross entropy-based losses to handle class imbalanced medical image segmentation. *Comput. Med. Imaging Graph.* **2022**, *95*, 102026. [CrossRef] [PubMed]
23. Khvedchenya, E. Pytorch Toolbelt. Available online: https://github.com/BloodAxe/pytorch-toolbelt (accessed on 28 April 2022)
24. Salehi, S.S.M.; Erdogmus, D.; Gholipour, A. Tversky loss function for image segmentation using 3D fully convolutional deep networks. In *Machine Learning in Medical Imaging, Proceedings of the 8th International Workshop, MLMI 2017, Held in Conjunction with MICCAI 2017, Quebec City, QC, Canada, 10 September 2017*; Springer: Cham, Switzerland, 2017. [CrossRef]
25. Bokhovkin, A.; Burnaev, E. Boundary Loss for Remote Sensing Imagery Semantic Segmentation. In *International Symposium on Neural Networks*; Springer: Cham, Switzerland, 2019; pp. 388-401.
26. Kervadec, H.; Bouchtiba, J.; Desrosiers, C.; Granger, E.; Dolz, J.; Ben Ayed, I. Boundary loss for highly unbalanced segmentation. In Proceedings of the 2nd International Conference on Medical Imaging with Deep Learning, London, UK, 8–10 July 2019; Volume 67, p. 101851. [CrossRef]
27. Pedregosa, F.; Varoquaux, G.; Gramfort, A.; Michel, V.; Thirion, B.; Grisel, O.; Blondel, M.; Prettenhofer, P.; Weiss, R.; Dubourg, V.; et al. Scikit-learn: Machine Learning in Python. *J. Mach. Learn. Res.* **2011**, *12*, 2825–2830.
28. Kim, H.; Ganslandt, T.; Miethke, T.; Neumaier, M.; Kittel, M. Deep Learning Frameworks for Rapid Gram Stain Image Data Interpretation: Protocol for a Retrospective Data Analysis. *JMIR Res. Protoc.* **2020**, *9*, e16843. [CrossRef] [PubMed]
29. Boehm, K.M.; Khosravi, P.; Vanguri, R.; Gao, J.; Shah, S.P. *Harnessing Multimodal Data Integration to Advance Precision Oncology*; Nature Publishing Group: Berlin, Germany, 2021; pp. 1–13.
30. Tizhoosh, H.R.; Pantanowitz, L. *Artificial Intelligence and Digital Pathology: Challenges and Opportunities*; Wolters Kluwer–Medknow Publications: Mumbai, India, 2018; Volume 9
31. World Health Organization. *WHO Classification of Tumours of the Lung, Pleura, Thymus and Heart*; World Health Organization: Geneva, Switzerland, 2015; Volume 7.

32. Kriegsmann, M.; Haag, C.; Weis, C.A.; Steinbuss, G.; Warth, A.; Zgorzelski, C.; Muley, T.; Winter, H.; Eichhorn, M.E.; Eichhorn, F.; et al. Deep Learning for the Classification of Small-Cell and Non-Small-Cell Lung Cancer. *Cancers* **2020**, *12*, 1604. [CrossRef] [PubMed]
33. Tschuchnig, M.E.; Oostingh, G.J.; Gadermayr, M. Generative adversarial networks in digital pathology: A survey on trends and future potential. *Patterns* **2020**, *1*, 100089. [CrossRef] [PubMed]
34. Peikari, M.; Salama, S.; Nofech-Mozes, S.; Martel, A.L. *A Cluster-Then-Label Semi-Supervised Learning Approach for Pathology Image Classification*; Nature Publishing Group: Berlin, Germany, 2018; Volume 8, pp. 1–13.
35. Janowczyk, A.; Basavanhally, A.; Madabhushi, A. Stain normalization using sparse autoencoders (StaNoSA): Application to digital pathology. *Comput. Med. Imaging Graph.* **2017**, *57*, 50–61. [CrossRef]
36. Song, T.H.; Sanchez, V.; Daly, H.E.; Rajpoot, N.M. Simultaneous cell detection and classification in bone marrow histology images. *IEEE J. Biomed. Health Inf.* **2018**, *23*, 1469–1476. [CrossRef]
37. Xu, J.; Xiang, L.; Liu, Q.; Gilmore, H.; Wu, J.; Tang, J.; Madabhushi, A. Stacked sparse autoencoder (SSAE) for nuclei detection on breast cancer histopathology images. *IEEE Trans. Med. Imaging* **2015**, *35*, 119–130. [CrossRef]
38. Pourian, N.; Karthikeyan, S.; Manjunath, B.S. Weakly supervised graph based semantic segmentation by learning communities of image-parts. In Proceedings of the IEEE International Conference on Computer Vision, Washington, DC, USA, 7–13 December 2015; pp. 1359–1367.
39. Wigness, M.; Rogers, J.G. Unsupervised semantic scene labeling for streaming data. In Proceedings of the IEEE Conference on Computer Vision and Pattern Recognition, Honolulu, HI, USA, 21–26 July 2017; pp. 4612–4621.
40. Mills, S. *Histology for Pathologists*; LWW Medical Book Collection; Lippincott Williams & Wilkins: Philadelphia, PA, USA, 2007.
41. Fletcher, C. *Diagnostic Histopathology of Tumors*; Elsevier Health Sciences: Amsterdam, The Netherlands, 2013.

Article

Deep Learning-Based Automatic Segmentation of Mandible and Maxilla in Multi-Center CT Images

Seungbin Park [1], Hannah Kim [2], Eungjune Shim [2], Bo-Yeon Hwang [3], Youngjun Kim [1,2], Jung-Woo Lee [3,*] and Hyunseok Seo [1,*]

[1] Center for Bionics, Korea Institute of Science and Technology, Seoul 02792, Korea; seungbin201803@gmail.com (S.P.); ceo@imagoworks.ai (Y.K.)
[2] Imagoworks, Inc., Seoul 06611, Korea; hannah.kim@imagoworks.ai (H.K.); ejshim@imagoworks.ai (E.S.)
[3] Department of Oral and Maxillofacial Surgery, School of Dentistry, Kyung Hee University, Seoul 02447, Korea; bo0426@hanmail.net
* Correspondence: omsace@khu.ac.kr (J.-W.L.); seo@kist.re.kr (H.S.)

Abstract: Sophisticated segmentation of the craniomaxillofacial bones (the mandible and maxilla) in computed tomography (CT) is essential for diagnosis and treatment planning for craniomaxillofacial surgeries. Conventional manual segmentation is time-consuming and challenging due to intrinsic properties of craniomaxillofacial bones and head CT such as the variance in the anatomical structures, low contrast of soft tissue, and artifacts caused by metal implants. However, data-driven segmentation methods, including deep learning, require a large consistent dataset, which creates a bottleneck in their clinical applications due to limited datasets. In this study, we propose a deep learning approach for the automatic segmentation of the mandible and maxilla in CT images and enhanced the compatibility for multi-center datasets. Four multi-center datasets acquired by various conditions were applied to create a scenario where the model was trained with one dataset and evaluated with the other datasets. For the neural network, we designed a hierarchical, parallel and multi-scale residual block to the U-Net (HPMR-U-Net). To evaluate the performance, segmentation with in-house dataset and with external datasets from multi-center were conducted in comparison to three other neural networks: U-Net, Res-U-Net and mU-Net. The results suggest that the segmentation performance of HPMR-U-Net is comparable to that of other models, with superior data compatibility.

Keywords: segmentation; mandible; craniomaxillofacial bone; deep learning; neural network; multi-center

Citation: Park, S.; Kim, H.; Shim, E.; Hwang, B.-Y.; Kim, Y.; Lee, J.-W.; Seo, H. Deep Learning-Based Automatic Segmentation of Mandible and Maxilla in Multi-Center CT Images. *Appl. Sci.* **2022**, *12*, 1358. https://doi.org/10.3390/app12031358

Academic Editor: Carmelo Militello

Received: 10 December 2021
Accepted: 21 January 2022
Published: 27 January 2022

1. Introduction

Segmentation of the craniomaxillofacial bones, such as the mandible and maxilla, in computed topography (CT) images is one of the crucial steps for generating three-dimensional (3D) models that are required for the diagnosis and treatment planning of craniomaxillofacial deformities, craniofacial tumor resection, or free flap reconstruction of the mandible [1,2]. Additionally, 3D segmentation of organs at risk (OARs) in head and neck (H&N) CT including the mandible is a critical step in radiotherapy planning for H&N cancer treatment [3].

The conventional segmentation task is performed manually using professional software, which is labor-intensive and time-consuming in clinical practice [4,5]. Additionally, manual segmentation has limitations such as low reproducibility and operator variability. Moreover, accurate segmentation of head CT is challenging owing to the complexity of the anatomical structures, the low contrast of soft tissue, artifacts caused by mental implants, and variations between individual patients [6]. In specific, weak and false edges of condyles appearing in CT images adversely affect the accurate segmentation of the mandible [7]. Figure 1 shows examples of the difficulties in segmenting the mandible and maxilla.

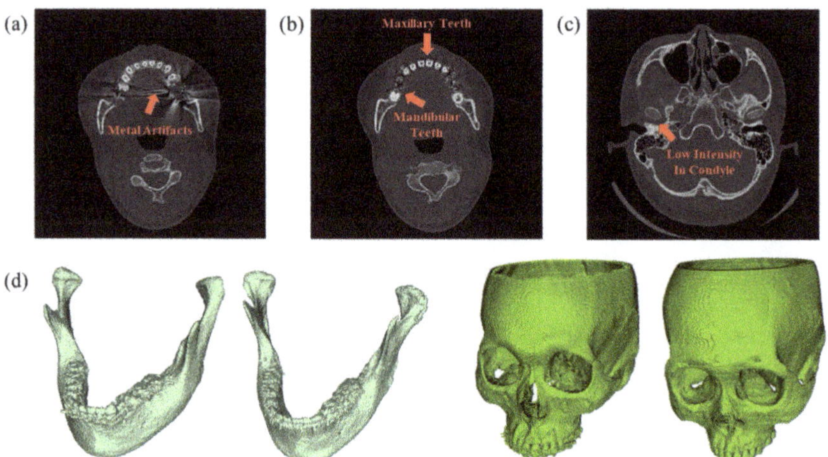

Figure 1. Difficulties in mandible and maxilla segmentation. (**a**) Metal artifacts caused by dental implants (**b**) Difficulty in distinguishing mandibular and maxillary teeth, or mandible and midface (**c**) Low intensity and thin edges in condyle (**d**) Inter-patient anatomical variance.

Automatic segmentation can improve efficiency and reliability, reducing segmentation time and clinician workload [7]. Numerous studies exist on automatic or semi-automatic segmentation of the mandible from CT scans, including OARs. In the Medical Image Computing and Computer-Assisted Intervention (MICCAI) 2015 Head and Neck Auto Segmentation Challenge [8], various approaches were proposed for the segmentation of OARs including the mandible. The use of public datasets, such as the Public Domain Database for Computational Anatomy (PDDCA) version 1.4.1, which was provided for the challenge, and how to evaluate the model performance have been a standard in head CT segmentation research. Most of these approaches utilize atlas-based methods [9] or model-based methods [10].

Atlas-based methods performs segmentation on novel data by image registration using the prior knowledge from the structures of interest [11]. Although atlas-based methods are popular and widely used for anatomy segmentation, they are sensitive to anatomical variations as they use a fixed set of atlases [12]. Moreover, they are computationally expansive and require many minutes to complete one registration task [13].

Statistical model-based methods utilize a statistical appearance model [14]. The models that best represent the shape or appearance variations in the structure of interest, which are obtained from training with a set of images and segmentations, are selected for a new patient image [15]. However, the shape or appearance described by the statistical model is limited to specific shapes, which gives it less flexibility unless large training sets are employed.

In some studies, atlas-based and statistical model-based methods have been combined with each other or with another method, leading to various other approaches for automatic mandible segmentation. Albrecht et al. [16] used a multi-atlas to obtain an initial segmentation of the OAR and an active shape model to refine the initial segmentation. Aghdasi et al. [17] employed anatomic landmarks and prior knowledge for segmentation. Chuang et al. [18] proposed a registration-based semi-automatic mandible segmentation pipeline that uses a nonlinear diffeomorphic method to register preprocessed test CT scans on the reference templates.

Recently, as convolutional neural networks (CNNs) have become more effective in computer vision, research on deep learning for medical image segmentation has increased exponentially [19]. The first deep learning-based algorithm utilizing a CNN for the segmentation of OARs in H&N CT was proposed by Ibragimov et al. [20], who employed a network with three convolution layers. Tong et al. [21] then incorporated a CNN with the pretrained shape representation model. Beyond simple CNNs, U-Net [22] has been one of the most popular CNNs for medical image segmentation. Compared with other CNNs, U-Net, with a simple and flexible structure, shows an outstanding performance in segmentation extracting image features by multi-scale recognition and fusion [23]. Several approaches have been developed by applying the U-Net structure as a baseline for mandible segmentation. Qiu et al. [1] used three U-Nets for orthogonal planes with dice loss to segment the mandible. AnatomyNet [13] was proposed to segment OARs from H&N CT, which was built on a 3D U-net architecture. A two-stage segmentation framework for OAR in CT was also proposed, which employs two 3D U-Nets for localization and segmentation [24].

Several studies have utilized U-Net with other structures together as well. Both a faster regional CNN and attention U-Net for localization and segmentation have been introduced by Lei et al. [24]. A recurrent segmentation CNN was proposed that embeds the CNN into a recurrent neural network for segmentation of the mandible from CT [7]. An attention mechanism, which has been advanced with deep learning models in computer vision, has been incorporated to U-Net for segmentation. Squeeze-and-excitation blocks were incorporated into U-Net for prostate zonal segmentation of multi-institutional MRI datasets, enhancing both intra- and cross-dataset generalization [25]. An attention gate model that can be integrated into CNN models was proposed to automatically learn to focus on target structures [26]. Focus U-Net with attention gate for spatial and channel-based attention was proposed for fast and accurate polyp segmentation [27].

However, there is an inevitable and considerable pitfall in data-driven methods including deep learning, which is the lack of data compatibility; that is, the method may fail to accurately segment images with varying properties, such as those acquired using different CT scanners and imaging protocols [28]. The compatibility of dataset in the models refers to the ability of models to inference the input images that have different distributions in the latent space from the multi-center training dataset [28]. In general, datasets are limited so that they cannot fully represent the general patient population in the clinic [29]. As a result, models trained on the specific center domain do not perform well on a different center domains with disparate data distribution [30]. This drawback is more significant when applying deep learning clinically on images from other institutions. For example, it is known that the Hounsfield units measurement varies between scanners [31]. The results of models targeted to CT can vary depending on the imaging parameters, the scanner type, calibration, or the scan date [29,32,33]. That is, multicenter data tend to have different data distributions, making trained neural network impractical. With consideration for this variability, it has been recently been required to test the artificial intelligence model with an external dataset [32]. From these limitations in clinical applications, data compatibility in deep learning for medical images has been an essential challenge to be addressed.

To solve this problem, research has been conducted to utilize multicenter data in neural network training [33,34]. Another potential solution to this problem is transfer learning [35–37], which trains with more easily obtained datasets from different domains to enhance performance [38]. However, these approaches have limitations for clinical use, as available medical data are scarce compared to natural images and are not sufficient for deep learning. Furthermore, labeling is more challenging with medical data.

In this study, we propose a framework for automated 3D segmentation of the mandible and maxilla using deep learning. We aim not only to accurately delineate the mandible and maxilla from CT, but also to improve the compatibility of multicenter data so that the model performs well on new domain data. To this end, we employed four multi-center datasets acquired by various conditions, with one used to train the models, and three used to evaluate the performance of the segmentation and the data compatibility. For the neural network, we applied residual connections [39] to U-Net, as it has been empirically and theoretically determined that the generalization is improved in residual networks compared with non-residual networks [40,41].

2. Materials and Methods

2.1. Data

We utilized four datasets: two of them from different centers (CenterA and CenterB) including mandible and maxilla segmentations and two public datasets (PDDCA and TCIA) for OAR segmentation in H&N CT. The CenterA dataset was randomly divided into training, validation, and test datasets, consisting of 146, 10, and 15 sets at the patient level, respectively. The training dataset was used to train the models, whereas the validation dataset was used to tune the hyperparameters of the models and check the validity of the training process. The test dataset from CenterA and other datasets were completely separated from the training and validation datasets, and were used for evaluating the performance of the models. Specifically, the PDDCA, TCIA, and CenterB datasets are external datasets that were used to evaluate the models for dataset compatibility. Detailed data characteristics of all datasets, including the number and size of slices, pixel spacing, and slice thicknesses, are presented in Table 1.

Table 1. Properties of the datasets.

Dataset	CenterA	PDDCA †	TCIA †	CenterB
No. of sets	171 (Train: 146, Validation: 10, Test: 15)	15	28	15
Acquisition type	MDCT	MDCT	MDCT	CBCT
Target structure	Mandible & Maxilla	OARs	OARs	Mandible & Maxilla
No. of slices	166–450, 208 ± 32	109–263, 154 ± 36	61–110, 93 ± 12	432
Slice size [pixel]	512	576	512	512
Pixel spacing [mm]	0.36–0.49, 0.44 ± 0.03	0.98–1.27, 1.11 ± 0.10	0.94–1.27, 1.04 ± 0.10	0.40
Slice thickness [mm]	0.50–1.04, 0.99 ± 0.07	2.0–3.0, 2.73 ± 0.31	2.50	0.40

† denotes the public dataset. 'No. of slices', 'Pixel spacing', and 'Slice thickness' are indicated as the range, the average, and the standard deviation across the cases or the exact value if they are all same.

CenterA and CenterB datasets include CT images and the corresponding segmentation of the mandible and maxilla provided by the clinical experts of oral and maxillofacial surgery department and orthodontic department, respectively. Targets in CenterA datasets were delineated manually by an expert surgeon (B.Y.H.) from Kyung Hee University Hospital, Seoul, Korea. Ethical approval was received from the institutional review board (IRB) (approval number KH-DT19033) for CenterA dataset. CenterB dataset was built with 15 sets of dental CBCT (i-CAT 17-19TM, Imaging Science International) from Chungang University Hospital, Seoul, Korea (approval number 1922-007-362). Those CT images were segmented by two well-trained biomedical engineers supervised by a clinical expert.

The PDDCA dataset is a public dataset for OAR segmentation in the H&N region of CT images released at the 2015 MICCAI H&N radiotherapy OAR segmentation challenge [8] provided and maintained by Dr. Sharp at Harvard Medical School. The CT scans in the dataset are available via the Cancer Imaging Archive (TCIA) and are originally from the radiation therapy oncology group (RTOG) 0522 study, which includes multi-institutional clinical studies from patients with stage III or IV H&N carcinoma [42]. The dataset consists of 48 H&N CT images with nine OAR structures manually re-segmented by experts for uniform quality and consistency. In the challenge, the dataset was divided into 25 training sets, 10 off-site test sets, and 5 on-site test sets. In this study, we employed 15 test sets with mandible annotation.

The TCIA dataset [43] contains 31 CT scans from TCIA [44] and segmentations for 21 OARs, in which we only used mandible segmentation. They were delineated by an experienced radiographer, with additional peer arbitration by another radiographer and a radiation oncologist. Both the PDDCA and TCIA datasets include a selected part of the Head–Neck Cetuximab open source dataset [45]; owing to different selection criteria and different train/validation/test set division, there are five scans present in both PDDCA and TCIA test sets.

Examples of all datasets are illustrated in Figure 2. CenterA dataset is different from the PDDCA and TCIA datasets in terms of pixel spacing, slice thickness, and the scan range of the CT images. Comparatively, the PDDCA and TCIA datasets include a wider range of bodies that target OARs. CenterB uses cone beam CT (CBCT), which is fundamentally different from multidetector CT (MDCT) datasets, meaning the performance of a model trained with MDCT may be hindered when inferencing CBCT. Generally, seg mentation of CBCT is more laborious and time-consuming than MDCT as the edge of the image is more blurred and noisy. Additionally, CenterB dataset includes many cases with orthognathic surgery or orthodontics, which makes segmentation more difficult owing to the noise caused by surgery plates or orthodontic appliances (Figure 3). By externally testing using datasets, including PDDCA, TCIA, and CenterB datasets, with various characteristics, it was possible to evaluate the compatibility of the models.

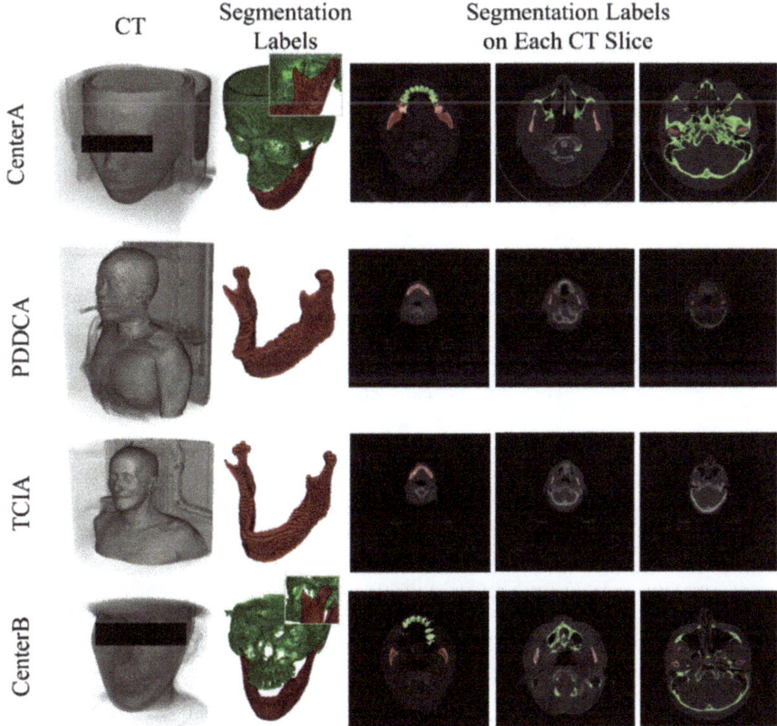

Figure 2. Example cases of the datasets.

All datasets were preprocessed using the same procedure. A threshold of −1000 and 2500 HU was employed for each scan and normalized between zero and one. Both CT scans and segmentation slices were cropped to fit the skull. All CT and segmentation volumes were resampled to be isotropic (512 × 512 × 512). For a fair evaluation, the predicted segmentations were conversely uncropped and resampled into the original spacing and thickness before the evaluation metrics were calculated. For the PDDCA and TCIA datasets, we only used the range of the mandible for the training dataset, while using the entire range of slices for the validation and testing.

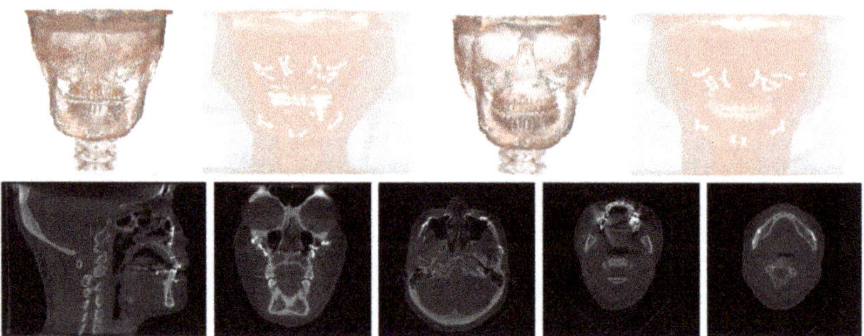

Figure 3. Examples of noises caused by surgical plates, orthodontic device, and dental implants in CenterB dataset, which make it more challenging to delineate the CT images.

Datasets were input to the models as 2.5D [46], in which the input was a volume of images consisting of the target slice and its adjacent slices, and the segmentation map corresponding to the center target slice was produced as an output of the model. This method enables the use of adjacent context information in 3D, whilst lowering the computational power required relative to 3D inputs. The proposed approach is applicable to mandible and maxilla segmentation as the adjacent upper and lower spatial information is important for distinguishing the mandible and maxilla in a slice that appears similar. In this study, the 11 slices, composed of one target slice and five upper and lower slices, were input for one slice of the segmentation map.

2.2. Framework and Network Architectures

The overall framework and detailed architectures of neural networks are displayed in Figure 4. Preprocessed CT scans are input to the neural network as 2.5D, which outputs one segmentation mask map for each target slice. This process was repeated for all slices in each patient scan. Afterwards, the segmented volume for each patient was post-processed.

For the neural network, we applied a hierarchical, parallel, and multi-scale residual (HPMR) block [47] to U-Net to enhance the data compatibility of the CNN model. This block was first designed to enhance the performance of a CNN for landmark localization with limited computational resources. The starting point of the architecture is a residual bottleneck block [39] that enables the stable optimization of a deeper model by assisting the propagation of information both forward and backward, improving performance. The other basis for the architecture is the inception block [48], which concatenates features from parallel paths with different receptive field sizes. Compared to the inception residual block, the HPMR block has a smaller number of parameters with the advantage of a parallel path. Compared to the existing research, we combined HPMR block to U-Net and showed its performance on the segmentation task. We used HPMR block for efficient learning to utilize advances of residual bottleneck block and parallel path with the lower number of parameters compared to using inception blocks.

We compared U-Net with HPMR blocks (HPMR-U-Net) to its base component architecture, U-Net, and U-Net with residual blocks (Res-U-Net) to verify the effects of HPMR blocks. Additionally, modified U-Net (mU-Net) [49] was selected as another state-of-the-art segmentation CNN model for comparison because it requires minimum increase of network parameters. Its residual block is composed of deconvolution and activation operations to pass features to the skip connection of the U-Net adaptively with the object size. mU-Net is designed not only to extract high-level features of large object edges, but also high-level global features of small objects. We hypothesized that the increase in the complexity of model, i.e., the increased number of parameters in the neural network, would hinder the data compatibility of the model. It is well known that overfitting, which impedes the data compatibility of a model, occurs when the number of parameters increases [50].

Therefore, we chose a simpler neural network with lower number of parameters than other state-of-the-art neural networks for comparison.

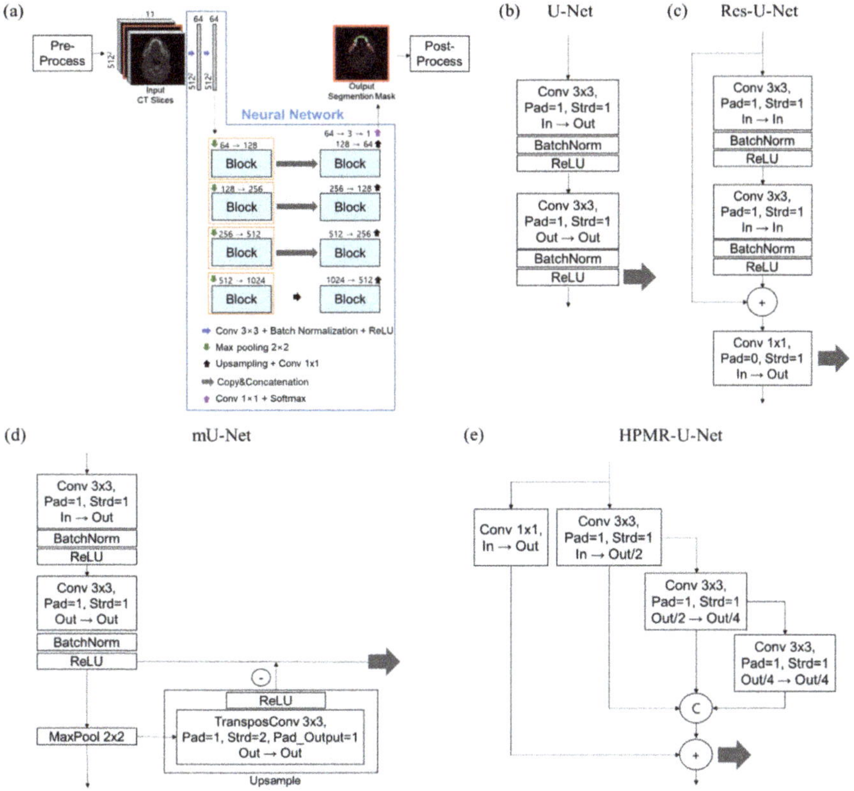

Figure 4. Overall framework and block architectures of neural networks. (**a**) Overall framework. The numbers above the boxes refer to the channel number of the feature maps. The orange boxes for each neural network are represented in (**b–e**), with (**b**) block architecture of U-Net, (**c**) block architecture of Res-U-Net, (**d**) block architecture of mU-Net, and (**e**) block architecture of HPMR-U-Net.

All networks were trained using PyTorch framework in Python under the same conditions for comparison. They were trained with a batch size of 10 for 30 epochs. We employed cross entropy loss as a loss function and Adam optimization with a learning rate of 10^{-5}. Training and evaluation were performed on the computer hardware resources of a Nvidia GeForce RTX 3090 with 24 GB memory and 16 of DIMM DDR4 Synchronous 2666 MHz with 32 GiB in a Linux environment.

2.3. Performance Evaluation

To evaluate the regular segmentation performance of the models, an in-house test was conducted with the separated test portion of CenterA dataset, with the ground truths and output segmentations from the models compared. Additionally, an external test was performed to evaluate the data compatibility in the models. Output segmentations for CT scans in external datasets (PDDCA, TCIA, and CenterB) were obtained and compared with the ground truths. The external test characterizes how the model can be utilized generally in varied data, which is common in clinical settings. In the absence of maxilla segmentations in the PDDCA and TCIA datasets, only mandible segmentations were considered. To quantitatively evaluate the segmentation performance of the models, we used the Dice coefficient (DC), 95% Hausdorff distance (95HD) and average surface distance (ASD) as evaluation metrics. Additionally, we qualitatively evaluated the segmentation results of the models by visualizing them in 3D.

The DC measures the degree of volumetric overlap between two volumes. It is defined as

$$\mathrm{DC} = \frac{2|\mathrm{GT} \cap \mathrm{OUT}|}{|\mathrm{GT}| + |\mathrm{OUT}|}, \tag{1}$$

where GT and OUT are the labeled voxel sets of the manual segmentation ground truth and output segmentation from the model, respectively.

The 95HD and ASD are distance-related metrics, with 95HD being the 95th percentile of the Hausdorff distance (HD) between the GT and OUT points. HD measures the distance of a point in the GT to the nearest point in the OUT. It is defined as

$$\max_{gt \in GT} \min_{out \in OUT} \|gt - out\|. \tag{2}$$

The 95th percentile is used to eliminate the impact of outliers from a small subset of inaccurate points when evaluating the overall segmentation performance. ASD measures the average distance between the GT and the OUT, defined as:

$$\mathrm{ASD} = \frac{1}{2}\left\{\frac{\sum_{out \in OUT} d(out,\ GT)}{|\mathrm{OUT}|} + \frac{\sum_{gt \in GT} d(gt,\ OUT)}{|\mathrm{GT}|}\right\}, \tag{3}$$

where $d(out, GT)$ is the minimum distance of a voxel on OUT to the voxels on GT, and $d(gt, OUT)$ is the minimum distance of voxel gt on GT to the voxels on OUT.

3. Results

Tables 2 and 3 display the calculated evaluation metrics between the ground truths and the model outputs for the in-house and external tests. In the in-house test with the CenterA dataset, although the scores of HPMR-U-Net were not the best among the models, the score differences were lower compared to those for the other datasets. From the result, it can be inferred that the performance of HPMR-U-Net for the CenterA dataset was comparable to that of the other models. In the external tests, the scores of HPMR-U-Net ranked first for all external datasets. The results indicate that HPMR-U-Net has the highest performance among the models in this study for the external datasets. Comparing results among external datasets, the differences in scores were the largest in the CenterB dataset, where CenterB dataset may have the largest characteristic difference in the image obtained by CBCT as compared to CenterA dataset acquired by MDCT.

Table 2. Results of in-house and external tests for mandible segmentation. The best case is bolded.

	In-House Test			**External Test**								
	CenterA			**PDDCA**			**TCIA**			**CenterB**		
	DC [%]	**95HD [mm]**	**ASD [mm]**	**DC [%]**	**95HD [mm]**	**ASD [mm]**	**DC [%]**	**95HD [mm]**	**ASD [mm]**	**DC [%]**	**95HD [mm]**	**ASD [mm]**
U-Net	98.3 ± 0.4	0.4 ± 0.1	**0.0 ± 0.0**	63.4 ± 20.2	7.3 ± 5.8	1.8 ± 3.5	62.8 ± 26.3	9.6 ± 11.8	3.2 ± 7.9	61.2 ± 17.9	33.7 ± 26.1	4.1 ± 4.0
Res-U-Net	98.2 ± 0.4	0.4 ± 0.1	0.1 ± 0.0	51.3 ± 20.1	13.5 ± 12.3	2.0 ± 1.8	46.3 ± 25.5	18.0 ± 19.8	6.5 ± 13.5	48.5 ± 13.1	28.8 ± 20.6	4.1 ± 3.3
mU-Net	**98.4 ± 0.3**	**0.4 ± 0.0**	**0.0 ± 0.0**	72.3 ± 21.6	5.6 ± 6.4	1.5 ± 3.5	71.4 ± 27.8	8.4 ± 12.9	2.5 ± 5.0	63.6 ± 14.7	22.5 ± 18.9	2.6 ± 2.2
HPMR-U-Net	97.4 ± 0.4	0.4 ± 0.1	0.1 ± 0.0	**86.5 ± 3.9**	**1.8 ± 1.3**	**0.2 ± 0.1**	**86.4 ± 6.2**	**2.8 ± 7.7**	**0.3 ± 0.7**	**77.7 ± 4.1**	**3.4 ± 0.6**	**0.7 ± 0.2**

Table 3. Results of in-house and external tests for maxilla segmentation. The best case is bolded.

	In-House Test			External Test		
	CenterA			CenterB		
	DC [%]	95HD [mm]	ASD [mm]	DC [%]	95HD [mm]	ASD [mm]
U-Net	**96.5 ± 0.8**	0.4 ± 0.1	0.1 ± 0.0	75.0 ± 5.7	9.0 ± 8.8	1.1 ± 0.8
Res-U-Net	96.2 ± 0.8	0.4 ± 0.1	0.1 ± 0.0	67.6 ± 12.3	17.1 ± 18.1	2.5 ± 3.4
mU-Net	96.5 ± 0.7	**0.4 ± 0.0**	0.1 ± 0.0	75.9 ± 5.1	8.6 ± 7.9	1.0 ± 0.7
HPMR-U-Net	90.2 ± 19.5	0.5 ± 0.1	0.1 ± 0.0	**82.8 ± 3.2**	**2.7 ± 1.6**	**0.4 ± 0.2**

Figures 5–8 show 3D rendered ground truths and the highest DC cases of the output segmentations converted to isosurfaces from volumes for each dataset. Corresponding to the results of the quantitative tests, the ground truth and the outputs for the CenterA dataset are similar for all models, as shown in Figure 5. By contrast, for the external datasets, there are visually noticeable differences in the output segmentations of HPMR-U-Net and other models.

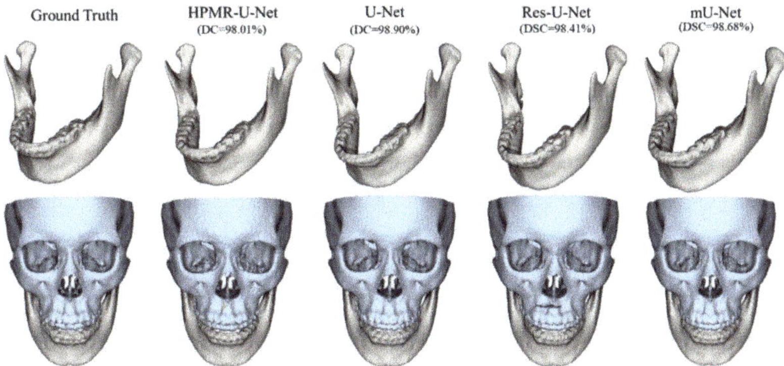

Figure 5. Sample case in the CenterA dataset. White and blue indicate the mandible and maxilla, respectively.

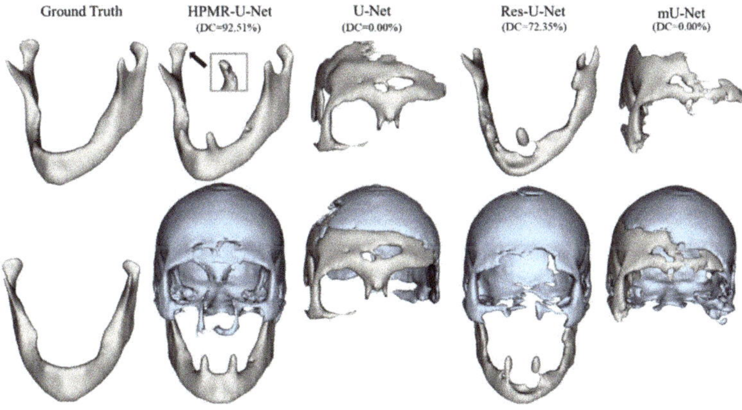

Figure 6. Sample case in the PDDCA dataset. White and blue indicate the mandible and maxilla, respectively. PDDCA has no maxilla ground truth.

There were prominent decreases in quality of segmentations from other models for the external datasets. For the PDDCA dataset in Figure 6, the DC scores for the mandible of U-Net and mU-Net were 0.0%, as the model could not find the mandible at all, that is, they were unable to distinguish between the mandible and the maxilla. There were also many losses in the segmentations of the mandible and maxilla in the outputs of Res-U-Net. By contrast, the outputs of HPMR-U-Net were more intact and closer to the ground truth. As the teeth were included in the CenterA dataset segmentations that were used in training, the teeth were also segmented, despite not being in the ground truth. The results for the TCIA dataset in Figure 7 are also similar to those of the PDDCA dataset. U-Net and Res-U-Net failed to segment the mandible, which resulted in a 0.0% DC. Additionally, mU-Net included many portions of the maxilla in the mandible output and lost a large portion of the segmentations. However, HPMR-U-Net exhibited high performance with a DC of 91.7%. As displayed in Figure 8 for the CenterB dataset, HPMR-U-Net also showed the highest performance among the models, with many lost sections in the other models. Furthermore, the other models were more unable to accurately separate the mandible and maxilla compared to HPMR-U-Net.

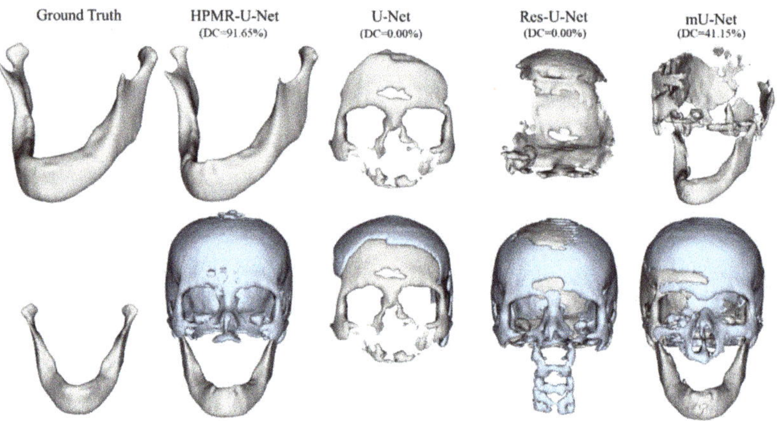

Figure 7. Sample case in the TCIA dataset. White and blue indicate mandible and maxilla, respectively. TCIA has no maxilla ground truth.

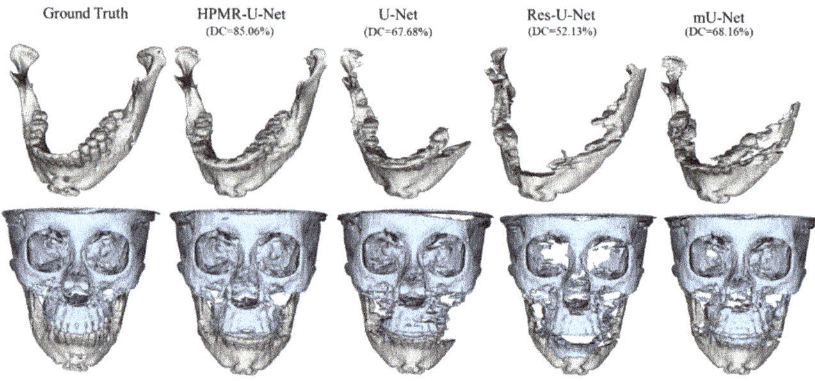

Figure 8. Sample case in CenterB dataset. White and blue indicate the mandible and maxilla, respectively.

Figure 9 shows rendered color maps in 3D for the distance from the ground truths to the output segmentations of the best DC case for the mandible in CenterA dataset to thoroughly examine the differences among the model outputs for this dataset. There were no significant differences, but the distances in the mandibular foramen were slightly different. This part is challenging to segment accurately owing to its small size, and the distance was less in the outputs of HPMR-U-Net than the other models.

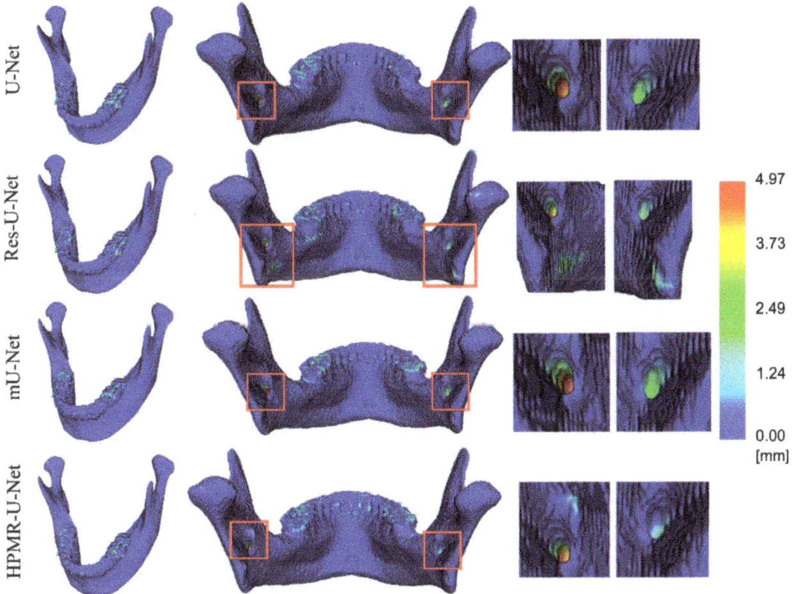

Figure 9. Color maps of surface distance from the ground truths to the output segmentations of the best dice coefficient case in the CenterA dataset for the mandible. The pieces of images on the right side are enlargement of mandibular foramen.

4. Discussion & Conclusions

The four neural networks compared in this research exhibited similar performance in the CenterA dataset, which was the domain used for training. Among other neural networks, U-Net and Res-U-Net were considered for the comparison because U-Net is a basic component of HPMR-U-Net, in which showing a difference would represent that HPMR block is effective compared to other basic architectures. Additionally, mU-Net was selected for the comparison as a state-of-the-art neural network for segmentation. We chose comparably simple neural networks because we hypothesized that the data compatibility of the more complex model with larger number of parameters would be worse because of overfitting. CenterA dataset was set as the train dataset because it was MDCT datasets in which easier to make ground truths than CBCT. With training with a dataset easier to constitute, we aimed to show the performance for other institutional MDCTs and CBCTs.

PDDCA and TCIA datasets were used for examples of MDCT and CenterB dataset for CBCT. For the external datasets of PDDCA, TCIA, and CenterB, HPMR-U-Net displayed significantly higher performance compared to the U-Net, Res-U-Net, and mU-Net models in both quantitative and qualitative evaluations.

All networks produced comparable results for data from the same cohort of the training dataset; however, they exhibited different results for data from out of the training dataset cohort. While the performances of other networks were degraded in the external datasets, HPMR-U-Net produced segmentation of the mandible and maxilla similar to the ground truths. From these results, HPMR-U-Net infers a high data compatibility for mandible and maxilla features in CT images.

The assumed differences in the data cohorts were reflected in the results. For the PDDCA and TCIA datasets, the performance degraded significantly, and the mandible and maxilla were not classified accurately. This is due to their slice thickness being different from that of CenterA dataset, even though they are MDCT. The inter-slice information is important to classify a pixel in a slice as the mandible or maxilla. The results for CenterB dataset were the worst among the external datasets for all models. The segmentation of CenterB dataset is more challenging as it is CBCT, which is not only different from the in-house dataset, but also contains more noise. Additionally, CenterB dataset contains variances in anatomical structure caused by surgeries and noise from surgical plates, orthodontic device, and dental implants (Figure 3). It is remarkable that the score difference between HPMR-U-Net and other models is significant for CenterB. For CBCT, which is a different image protocol than MDCT that was used to train, there was a significant degradation of performance in other models, but minimal degradation in HPMR-U-Net. This demonstrates that HPMR-U-Net is more robust than other models to various data domains that may be different from the training data.

We assume that one of the reasons for the better performance of HPMR-U-Net compared to Res-U-Net is the number of parameters. The higher the complexity of the hypothesis space of the deep neural network, the worse is the generalizability, according to the principle of Occam's razor [51]. The number of parameters in HPMR-U-Net is 12,042,179, which is smaller than Res-U-Net with 17,118,019, U-Net with 28,959,299, and mU-Net with 35,230,019. The HPMR block could efficiently decrease the overall number of parameters, which as a result could enhance the generalizability of the model while maintaining its segmentation performance.

In future work, an attempt will be made to improve the performance in external datasets for actual clinical applications when the neural network is trained with only one data domain. Additionally, the structure of the neural network with residual connections and HPMR block can be analyzed theoretically to establish the reason for the greater generalizability, which may lead to the design of a stronger neural network for generalization.

In this study, we applied deep learning to accurately segment the mandible and maxilla from CT and improve the compatibility in the segmentation model. To achieve this, we utilized HPMR-U-Net and compared its results with those of U-Net, Res-U-Net, and mU-Net with in-house and external tests. The results show that the segmentation performance of HPMR-U-Net in the in-house test dataset was comparable to that of the other models. In particular, the data compatibility of HPMR-U-Net was superior to other models in the external datasets of PDDCA, TCIA, and CenterB, which have varying properties such as image protocol, pixel spacing, slice thickness, and target range.

Author Contributions: Conceptualization, S.P., E.S., Y.K., J.-W.L. and H.S.; methodology, S.P., H.K. and H.S.; software, S.P., H.K. and E.S.; validation, S.P.; formal analysis, S.P.; investigation, S.P.; resources, Y.K. and H.S.; data curation, B.-Y.H. and J.-W.L.; writing—original draft preparation, S.P.; writing—review and editing, S.P., H.K., Y.K., J.-W.L. and H.S.; visualization, S.P. and H.K.; supervision, Y.K., J.-W.L. and H.S.; project administration, Y.K., J.-W.L. and H.S.; funding acquisition, Y.K., J.-W.L. and H.S. All authors have read and agreed to the published version of the manuscript.

Funding: This research was supported by KIST Institutional Program (grant number: 2E31158) and the Korea Medical Device Development Fund grant funded by the Korea government (the Ministry of Science and ICT, the Ministry of Trade, Industry and Energy, the Ministry of Health & Welfare, the Ministry of Food and Drug Safety) (Project Number: 9991006675, 202011A02, KMDF_PR_20200901_0002). In addition, this research was supported by a grant of the Korea Health Technology R&D Project through the Korea Health Industry Development Institute (KHIDI), funded by the Ministry of Health & Welfare, Republic of Korea (grant number: HI18C1224).

Institutional Review Board Statement: The study was approved by the Institutional Review Board of Chungang University Hospital, Seoul, Korea (1922-007-362) and Kyung Hee University Hospital, Seoul, Korea (KH-DT19033).

Informed Consent Statement: Patient consent was waived because it was stated that the data can be used retrospectively without DICOM tags in IRB approval.

Data Availability Statement: Data sharing is not applicable to this article.

Acknowledgments: The authors wish to express their thanks for the support of Uilyong Lee (Department Oral and maxillofacial Surgery, Chung-Ang University) in the collection of datasets.

Conflicts of Interest: The authors declare no conflict of interest.

References

1. Qiu, B.; Guo, J.; Kraeima, J.; Glas, H.H.; Borra, R.J.; Witjes, M.J.; van Ooijen, P.M. Automatic Segmentation of the Mandible from Computed Tomography Scans for 3D Virtual Surgical Planning Using the Convolutional Neural Network. *Phys. Med. Biol.* **2019**, *64*, 175020. [CrossRef] [PubMed]
2. Wang, L.; Chen, K.C.; Gao, Y.; Shi, F.; Liao, S.; Li, G.; Shen, S.G.; Yan, J.; Lee, P.K.; Chow, B. Automated Bone Segmentation from Dental CBCT Images Using Patch-based Sparse Representation and Convex Optimization. *Med. Phys.* **2014**, *41*, 043503. [CrossRef] [PubMed]
3. Kodym, O.; Španěl, M.; Herout, A. Segmentation of Head and Neck Organs at Risk Using Cnn with Batch Dice Loss. In Proceedings of the German Conference on Pattern Recognition; Springer: Stuttgart, Germany, 2018; pp. 105–114.
4. Byrne, N.; Velasco Forte, M.; Tandon, A.; Valverde, I.; Hussain, T. A Systematic Review of Image Segmentation Methodology, Used in the Additive Manufacture of Patient-Specific 3D Printed Models of the Cardiovascular System. *JRSM Cardiovasc. Dis.* **2016**, *5*, 2048004016645467. [CrossRef] [PubMed]
5. Huff, T.J.; Ludwig, P.E.; Zuniga, J.M. The Potential for Machine Learning Algorithms to Improve and Reduce the Cost of 3-Dimensional Printing for Surgical Planning. *Expert Rev. Med. Devices* **2018**, *15*, 349–356. [CrossRef]
6. Wang, Z.; Wei, L.; Wang, L.; Gao, Y.; Chen, W.; Shen, D. Hierarchical Vertex Regression-Based Segmentation of Head and Neck CT Images for Radiotherapy Planning. *IEEE Trans. Image Process.* **2017**, *27*, 923–937. [CrossRef]
7. Qiu, B.; Guo, J.; Kraeima, J.; Glas, H.H.; Borra, R.J.; Witjes, M.J.; Ooijen, P.M.V. Recurrent Convolutional Neural Networks for Mandible Segmentation from Computed Tomography. *arXiv* **2020**, arXiv:2003.06486.
8. Raudaschl, P.F.; Zaffino, P.; Sharp, G.C.; Spadea, M.F.; Chen, A.; Dawant, B.M.; Albrecht, T.; Gass, T.; Langguth, C.; Lüthi, M. Evaluation of Segmentation Methods on Head and Neck CT: Auto-segmentation Challenge 2015. *Med. Phys.* **2017**, *44*, 2020–2036. [CrossRef]
9. Chen, A.; Dawant, B. A Multi-Atlas Approach for the Automatic Segmentation of Multiple Structures in Head and Neck CT Images. *MIDAS J.* **2015**. [CrossRef]
10. Mannion-Haworth, R.; Bowes, M.; Ashman, A.; Guillard, G.; Brett, A.; Vincent, G. Fully Automatic Segmentation of Head and Neck Organs Using Active Appearance Models. *MIDAS J.* **2015**. [CrossRef]
11. Han, X.; Hoogeman, M.S.; Levendag, P.C.; Hibbard, L.S.; Teguh, D.N.; Voet, P.; Cowen, A.C.; Wolf, T.K. *Atlas-Based Auto-Segmentation of Head and Neck CT Images*; Springer: Berlin/Heidelberg, Germany, 2008; pp. 434–441.
12. Linares, O.C.; Bianchi, J.; Raveli, D.; Neto, J.B.; Hamann, B. Mandible and Skull Segmentation in Cone Beam Computed Tomography Using Super-Voxels and Graph Clustering. *Vis. Comput.* **2019**, *35*, 1461–1474.
13. Zhu, W.; Huang, Y.; Zeng, L.; Chen, X.; Liu, Y.; Qian, Z.; Du, N.; Fan, W.; Xie, X. AnatomyNet: Deep Learning for Fast and Fully Automated Whole-volume Segmentation of Head and Neck Anatomy. *Med. Phys.* **2019**, *46*, 576–589. [CrossRef] [PubMed]
14. Cootes, T.F.; Edwards, G.J.; Taylor, C.J. Active Appearance Models. *IEEE Trans. Pattern Anal. Mach. Intell.* **2001**, *23*, 681–685. [CrossRef]
15. Fritscher, K.D.; Peroni, M.; Zaffino, P.; Spadea, M.F.; Schubert, R.; Sharp, G. Automatic Segmentation of Head and Neck CT Images for Radiotherapy Treatment Planning Using Multiple Atlases, Statistical Appearance Models, and Geodesic Active Contours. *Med. Phys.* **2014**, *41*, 051910. [CrossRef] [PubMed]
16. Albrecht, T.; Gass, T.; Langguth, C.; Lüthi, M. Multi Atlas Segmentation with Active Shape Model Refinement for Multi-Organ Segmentation in Head and Neck Cancer Radiotherapy Planning. *MIDAS J.* **2015**. [CrossRef]
17. Aghdasi, N.; Li, Y.; Berens, A.; Moe, K.; Hannaford, B. Automatic Mandible Segmentation on CT Images Using Prior Anatomical Knowledge. *MIDAS J.* **2016**. [CrossRef]
18. Chuang, Y.J.; Doherty, B.M.; Adluru, N.; Chung, M.K.; Vorperian, H.K. A Novel Registration-Based Semi-Automatic Mandible Segmentation Pipeline Using Computed Tomography Images to Study Mandibular Development. *J. Comput. Assist. Tomogr.* **2018**, *42*, 306. [CrossRef]
19. Shen, D.; Wu, G.; Suk, H.-I. Deep Learning in Medical Image Analysis. *Annu. Rev. Biomed. Eng.* **2017**, *19*, 221–248. [CrossRef]
20. Ibragimov, B.; Xing, L. Segmentation of Organs-at-risks in Head and Neck CT Images Using Convolutional Neural Networks. *Med. Phys.* **2017**, *44*, 547–557. [CrossRef]

21. Tong, N.; Gou, S.; Yang, S.; Ruan, D.; Sheng, K. Fully Automatic Multi-organ Segmentation for Head and Neck Cancer Radiotherapy Using Shape Representation Model Constrained Fully Convolutional Neural Networks. *Med. Phys.* **2018**, *45*, 4558–4567. [CrossRef]
22. Ronneberger, O.; Fischer, P.; Brox, T. *U-Net: Convolutional Networks for Biomedical Image Segmentation*; Springer: Berlin/Heidelberg, Germany, 2015; pp. 234–241.
23. Liu, L.; Cheng, J.; Quan, Q.; Wu, F.-X.; Wang, Y.-P.; Wang, J. A Survey on U-Shaped Networks in Medical Image Segmentations. *Neurocomputing* **2020**, *409*, 244–258. [CrossRef]
24. Wang, Y.; Zhao, L.; Wang, M.; Song, Z. Organ at Risk Segmentation in Head and Neck Ct Images Using a Two-Stage Segmentation Framework Based on 3D U-Net. *IEEE Access* **2019**, *7*, 144591–144602. [CrossRef]
25. Rundo, L.; Han, C.; Nagano, Y.; Zhang, J.; Hataya, R.; Militello, C.; Tangherloni, A.; Nobile, M.S.; Ferretti, C.; Besozzi, D.; et al. USE-Net: Incorporating Squeeze-and-Excitation Blocks into U-Net for Prostate Zonal Segmentation of Multi-Institutional MRI Datasets. *Neurocomputing* **2019**, *365*, 31–43. [CrossRef]
26. Schlemper, J.; Oktay, O.; Schaap, M.; Heinrich, M.; Kainz, B.; Glocker, B.; Rueckert, D. Attention Gated Networks: Learning to Leverage Salient Regions in Medical Images. *Med. Image Anal.* **2019**, *53*, 197–207. [CrossRef] [PubMed]
27. Yeung, M.; Sala, E.; Schönlieb, C.-B.; Rundo, L. Focus U-Net: A Novel Dual Attention-Gated CNN for Polyp Segmentation during Colonoscopy. *Comput. Biol. Med.* **2021**, *137*, 104815. [CrossRef] [PubMed]
28. Liang, X.; Nguyen, D.; Jiang, S.B. Generalizability Issues with Deep Learning Models in Medicine and Their Potential Solutions: Illustrated with Cone-Beam Computed Tomography (CBCT) to Computed Tomography (CT) Image Conversion. *Mach. Learn. Sci. Technol.* **2020**, *2*, 015007. [CrossRef]
29. Qiu, B.; van der Wel, H.; Kraeima, J.; Glas, H.H.; Guo, J.; Borra, R.J.H.; Witjes, M.J.H.; van Ooijen, P.M.A. Automatic Segmentation of Mandible from Conventional Methods to Deep Learning—A Review. *J. Pers. Med.* **2021**, *11*, 629. [CrossRef]
30. Hesse, L.S.; Kuling, G.; Veta, M.; Martel, A.L. Intensity Augmentation to Improve Generalizability of Breast Segmentation Across Different MRI Scan Protocols. *IEEE Trans. Biomed. Eng.* **2021**, *68*, 759–770. [CrossRef]
31. Bosniak, M.A. The Current Radiological Approach to Renal Cysts. *Radiology* **1986**, *158*, 1–10. [CrossRef]
32. Bluemke, D.A.; Moy, L.; Bredella, M.A.; Ertl-Wagner, B.B.; Fowler, K.J.; Goh, V.J.; Halpern, E.F.; Hess, C.P.; Schiebler, M.L.; Weiss, C.R. Assessing Radiology Research on Artificial Intelligence: A Brief Guide for Authors, Reviewers, and Readers—From the *Radiology* Editorial Board. *Radiology* **2020**, *294*, 487–489. [CrossRef]
33. Kim, H.; Shim, E.; Park, J.; Kim, Y.-J.; Lee, U.; Kim, Y. Web-Based Fully Automated Cephalometric Analysis by Deep Learning. *Comput. Methods Programs Biomed.* **2020**, *194*, 105513. [CrossRef]
34. Tao, Q.; Yan, W.; Wang, Y.; Paiman, E.H.M.; Shamonin, D.P.; Garg, P.; Plein, S.; Huang, L.; Xia, L.; Sramko, M.; et al. Deep Learning–Based Method for Fully Automatic Quantification of Left Ventricle Function from Cine MR Images: A Multivendor, Multicenter Study. *Radiology* **2019**, *290*, 81–88. [CrossRef] [PubMed]
35. B, S.; R, N. Transfer Learning Based Automatic Human Identification Using Dental Traits- An Aid to Forensic Odontology. *J. Forensic Leg. Med.* **2020**, *76*, 102066. [CrossRef] [PubMed]
36. Ghafoorian, M.; Mehrtash, A.; Kapur, T.; Karssemeijer, N.; Marchiori, E.; Pesteie, M.; Guttmann, C.R.G.; de Leeuw, F.-E.; Tempany, C.M.; van Ginneken, B.; et al. Transfer Learning for Domain Adaptation in MRI: Application in Brain Lesion Segmentation. In *International Conference on Medical Image Computing and Computer-Assisted Intervention*; Springer: Berlin/Heidelberg, Germany, 2017; pp. 516–524. [CrossRef]
37. Lee, K.-S.; Jung, S.-K.; Ryu, J.-J.; Shin, S.-W.; Choi, J. Evaluation of Transfer Learning with Deep Convolutional Neural Networks for Screening Osteoporosis in Dental Panoramic Radiographs. *J. Clin. Med.* **2020**, *9*, 392. [CrossRef] [PubMed]
38. Weiss, K.; Khoshgoftaar, T.M.; Wang, D. A Survey of Transfer Learning. *J. Big Data* **2016**, *3*, 9. [CrossRef]
39. He, K.; Zhang, X.; Ren, S.; Sun, J. Deep Residual Learning for Image Recognition. In Proceedings of the IEEE Conference on Computer Vision and Pattern Recognition, Las Vegas, NV, USA, 27–30 June 2016; pp. 770–778.
40. Frei, S.; Cao, Y.; Gu, Q. Algorithm-Dependent Generalization Bounds for Overparameterized Deep Residual Networks. *arXiv* **2019**, arXiv:1910.02934.
41. Huang, K.; Tao, M.; Wang, Y.; Zhao, T. Why Do Deep Residual Networks Generalize Better than Deep Feedforward Networks? —A Neural Tangent Kernel Perspective. 2020, 12. *arXiv* **2020**, arXiv:2002.06262.
42. Ang, K.K.; Zhang, Q.; Rosenthal, D.I.; Nguyen-Tan, P.F.; Sherman, E.J.; Weber, R.S.; Galvin, J.M.; Bonner, J.A.; Harris, J.; El-Naggar, A.K. Randomized Phase III Trial of Concurrent Accelerated Radiation plus Cisplatin with or without Cetuximab for Stage III to IV Head and Neck Carcinoma: RTOG 0522. *J. Clin. Oncol.* **2014**, *32*, 2940. [CrossRef]
43. Nikolov, S.; Blackwell, S.; Zverovitch, A.; Mendes, R.; Livne, M.; De Fauw, J.; Patel, Y.; Meyer, C.; Askham, H.; Romera-Paredes, B. Deep Learning to Achieve Clinically Applicable Segmentation of Head and Neck Anatomy for Radiotherapy. *arXiv* **2018**, arXiv:1809.04430.
44. Clark, K.; Vendt, B.; Smith, K.; Freymann, J.; Kirby, J.; Koppel, P.; Moore, S.; Phillips, S.; Maffitt, D.; Pringle, M. The Cancer Imaging Archive (TCIA): Maintaining and Operating a Public Information Repository. *J. Digit. Imaging* **2013**, *26*, 1045–1057. [CrossRef]
45. Bosch, W.R.; Straube, W.L.; Matthews, J.W.; Purdy, J.A. Data from Head-Neck_cetuximab. *Cancer Imaging Arch.* **2015**, *10*, K9.
46. Han, X. Automatic Liver Lesion Segmentation Using A Deep Convolutional Neural Network Method. *Med. Phys.* **2017**, *44*, 1408–1419. [CrossRef] [PubMed]

47. Bulat, A.; Tzimiropoulos, G. Binarized Convolutional Landmark Localizers for Human Pose Estimation and Face Alignment with Limited Resources. In Proceedings of the IEEE International Conference on Computer Vision, Venice, Italy, 22–29 October 2017; pp. 3706–3714.
48. Szegedy, C.; Liu, W.; Jia, Y.; Sermanet, P.; Reed, S.; Anguelov, D.; Erhan, D.; Vanhoucke, V.; Rabinovich, A. Going Deeper with Convolutions. In Proceedings of the Proceedings of the IEEE Conference on Computer Vision and Pattern Recognition, Boston, MA, USA, 7–12 June 2015; pp. 1–9.
49. Seo, H.; Huang, C.; Bassenne, M.; Xiao, R.; Xing, L. Modified U-Net (MU-Net) with Incorporation of Object-Dependent High Level Features for Improved Liver and Liver-Tumor Segmentation in CT Images. *IEEE Trans. Med. Imaging* **2019**, *39*, 1316–1325. [CrossRef] [PubMed]
50. Gupta, S.; Gupta, R.; Ojha, M.; Singh, K.P. A Comparative Analysis of Various Regularization Techniques to Solve Overfitting Problem in Artificial Neural Network. In Proceedings of the Data Science and Analytics; Panda, B., Sharma, S., Roy, N.R., Eds.; Springer: Singapore, 2018; pp. 363–371.
51. He, F.; Liu, T.; Tao, D. Why ResNet Works? Residuals Generalize. *arXiv* **2019**, arXiv:1904.01367.

MDPI

Article

On Unsupervised Methods for Medical Image Segmentation: Investigating Classic Approaches in Breast Cancer DCE-MRI

Carmelo Militello [1,*], Andrea Ranieri [2], Leonardo Rundo [3,4,5], Ildebrando D'Angelo [6,7], Franco Marinozzi [2], Tommaso Vincenzo Bartolotta [6,8], Fabiano Bini [2] and Giorgio Russo [1]

1 Institute of Molecular Bioimaging and Physiology, Italian National Research Council (IBFM-CNR), 90015 Cefalu, Italy; giorgio.russo@ibfm.cnr.it
2 Department of Mechanical and Aerospace Engineering, Sapienza University of Rome, 00184 Roma, Italy; ranieri.1837003@studenti.uniroma1.it (A.R.); franco.marinozzi@uniroma1.it (F.M.); fabiano.bini@uniroma1.it (F.B.)
3 Department of Radiology, University of Cambridge, Cambridge CB2 0QQ, UK; lr495@cam.ac.uk
4 Cancer Research UK Cambridge Centre, Cambridge CB2 0RE, UK
5 Department of Information and Electrical Engineering and Applied Mathematics (DIEM), University of Salerno, 84084 Fisciano, Italy
6 Department of Radiology, Fondazione Istituto "G. Giglio", 90015 Cefalu, Italy; ildebrando.dangelo@hsrgiglio.it (I.D.); tommasovincenzo.bartolotta@unipa.it (T.V.B.)
7 Breast Unit, Fondazione Istituto "G. Giglio", 90015 Cefalu, Italy
8 Section of Radiology—Department of Biomedicine, Neuroscience and Advanced Diagnostics (BiND), University Hospital "Paolo Giaccone", 90127 Palermo, Italy
* Correspondence: carmelo.militello@ibfm.cnr.it

Abstract: Unsupervised segmentation techniques, which do not require labeled data for training and can be more easily integrated into the clinical routine, represent a valid solution especially from a clinical feasibility perspective. Indeed, large-scale annotated datasets are not always available, undermining their immediate implementation and use in the clinic. Breast cancer is the most common cause of cancer death in women worldwide. In this study, breast lesion delineation in Dynamic Contrast Enhanced MRI (DCE-MRI) series was addressed by means of four popular unsupervised segmentation approaches: Split-and-Merge combined with Region Growing (SMRG), k-means, Fuzzy C-Means (FCM), and spatial FCM (sFCM). They represent well-established pattern recognition techniques that are still widely used in clinical research. Starting from the basic versions of these segmentation approaches, during our analysis, we identified the shortcomings of each of them, proposing improved versions, as well as developing ad hoc pre- and post-processing steps. The obtained experimental results, in terms of area-based—namely, Dice Index (DI), Jaccard Index (JI), Sensitivity, Specificity, False Positive Ratio (FPR), False Negative Ratio (FNR)—and distance-based metrics—Mean Absolute Distance (MAD), Maximum Distance (MaxD), Hausdorff Distance (HD)—encourage the use of unsupervised machine learning techniques in medical image segmentation. In particular, fuzzy clustering approaches (namely, FCM and sFCM) achieved the best performance. In fact, for area-based metrics, they obtained DI = 78.23% ± 6.50 (sFCM), JI = 65.90% ± 8.14 (sFCM), sensitivity = 77.84% ± 8.72 (FCM), specificity = 87.10% ± 8.24 (sFCM), FPR = 0.14 ± 0.12 (sFCM), and FNR = 0.22 ± 0.09 (sFCM). Concerning distance-based metrics, they obtained MAD = 1.37 ± 0.90 (sFCM), MaxD = 4.04 ± 2.87 (sFCM), and HD = 2.21 ± 0.43 (FCM). These experimental findings suggest that further research would be useful for advanced fuzzy logic techniques specifically tailored to medical image segmentation.

Keywords: medical image segmentation; breast cancer; pattern recognition; machine learning; clinical feasibility; magnetic resonance imaging; computer-assisted segmentation

Citation: Militello, C.; Ranieri, A.; Rundo, L.; D'Angelo, I.; Marinozzi, F.; Bartolotta, T.V.; Bini, F.; Russo, G. On Unsupervised Methods for Medical Image Segmentation: Investigating Classic Approaches in Breast Cancer DCE-MRI. *Appl. Sci.* **2022**, *12*, 162. https://doi.org/10.3390/app12010162

Academic Editor: Syoji Kobashi

Received: 10 November 2021
Accepted: 19 December 2021
Published: 24 December 2021

1. Introduction

The use of advanced imaging technologies has significantly improved the quality of medical care delivered to patients, allowing medical imaging to be an essential part of today's healthcare system [1]. In fact, medical imaging comprises techniques for acquiring images that convey detailed information about the anatomy and physiology of the imaged organs [2]. Moreover, many imaging-enabled tools were developed, supporting clinicians in several tasks of the care process: assisted segmentation [3,4], diagnosis support [5], treatment response assessment [6], radiomic analyses [7,8]. For these reasons, computer-assisted image analysis is considered an essential instrument in the clinical workflow [9,10].

Despite the technological progress characterizing the modern era, some medical tasks, such as image annotation, are still performed manually, often via time-consuming and operator-dependent procedures. As an example, the Gross Tumor Volume (GTV) segmentation for radiotherapy treatments is usually delineated by means of a fully manual procedure [11,12]. Considering that dozens of slices have to be contoured, this manual process is extremely time-consuming. Moreover, operator dependence is critical in terms of result reproducibility. Indeed, these manual procedures are strongly dependent on clinician's knowledge and experience: this means that a remarkable intra- and inter-operator variability can seriously affect the segmentation and quantification results.

In this scenario, computer-assisted approaches (automatic or semi-automatic) allow us to mitigate some of the typical drawbacks of manual procedures. Semi-automatic segmentation techniques involve a minimal level of user interactions and exploit automated algorithms to produce accurate and repeatable results. User interaction, for example, may involve the selection of an approximate initial ROI, which is subsequently used to segment the image [13]. As a consequence, semi-automatic approaches provide more reproducible measurements—compared to fully-manual ones—with a significant reduction of the segmentation time. In particular, fully-automatic segmentation approaches do not require any user interaction. Most existing fully-automatic approaches exploit machine learning or deep learning techniques—such as Support Vector Machines (SVMs) or deep Convolutional Neural Networks (CNNs)—to successfully handle the variability characterizing biomedical data [14,15]. In this scenario, supervised learning techniques are, in general, more complex since they require high computation times and a large amount of labeled data for training. Furthermore, it is important to point out that—from a clinical feasibility perspective—machine learning approaches, which do not require training, are advantageous: the amount of labeled data needed to adequately train and evaluate the approaches based on deep learning is not always available [16,17].

The aim of this work is to show the potential of unsupervised pattern recognition techniques, which do not require training and can be more easily integrated into care routine, especially from a clinical feasibility perspective. For this reason, supervised approaches—such as CNNs—were not treated here. As a relevant case study, we consider the segmentation of contrast-enhancing masses on DCE-MRI. It is worth noting that, in [18], classical unsupervised techniques, both automatic and semi-automatic, allowed us to obtain results comparable or superior to the deep learning approaches.

From experimental evidence, it is possible to observe that fuzzy clustering techniques significantly outperformed direct region detection approaches (i.e., split-and-merge and region growing) and crisp k-means. FCM and sFCM obtained comparable results, although the integration of spatial information into the sFCM allowed for the best performance. Therefore, the explicit management of segmentation uncertainty via multiple degrees of class memberships, along with spatial information, represented the best computational framework for the problem at hand. The main contributions of this study are:

- unsupervised segmentation methods, based on classic pattern recognition techniques, can still provide an effective solution despite the prevalence of supervised approaches, such as deep CNNs;
- the use of traditional unsupervised approaches—requiring no training—is an advantage in terms of of immediate clinical feasibility;
- fuzzy clustering techniques significantly outperformed direct region detection approaches and crisp k-means;
- the integration of spatial information into the FCM algorithm achieved the best performance;

- with the goal of providing a guide for beginners, as well as possibly enabling new future extensions from other researchers, this study provides all the technical information needed to understand both the functioning of each of the studied algorithms and the implemented workflow.

The remainder of this work is organized as follows. Section 2 introduces the theoretical background about unsupervised segmentation approaches, focusing on the algorithms used in this study. A detailed description of the performed analysis and of the implemented processing steps is proposed in Section 3, where the exploited DCE-MRI dataset is also described. Section 4 formulates the metrics used to evaluate the performance of the analyzed approaches. Section 5 illustrates the experimental results, including a discussion about the comparison of the proposed techniques. Finally, discussion and conclusions are provided in Section 6.

2. Theoretical Background

A number of algorithms and techniques for image segmentation have been developed and implemented over the years, and a large amount of literature papers about this non-trivial task were proposed. The aim of this section is to present a brief overview about theoretical notions of literature approaches from which this work drawn inspiration. For this reason, we decided for a comprehensive description, to provide the reader with all the technical information needed to understand the functioning of each of the investigated algorithms, as well as the implemented workflow.

The segmentation involves the image partitioning into homogeneous and meaningful sub-regions. By a formal point of view, the segmentation of an image $\mathcal{I}$ involves the identification of a finite set of regions $\mathcal{R}_1, \mathcal{R}_2, \ldots, \mathcal{R}_N$ as in Equation (1):

$$\bigcup_{i=1}^{N} \mathcal{R}_i = \mathcal{I}' \tag{1}$$

with the following constraints:

$$\mathcal{R}_i \cap \mathcal{R}_j = \emptyset, \quad \text{for} \quad i \neq j, \tag{2}$$

$$\mathsf{P}(\mathcal{R}_i) = \mathsf{TRUE}, \quad \text{for} \quad i = 1, 2, \ldots, N, \tag{3}$$

$$\mathsf{P}(\mathcal{R}_i \cup \mathcal{R}_j) = \mathsf{FALSE}, \quad \text{for} \quad i \neq j. \tag{4}$$

With more details, P is an appropriate logical predicate leading the segmentation process. Equation (1) states that the union of all the sub-regions resulting from segmentation process. Equation (2) points out that the intersection of two different sub-regions is the empty set: this means that the segmented sub-regions do not overlap each other. According to Equation (3), the result of the logical predicate P on all the pixels belonging to the same sub-region is always TRUE; in other words, all the pixels belonging to the same region share the same characteristics according to the predicate P. As a consequence, Equation (4) states that the result of P on the union of two distinct sub-regions is FALSE.

It is important to clarify that each type of medical image has a specific set of features reflecting its own properties: in fact, each image is the result of a complex interaction between the human body and the scanner (i.e., X-rays for CT, magnetic fields for MRI, radioactive decay for nuclear medicine exams) [19]. As a consequence, not all the segmentation techniques obtain the same results on all image types, but some algorithms yield better results when applied on a specific kind of image. Furthermore, it is necessary to point out that, typically, an approach that works very well with one type of image does not mean that it continues to perform well even on different images. As a matter of fact, bioimages exhibit a very high variability, as they depend on various factors, both intrinsic (e.g., patients) and extrinsic (e.g., imaging modalities, acquisition parameters). Accordingly, ad hoc modifications might be required at the level of both the segmentation approach and the pre-/post-processing phases.

In what follows, we outline the techniques investigated and compared for contrast-enhancing mass segmentation on DCE-MRI.

2.1. Split-and-Merge Combined with Region Growing

The simplest segmentation approaches use a global threshold applied on pixel intensity to partition the original image: pixels with an intensity greater than threshold T are assigned to one region, while those below the threshold T are assigned to another one [20–23]. In this way, a binary image is created providing the segmentation of the original image with respect to the chosen threshold value. Split-and-Merge and Region Growing (SMRG) is basically a threshold-based approach that combines Split-and-Merge (SM)—composed of a first split (top-down) phase of the image followed by a merge (bottom-up) phase—with RG for the refinement of the identified regions [23].

The SM algorithm represents a valid alternative to thresholding, because it can find homogeneous regions in terms of uniformity criteria [24–27]. Unlike SM, the RG algorithm, starting from one or more seed-points, identifies an ROI through a growing procedure guided by appropriate similarity properties that describe ROI intensity features. Generally SR and RG are used individually for image segmentation, but the use of both together allows us to exploit the overall potential [28,29].

With more details, the idea behind the SM algorithm involves successive splits of the whole image into disjoint regions lead by a logical predicate P. The algorithm starts with an arbitrary partition $\mathcal{R}$ of the original image (i.e., the whole image) and yields an output composed of uniform sub-regions $\mathcal{R}_i$, for $i = 1, 2, \ldots, n$, according to the logical criterion expressed by P. At the generic step t, if $\mathsf{P}(\mathcal{R}_i) = \mathsf{FALSE}$, each region $\mathcal{R}_i$ is split into four sub-regions (also called 'quad-regions'): this process iteratively continues until a quad-region such that $\mathsf{P}(\mathcal{R}_i) = \mathsf{TRUE}$ or with an area smaller than a certain threshold is found. The logical predicate chosen for this work allows us to find the quad-regions with a mean intensity that is greater than the threshold value yielded by the Otsu's method [30]. If only splitting is used, the final partition contains adjacent regions with identical properties: this drawback can be overcome by merging only adjacent regions where the combined pixels satisfy the predicate P.

After the initial rough ROI identification obtained by means of SM, the RG algorithm expands this initial region to properly identify the lesion. According to the classic RG algorithm, each region begins its own growth from a single pixel (seed-point). Instead, when the SM is exploited, the seed-point can be obtained by using the ROI yielded by the SM algorithm (more correctly, a seed-region) [23]: the seed-region is iteratively grown by evaluating, for each pixel on the boundary, its 8-neighborhood as candidate for the growth. A stopping rule is necessary for interrupting the growing procedure if no more pixels match the membership criterion, which often refers to the proximity of the pixel intensities.

2.2. K-Means

Classification refers to data labeling into disjoint sets according to a common set of features. Among these, clustering algorithms can be used to determine the natural structures in the data. Clustering algorithms can use more sophisticated properties of the image: in digital imaging, this means that spatial and/or spectral features concerning pixels can be exploited [31].

K-means is an unsupervised clustering technique that aims at partitioning an input set of N observations into k clusters [32]. Let $\mathcal{X} = \{x_1, x_2, \ldots, x_N\}$ be a set of vector observations such that $x_i \in \mathbb{R}^n$ for $i = 1, 2, \ldots, N$. In image segmentation, each component of a vector x represents a numerical pixel attribute: if segmentation is based on gray-scale intensity alone, the n-dimensional observation $x_i \in \mathbb{R}^n$ degenerates into the scalar value $x_i \in \mathbb{R}$ representing the intensity of the i-th pixel. The final purpose of k-means is to partition the N observations into $k < N$ disjoint cluster sets $\mathcal{C} = \{\mathcal{C}_1, \mathcal{C}_2, \ldots, \mathcal{C}_k\}$ such that the sum of the distances from each point in a set to the mean of that set is minimum. From a mathematical point of view, k-means turns classification into an optimization problem with the following cost function in Equation (5):

$$\mathcal{J}(\mathcal{X},\mathcal{V}) = \underset{c}{\operatorname{argmin}}\Big(\sum_{i=1}^{k}\sum_{x\in\mathcal{C}_i}|x - v_i|^2\Big), \tag{5}$$

where v_i is the centroid of the samples in the set $\mathcal{C}_i$ for $i = 1,2,\ldots,k$. The function $\mathcal{J}(\mathcal{X},\mathcal{V})$ has no analytical solution: as a result, k-means proceeds by iteratively finding the minimum of its cost function. In particular, at each iteration t the centroid values of the prototype set $\hat{\mathcal{V}}^{(t)}$) are updated according to Equation (6).

$$\hat{v}_i^{(t)} = \frac{1}{|\mathcal{C}_i^{(t)}|}\sum_{x\in\mathcal{C}_i} x', \tag{6}$$

where $|\mathcal{C}_i^{(t)}|$ is the number of objects belonging to the i-th at the step t.

2.3. Fuzzy C-Means

Biomedical images are characterized by an intrinsic uncertainty, thus causing not well-defined regions (e.g., blurry boundaries or poor anatomical details). This aspect makes thresholding approaches and crisp approaches (such as k-means) not always suitable. Therefore, the natural fuzziness characterizing the Fuzzy C-Means (FCM) clustering approach allows to reach better segmentation results than the hard partitioning offered by k-means [33–36].

The FCM algorithm is an unsupervised clustering technique that searches for the optimal partition of an input data set. The idea leading the FCM classification process is that of minimizing the intra-cluster variance as well as maximizing the inter-cluster variance, in terms of a distance metrics between the feature vectors. Formally, the FCM technique searches for the optimal partition of an input data set $\mathcal{X} = \{x_1, x_2, \ldots, x_N\}$ of N objects into C clusters. With respect to the k-means algorithm in which each point is assigned to one cluster only, FCM allows each object to belong to multiple clusters with different degrees of membership. This *'soft'* classification allows us to define a fuzzy partition, $\mathcal{P}$ defined as a fuzzy set family $\mathcal{P} = \{Y_1, Y_2, \ldots, Y_C\}$ such that each point can have a partial membership to multiple clusters. In mathematical terms, the matrix $U = [u_{ik}] \in R^{C\times N}$ denotes a fuzzy C-partition of the data set X by means of C membership functions $u_i : X \to [0,1]$, whose values $u_{ik} := u_i(x_k) \in [0,1]$ represent membership grades of each element x_k to the i-th fuzzy cluster Y_i, and have to hold the constraints in Equation (7):

$$\begin{cases} 0 \leq u_{ik} \leq 1 \\ \sum\limits_{i=1}^{C} u_{ik} = 1, \forall k \in 1,2,\ldots,N \\ 0 < \sum\limits_{i=1}^{N} u_{ik} < N, \forall i \in 1,2,\ldots,C \end{cases} . \tag{7}$$

Although hard clustering works well on compact and well-separated groups of data, in many real-world situations clusters overlap each other: as a result, assigning them with gradual memberships by exploiting a soft computing approach may be more appropriate. Computationally, the FCM algorithm assigns to the sample x_k the membership function values using the relative distance (i.e., intensity value similarity) of x_k from the C prototype points $\mathcal{V} = \{v_1, v_2, \ldots, v_C\}$ identifying the centroids of the C clusters. Such as k-means algorithm, FCM may be rewritten as an optimization problem with respect to the objective function in Equation (8).

$$\mathcal{J}_m(U,\mathcal{V};\mathcal{X}) = \sum_{i=1}^{C}\sum_{k=1}^{N}(u_{ik})^m|x_k - v_i|^2, \tag{8}$$

where:

- m is the fuzzification constant (i.e., a weighting exponent such that $1 \leq m < \infty$) controlling the fuzziness of the classification process. If $m = 1$, the FCM algorithm degenerates to a k-means clustering: in general, the higher the m value the greater will be the fuzziness degree (the most common value is $m = 2$);
- U is the fuzzy C-partition of the set $\mathcal{X}$;
- $|x_k - v_i|^2$ is the Euclidean distance between the elements x_k and the centroid $v_i \in \mathcal{V}$.

Considering that the optimization problem described by FCM does not have a closed-form solution, the minimum of the cost function $\mathcal{J}_m(\mathcal{U}, \mathcal{V}; \mathcal{X})$ has to be found iteratively. In particular, at each iteration t, the centroid values of the prototype set $\hat{\mathcal{V}}^{(t)}$ and the elements of the matrix $\hat{\mathcal{U}}^{(t)}$ are updated according to Equations (9) and (10), respectively.

$$\hat{v}_i^{(t)} = \frac{\sum_{j=1}^{N} (\hat{u}_{ij}^{(t)})^m x_j}{\sum_{j=1}^{N} (\hat{u}_{ij}^{(t)})^m}, \tag{9}$$

$$\hat{u}_{ik}^{(t)} = \left(\sum_{j=1}^{C} \left(\frac{|x_k - \hat{v}_i^{(t)}|}{|x_k - \hat{v}_j^{(t)}|} \right)^{\frac{2}{m-1}}, \right)^{-1}, \tag{10}$$

with $m > 1$ and $x_k \neq \hat{v}_j^{(t)}, \forall j, k$.

At each iteration, each object x_k is compared with the elements of the centroid vector and is assigned to the nearest cluster. The process stops when the convergence condition (i.e., the matrix norm distance between $\hat{\mathcal{U}}^{(t+1)}$ and $\hat{\mathcal{U}}^{(t)}$ is less than a fixed value (i.e., minimum improvement in the objective function $\mathcal{J}$ between two consecutive iterations ϵ) or the maximum number of iterations $T_{\max}$ is reached. After the convergence, a defuzzification is applied to assign each pixel to the cluster with the highest membership degree, thus achieving a binary classification.

2.4. Spatial Fuzzy C-Means

The traditional FCM clustering does not take into account spatial relationship among neighboring pixels, making it sensitive to noise and other imaging artifacts [37,38]. Breast lesions generally tend to grow in an isotropic way, preserving a pseudo-spherical appearance [39]. Relying on those features, it is expected that neighbouring pixels in a digital image are highly correlated and that the probability that they belong to the same cluster is great. Therefore, the use of the sFCM, taking advantage of spatial relationship of neighbouring pixels, can help image segmentation [40,41]. The spatial function used by sFCM is defined in Equation (11):

$$h_{ij} = \sum_{k \in \mathcal{N}(x_j)} u_{ik}, \tag{11}$$

where $\mathcal{N}(x_j)$ represents a square neighborhood (in the spatial domain) around the pixel x_j. The term h_{ij} represents the probability that the pixel x_j belongs to i-th cluster: as a result, the spatial function of a pixel for a cluster is large if the majority of its neighbourhood belongs to the same cluster. The contribution of the spatial function modifies the classic FCM membership function according to Equation (12):

$$u'_{ij} = \frac{u_{ij}^p h_{ij}^q}{\sum_{k=1}^{C} u_{kj}^p h_{kj}^q}, \tag{12}$$

where p and q control the relative importance of both functions. In a homogeneous region, the spatial functions simply emphasizes the original membership and the clustering remains unchanged. On the other hand, for a noisy pixel this formula reduces the weighting of a noisy cluster by the labels of its neighboring pixels: as a result, misclassified pixels from noisy regions can easily be corrected. The sFCM clustering process involves two steps at each iteration: the former, which is the same as that in standard FCM, allows for calculating the membership function in the features domain, while the latter maps the membership information of each pixel into the spatial domain and allows to calculate the spatial functions for each pixel of the image. At this point, the FCM iteration proceeds with the new membership that is incorporated with the spatial function and stops when the maximum difference between cluster centers at two successive iterations is lower than a certain threshold. After the convergence, a defuzzification scheme (i.e., maximum membership) is applied.

3. Materials and Methods

As introduced in Section 1, the purpose of this work is to present an in-depth analysis of classical unsupervised segmentation algorithms, developing them appropriately to adapt them to the clinical case addressed (i.e., breast lesions detection), thus improving the overall performance.

This section, along with showing the characteristics of the DCE-MRI dataset analyzed, describes the processing pipeline of the proposed analysis. It is necessary to point out that particular attention was paid to the optimization of each step for the specific case study, basically on three different levels:

- *pre-processing*: obtaining images with similar characteristics for the downstream processing steps;
- *segmentation*: optimization of the parameters of the investigated segmentation algorithms;
- *post-processing*: definition of appropriate region properties (based on connected-components) to refine the segmentation results.

3.1. MRI Dataset Description

The analysis of this study was performed on a clinical DCE-MRI dataset composed of 50 patients with breast cancer: a total of 599 slices were processed. The main details on MRI acquisition parameters are reported in Table 1 , while in Figure 1 the phases related to some benign and malignant lesions are shown. The dataset includes patients with different stages of breast cancer, allowing us to cover a wide clinical scenario: as matter of fact, this dataset contains various levels of segmentation difficulty with some scans showing low contrast and large inhomogeneities. Lesions with a non-homogeneous enhancement region, masses with an irregular shape or necrotic core are also included.

Table 1. Some characteristics of the DCE-MRI dataset used for this work.

Characteristic	Value
Scanner manufacturer	General Electric (GE)
Scanner model	Signa HDxt
MRI sequence	DCE-MRI
Time Repetition [ms]	37.720
Time Echo [ms]	17.640
Slice thickness [mm]	2
Interslice spacing [mm]	1
Slice pixel spacing [mm]	0.6875
Matrix size [pixels]	512 × 512

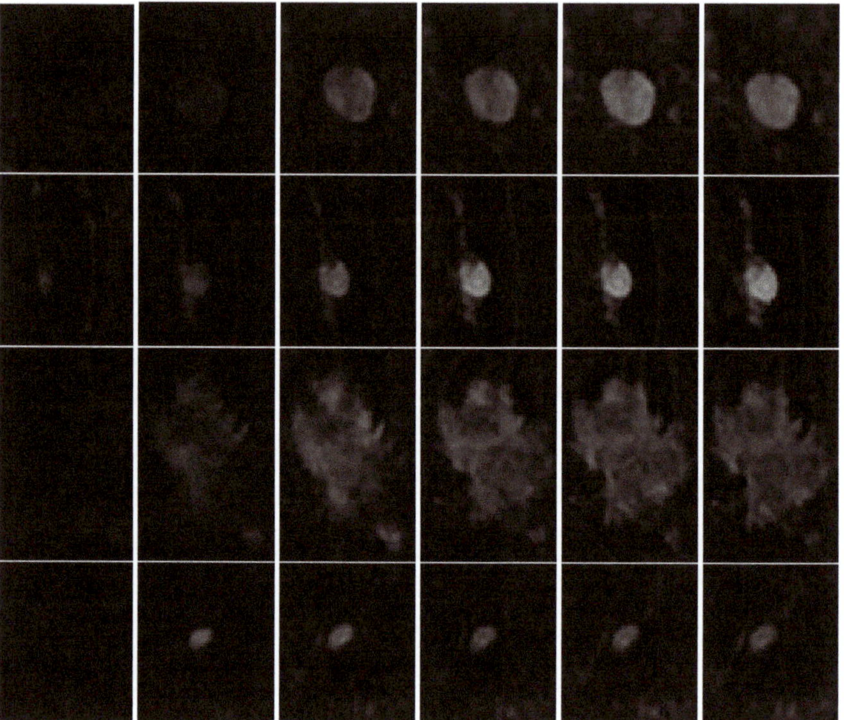

Figure 1. Examples of DCE-MRI phases related to four breast cancer patients (one per row): malignant lesions are shown on rows 1 and 3, while benign lesions are shown on rows 2 and 4. Among phases, for segmentation purposes the clinician selected the strongest one. In particular, for these lesions, the 4th, 3rd, 4th, and 3rd phase was chosen, respectively.

3.2. The Proposed Analysis

Clinical knowledge points out that breast lesions in DCE-MRI appear hyperintense compared to the adipose and muscle surrounding tissues. Furthermore, anatomical atlases and diagnosis reports also refer that breast lesions have the tendency to grow in an isotropic manner, preserving a pseudo-spherical shape [39]. Relying on these morphological features, in this study we decided to start the segmentation from the central slice, which should be the one with the largest area: adjacent slices were processed successively. The idea behind this choice is to segment initially the slice in which the lesion appears more evident, finding the centroid of the connected-component identifying the lesion. On slices different from the central one, the lesion ROI was found by identifying the connected-component with the centroid that has the minimum distance from the centroid found for the central slice. The whole segmentation process can be summarized as follows:

- *dataset loading*: all DCE-MRI images (belonging to the selected patient) and its ground-truth are loaded;
- *ROI selection*: a bounding box containing the lesion is manually drawn in the central image: in this way, the algorithm analyzes only the pixels within the box, reducing computational times and avoiding classification mismatches;
- *pre-processing*: necessary to reduce noise and provide images with similar characteristics for the next processing steps;
- *lesion segmentation*: once the images are pre-processed, the MR image stack of each patient is segmented by all the analyzed methods (i.e., SMRG, k-means, FCM, and sFCM);
- *post-processing*: performed to refine the segmentation and to properly identify the ROI from each binary mask obtained in the previous step;

- *performance evaluation*: the final masks are stored and performance metrics (i.e., area-based and distance-based metrics) are calculated comparing the masks against the ground-truth.

In addition to the well-known segmentation techniques, also the exploited pre- and post-processing operations are widely used by the international community in the field of medical imaging. In fact, the literature offers many works that use the same pre-processing [42,43] and post-processing [44,45] steps in the image analysis phases (before segmentation), as well as in the refinement phases (after segmentation), with excellent results.

Figure 2 shows the flow diagram of the implemented and proposed analysis. Each processing block is described in the following subsections. All the methods were developed using the MatLab environment (Natick, MA, USA). The code is available via GitHub: https://github.com/carmilitello/UnsupervisedSegmentation.git (accessed on 9 November 2021).

3.2.1. Dataset Loading and ROI Selection

This step selects the MRI series to be analyzed. The data loading step, even though does not represent a real processing step, is essential for each algorithm—both supervised and unsupervised—to obtain the data to process. Moreover, the concept of 'supervision' refers generally to the training phase of an algorithm by requiring labeled data. For these reasons, the 'unsupervised' nature of the algorithm used here is preserved. In order to reduce computational times and improve segmentation performance, a minimal user interaction is needed. In fact, by manually tracing a rectangular bounding-box on the image, the operator provides to approximately select the initial ROI containing the breast lesion [13]. Once traced, this ROI is used to crop all the slices of the MR image stack

The manual selection of the initial ROI containing the lesion is useful to reduce the processing time (since only a portion of the whole image is processed), as well as to exclude regions containing pixels with characteristics similar to the lesion that could complicate the algorithm performance, thus invalidating the final result. Moreover, the initial, interactive input provides the clinician with the ability of confidently controlling the entire segmentation process, which is generally preferred compared to a fully automatic process [18,46]. Moreover, considering that the method works on an ROI—selected by the clinician—it is assumed that the tumor is always present within the ROI.

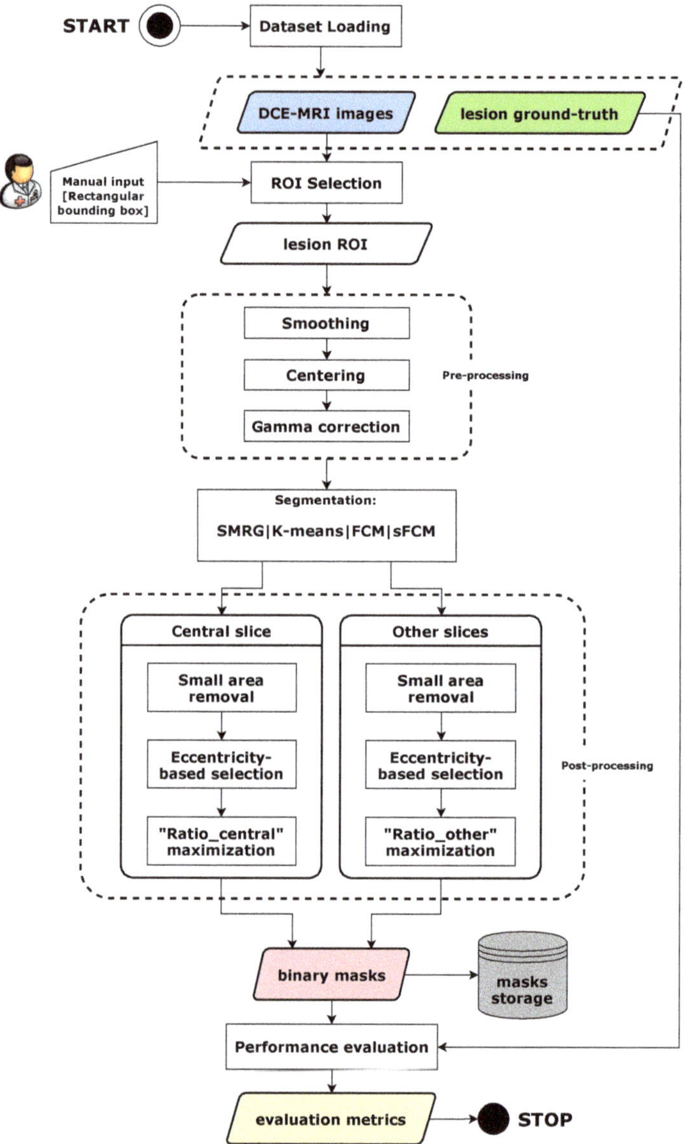

Figure 2. Flow diagram of the proposed analysis for breast lesion segmentation.

3.2.2. Pre-Processing

In order to improve segmentation results, some pre-processing operations are applied after the ROI selection (Figure 3). In particular, the aim of this preliminary phase is to perform denoising and data pre-processing, allowing us to achieve reliable results during the segmentation phase.

The first pre-processing operation deals with noise reduction [47]. The use of median or average (with Gaussian or flat kernels) filters is a common choice in MR image processing. For instance, median filtering was applied to facilitate the ROI identification by reducing the outlier introduced by anatomical peculiarities [48] or also deal with small patient shifts [49]. We used Gaussian kernels having the form in Equation (13), which are the only circular symmetric kernels that are also separable and that allow you to reduce noise by altering the image less than the average filter.

$$G(r) = ke^{-\frac{r^2}{2\sigma^2}}, \quad (13)$$

where r, σ and k represent the radius, the standard deviation and the normalization factor of the Gaussian function G, respectively. Kernel normalization, obtained by multiplying its coefficients by k—obtained as the inverse of the sum of all kernel coefficients—has two purposes: *(i)* the average value of an area of constant intensity would equal that intensity in the filtered image (as it should) and *(ii)* it prevents the introduction of biases during filtering (i.e., the sum of the pixels in the original and filtered images will be the same). For the purpose of this work, a Gaussian kernel with $r = 5$ and $\sigma = 0.6$ was chosen: this standard deviation value performed denoising without an excessive blurring of the original image.

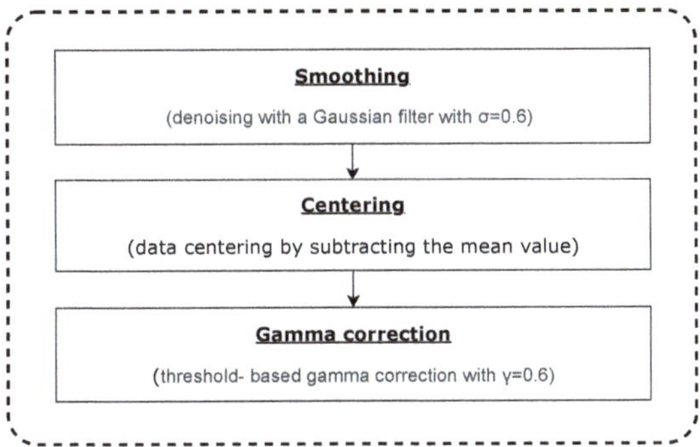

Figure 3. Flow diagram of the pre-processing steps approach.

The next pre-processing step subtracts the mean value from the original image. This is a very helpful step in signal processing because it emphasizes signal variations by shifting its mean to zero. For this reason, during the pre-processing phase, the algorithm removed this bias from the original image amplifying gray level variations. Once the mean value subtraction, the pixels resulting in a negative value have to be clipped to zero in order to avoid visualization problems in the following steps.

After noise reduction and mean subtraction, the last step performs a modified gamma-transformation to stretch the original image histogram. The general form of a gamma transformation is defined in Equation (14):

$$s = cr^{\gamma}, \quad (14)$$

where c and γ are positive constants. Power-law curves with fractional values of γ map a narrow range of dark input values into a wider range of output values, and vice versa for bright input values. When $c = \gamma = 1$ the gamma transformation reduces to the identity transformation. The modified gamma transformation proposed in this work used $c = 1$ and $\gamma = 0.7$, by means a piece-wise function in which the final value of each pixel depends on a certain threshold value θ. In particular, the function mapping the initial value of a pixel $r(x, y)$ into its final value $s(x, y)$ is defined in Equation (15). The value of θ, used for the piece wise gamma selection, is the one suggested by the Otsu's thresholding method [30]. Figure 4 shows the results of each pre-processing step applied on two (one per row) DCE-MRI breast lesions.

$$s(x,y) = \begin{cases} r(x,y) & if \quad r(x,y) \leq \theta) \\ r(x,y)^{\gamma} & if \quad r(x,y) > \theta) \end{cases} . \quad (15)$$

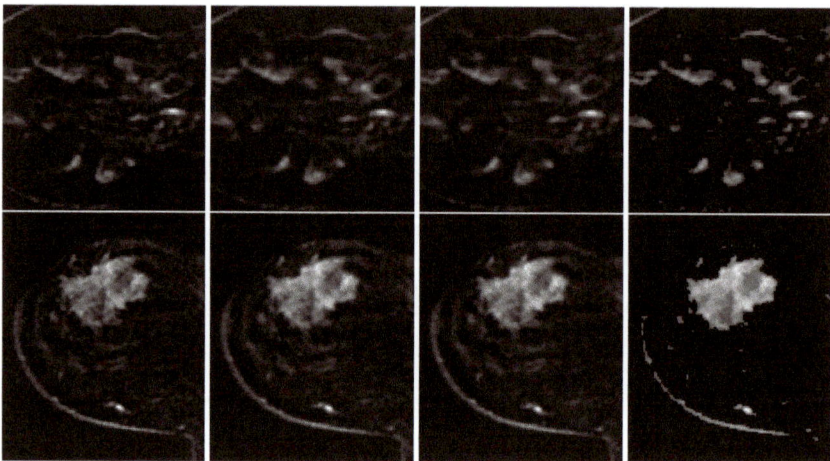

Figure 4. Examples of pre-processing steps on two breast lesions (one per row): (**1st column**) DCE-MRI breast image (after cropping); (**2nd column**) Gaussian smoothing result; (**3rd column**) centering result; and (**4th column**) gamma correction result. With respect to the position in the DCE-MRI sequence, all images have been rotated 90° clockwise, in order to improve the graphic representation and allow all the images relating to a specific lesion to be displayed in a single line.

3.2.3. Segmentation

After pre-processing, the segmentation involves the partitioning of a digital image into multiple sets of pixels, according to the specific clinical purpose. Nowadays, segmentation of digital images still remains one of the most challenging topics in image processing: in fact, medical image segmentation is often still performed manually, via time-consuming and operator-dependent procedures.

As indicated in Section 3.2.1, to start the segmentation process, it is necessary to select the ROI containing the lesion in the central image. The central slice—which is generally the one with the largest tumor section—is determined automatically after the operator sets the range of the initial and final slices containing the tumor. Starting from the selected ROI in the central slice, the ROIs in the adjacent slices are determined starting from the centroid of the segmented lesion in the previous slide. By doing so, the ROIs (set automatically) in the adjacent slices are centered (or almost) with the lesion section they contain. After the segmentation, a check is made to verify that the lesion does not touch the ROI boundary. In fact—considering that the ROI size is set equal to that of the ROI in the previous slice—it could happen that the selected ROI is too small and, consequently, the lesion was cropped. If so, the ROI is enlarged in order to fully contain the lesion section and repeated segmentation.

The segmentation is sequential: starting from the central slice, all the upper slices are processed first and then the lower ones. All the segmentation steps are the same for both the central slice and the other slices. The only difference lies in the different definition of the ratios—defined in Equations (18) and (19)—used in the post-processing steps, differentiated with the goal of optimizing the segmentation result.

SMRG Setting

There are several parameters controlling the SMRG behavior.

First of all, the splitting predicate P, a logical predicate fundamental to achieve satisfactory outputs. In order to identify breast lesions, homogeneity criteria are defined in terms of the mean value μ of each quad-regions. Here, the logical predicate in Equation (16) was used:

$$\mathsf{P} := \begin{cases} \mathsf{TRUE} & \text{if} \quad (\mu > 0.1) \wedge (\mu < 0.6) \\ \mathsf{FALSE} & \text{otherwise} \end{cases} . \tag{16}$$

The minimum block dimension $\rho_{\min}$ sets the minimum quad-region size beyond which no further splitting is carried out: the best results are found with minimum block dimensions of 4×4 pixels, because small regions are detected too. After the splitting phase, the final partition will contain adjacent regions with identical properties: this drawback can be addressed by merging only adjacent regions whose combined pixels satisfy the predicate P.

Finally, a stopping rule interrupts the growing procedure if no more pixels match the membership criterion. For the aim of this work, a reasonable stopping rule uses a similarity criterion between the candidate pixel to be incorporated and the pixels already belonging to the region. The criterion for the stopping rule is defined in terms of absolute distance between the regional mean of each quad-region and the threshold provided by the Otsu's method [30] on the original cropped image, both of them calculated without the contribution of null pixels.

Clustering Setting

Here, the optimal setting used for k-means, FCM and sFCM clustering algorithms is reported, in terms of the following parameters: *(i)* number of clusters; *(ii)* maximum number of iterations; *(iii)* minimum improvement in objective function between two consecutive iterations; *(iv)* exponent for the fuzzy partition matrix (only for fuzzy approaches); *(v)* $< p, q >$ parameters (only for fuzzy approaches).

Regarding the number of clusters, the lower the partition fuzziness, the better the segmentation result: the best clustering is achieved when V_{pe} is minimal. If the partition entropy for $C = 2$ is lower than the one obtained for $C = 3$, the lesion does not include any necrotic region: if so, pixels belonging the brightest cluster are turned to 1 and the others to 0. Otherwise, the two brightest clusters are fused together and their pixels are turned to 1, allowing for the inclusion of necrotic cores into the preliminary mask. Considering that, sometimes breast lesions can include an inner necrotic region that appears darker (i.e., hypo-intense) with respect to the rest of the lesion. In those cases where a binary clustering ($C = 2$) does not allow for properly detecting a necrosis, the number of clusters might be increased ($C = 3$). The choice of the optimal number of clusters to be used in k-means, FCM and sFCM is automatically set by evaluating the partition entropy for both cases $C = 2$ and $C = 3$. The partition entropy V_{pe} is a cluster validity function defined as in Equation (17):

$$V_{pe} = \frac{-\sum\limits_{j=1}^{N} \sum\limits_{i=1}^{C} u_{ij} \log u_{ij}^2}{N}. \tag{17}$$

The $\langle p, q \rangle$ parameters, dealing with FCM and sFCM clustering, control the relative importance of both functions defined in Equations (10) and (11), respectively. In particular, with $\langle p, q \rangle = \langle 1, 1 \rangle$, we sFCM is equivalent to the traditional FCM. As a matter of fact, to properly weight spatial information, the values $\langle p, q \rangle >= \langle 1, 2 \rangle$ were used. The choice was guided by the analysis and results obtained in [18], tackling a similar problem on DCE-MRI images. All these parameters are reported in Table 2.

At the end of the segmentation of each slice, the algorithm also verifies whether or not the contour of the detected lesion touches the boundaries of the initial rectangular bounding-box: if so, the shape of the lesion results in a cut version of the original one. For this reason, when the lesion intersects the boundaries of the original $m \times n$ rectangular crop, the algorithm shifts and expands the bounding-box itself until the boundaries of both the lesion and the crop do not intersect each other.

Table 2. Parameters setting for k-means, FCM and sFCM clustering algorithms.

Unsupervised Algorithm	Number of Clusters	Maximum Number of Iterations	Minimum Tolerance	Exponent of Fuzzy Partition Matrix	$\langle p, q \rangle$ Parameters
k-means	2 or 3	100	N.A.	N.A.	N.A.
FCM	2 or 3	100	1×10^{-5}	2.0	<1,1>
sFCM	2 or 3	100	1×10^{-2}	2.0	<1,2>

3.2.4. Post-Processing

The result of the segmentation process is a binary image in which a label is assigned to each pixel: 0 if the pixel does not belong to the mask, and 1 otherwise. The aim of the post-processing phase is to properly choose, among the connected-components resulting from the previous step, the only one representing the lesion. The post-processing itself consists of several sub-steps allowing for removing connected-components that do not meet specific morphological criteria (Figure 5).

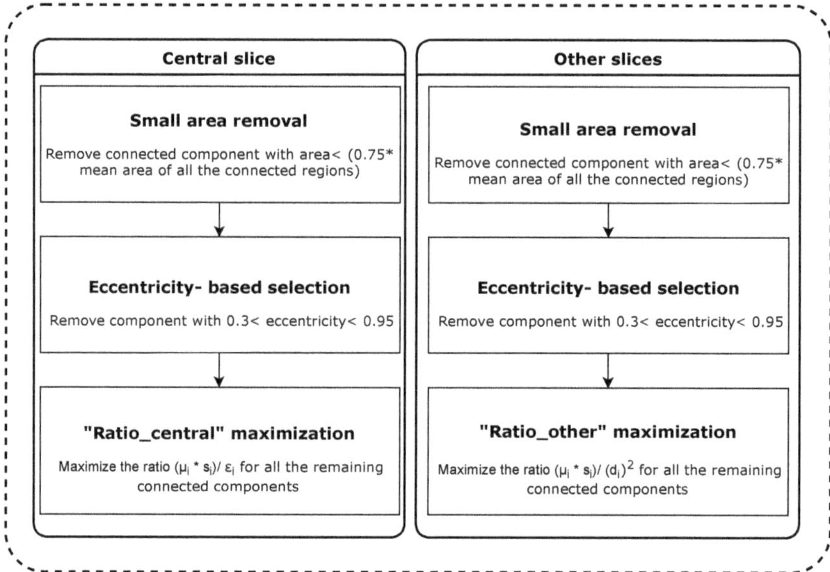

Figure 5. Diagram of the post-processing steps for both central slice and other slices.

The first post-processing phase allows us to extract all the connected-components from the raw binary mask. For each connected-component, a list of morphological features (i.e., extreme points, area, centroid, eccentricity, mean intensity and solidity) are computed that the algorithm is going to exploit in the following steps. In order to remove spurious connected-components with an area that is too small to be considered as a lesion, an area-based selection is performed: all the connected-components with an area smaller than $0.75 \times$ meanArea of all the connected-components. In such a way, spurious regions with a very small area are removed, reducing the number of valid candidates to be analyzed in the following steps.

Clinical evidence allows us to state that breast lesions generally grows in an isotropic way preserving a pseudo-spherical appearance [39]. As a consequence, their particular shape can be described in terms of eccentricity, which is defined as the ratio of the distance between the *foci* of an ellipse and its major axis length. Eccentricity values lay in the range $[0, 1]$: the extreme values represent degenerate cases identifying a circle and a line segment, respectively. The aim of this post-processing step is to delete from the list of the lesion candidates the regions with eccentricity lower than 0.3 and higher than 0.9 (these threshold values were determined experimentally). Eccentricity values close to 1 relate to very elongated lesions, while eccentricity values close to 0 relate to almost perfectly round shapes. In the case of breast lesions, even if overall there are rounded lesions, the lesion has a 'lobed' trend for which the final value of the eccentricity never takes values below 0.2–0.4. Using 0.3 as eccentricity lower limit in the post-processing phase, allow us to eliminate those (almost perfectly) circular connected components with pixel values similar to the lesion (and which are therefore incorrectly selected), but which are instead part of the background. As a consequence of their tendency to be round-like shaped, breast lesions exhibit an high solidity, which is defined as the ratio between the region area and the (including the region) convex polygon area. Considering this morphological information, the lesion ROI identification can be turned into an optimization problem with the ultimate goal of maximizing ratioCentral, defined in Equation (18):

$$\text{ratioCentral}_i = \frac{\mu_i s_i}{\epsilon_i}, \tag{18}$$

where μ_i is the mean intensity, s_i is the solidity and ϵ_i denote the eccentricity of the i-th connected-component, respectively.

The post-processing in the other slices differs from the central slice. In this case, the parameter to maximize ratioOther is defined according to Equation (19):

$$\text{ratioOther}_i = \frac{\mu_i s_i}{d_i^2}, \tag{19}$$

where d_i^2 represents the square of the distance between the centroid of the lesion region in the central slice and the centroids of all the connected-components in the current slice. In order to reduce false positives identification, ratioOther maximization searches for the connected-component with the centroid that is the closest to the one of the central slice mask. Recalling that breast lesions usually preserve a pseudo-spherical appearance, as the slices move away from the central one, the cross section identifying the lesion is reduced. In Equation (19) the replacement of ϵ_i with d_i^2 aims to avoid misclassification of spurious regions in slices different from the central one.

Figure 6 shows the results of each post-processing step applied to a segmentation mask obtained by means of sFCM clustering. Unfortunately, breast lesions might be characterized by different scenario—in terms of uniformity, contrast and well-defined boundaries—which a simple thresholding cannot properly manage. Figure 7 shows two segmentation results that allow us to appreciate the lesion non-homogeneity that only clustering approaches can properly manage, thus maximizing the result accuracy.

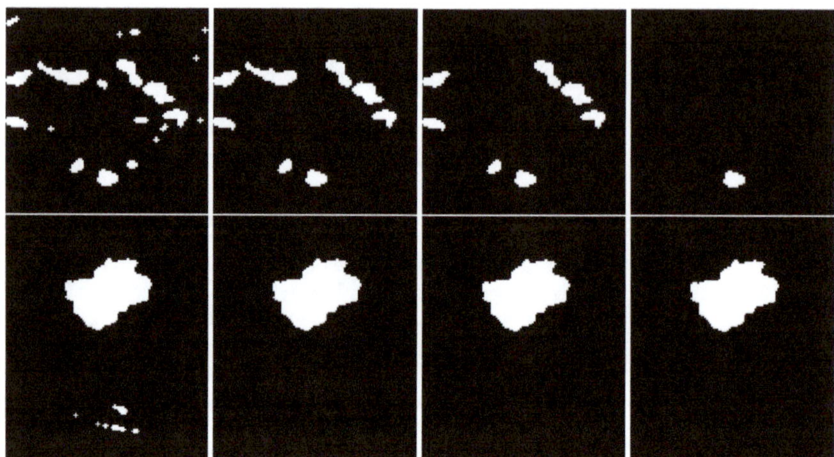

Figure 6. Examples of post-processing steps on two breast lesions (one per row): (**1st column**) binary mask obtained after sFCM clustering; (**2nd column**) mask after the small-area removal; (**3rd column**) mask after both the small-area removal and the eccentricity-based selection; and (**4th column**) mask after ratio criteria, defined in Equations (18) and (19). With respect to the position in the DCE-MRI sequence, all images have been rotated 90° clockwise, in order to improve the graphical representation and allow all the images, related to aspecific lesions, to be displayed in a single line.

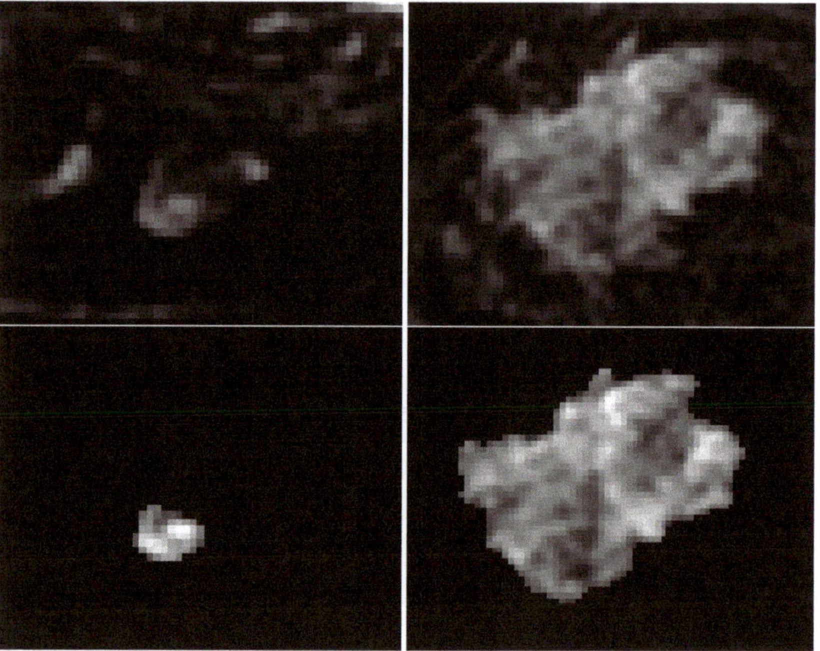

Figure 7. Two examples of segmentation obtained by sFCM: (**1st row**) the original DCE-MRI images; (**2nd row**) the corresponding segmentation. Compared to the images shown in Figures 4 and 6, a $2\times$ zoom factor was used, that allows us to better appreciate the lesion non-homogeneity that only clustering approaches can properly manage, thus maximizing the final result accuracy.

4. Segmentation Performance Evaluation

Several evaluation measures are computed to quantify segmentation performances by comparing masks—obtained from analyzed unsupervised methods—and the ground-truth—provided by a radiologist, with more than 5-year experience on breast MRI, in consensus with a consultant breast radiologist (with more than 30-year experience on breast imaging). To obtain an accurate and detailed quantification, both area-based and distance-based metrics were used. The reason behind this choice is that area-based metrics strongly depend on region size and are not always able to evaluate the precision of a segmentation approach. On the other hand, distance-based metrics take into account the distance between the boundaries of the two segmentations to be compared, ignoring the actual volume difference between the two masks.

4.1. Spatial Area-Based Metrics

Spatial area-based metrics compare the semi-automatic segmented regions with the manually segmented ones ($\mathcal{R}_A$ and $\mathcal{R}_T$, respectively) by calculating the overlapping percentage of area between the two masks obtained from the segmentation of the image $\mathcal{I}$. Recalling some basics on statistical decision theory measures, the regions containing 'true positives' (TP), 'false positive' (FP), 'false negatives' (FN), and 'true negatives' (TN) are defined as:

$$\begin{aligned} \mathcal{R}_{\text{TP}} &= \mathcal{R}_A \cap \mathcal{R}_T \\ \mathcal{R}_{\text{FP}} &= \mathcal{R}_A - \mathcal{R}_T \\ \mathcal{R}_{\text{FN}} &= \mathcal{R}_T - \mathcal{R}_A \\ \mathcal{R}_{\text{TN}} &= \mathcal{I} - \mathcal{R}_T - \mathcal{R}_A \end{aligned}$$

When validating the segmentation results, the two most used area-based metrics are the Dice Index (DI) and the Jaccard Index (JI), defined in Equations (20) and (21), respectively. DI and JI are used to describe how much similar the manual (ground-truth) and the semi-automatic segmentations are: the greater they are, the higher is the overlapping percentage between the two masks.

$$\text{DI} = \frac{2 \cdot \mathcal{R}_{\text{TP}}}{\mathcal{R}_A + \mathcal{R}_T} \tag{20}$$

$$\text{JI} = \frac{\mathcal{R}_A \cap \mathcal{R}_T}{\mathcal{R}_A \cup \mathcal{R}_T} \tag{21}$$

Sensitivity and specificity—defined in Equations (22) and (23)—represent the portion of positive pixels (foreground) and negative pixels (background) correctly detected by a segmentation method with respect to the ground-truth, respectively.

$$\text{Sensitivity} = \frac{\mathcal{R}_{\text{TP}}}{\mathcal{R}_{\text{TP}} + \mathcal{R}_{\text{FN}}} \tag{22}$$

$$\text{Specificity} = \frac{\mathcal{R}_{\text{TN}}}{\mathcal{R}_{\text{TN}} + \mathcal{R}_{\text{FP}}} \tag{23}$$

False Positive Ratio (FPR) and False Negative Ratio (FNR)—defined in Equations (24) and (25)—denote the presence of false positives and false negative compared to the reference region, respectively.

$$\text{FPR} = \frac{\mathcal{R}_{\text{FP}}}{\mathcal{R}_{\text{FP}} + \mathcal{R}_{\text{TN}}}, \tag{24}$$

$$\mathrm{FNR} = \frac{\mathcal{R}_{\mathrm{FN}}}{\mathcal{R}_{\mathrm{FN}} + \mathcal{R}_{\mathrm{TP}}}. \tag{25}$$

4.2. Spatial Distance-Based Metrics

Area-based metrics are susceptible to differences between the positions of segmented regions and strongly dependent on their own size. To take into account the spatial position of the pixels, it is necessary to quantify the distance between the boundaries computed by the semi-automatic methods and the ground-truth delineated by the expert. Let $A = \{a_i : i = 1, 2, \ldots, K\}$ be the set of vertices belonging to the semi-automatic mask and $T = \{t_j : i = 1, 2, \ldots, N\}$ the set of vertices belonging to the ground-truth, the distance between the i-th pixel in A and the set T is defined as:

$$d(a_i, T) = \min_{j \in \{1,2,\ldots,N\}} ||a_i - t_j||_2, \tag{26}$$

where $||a_i - t_j||_2$ denotes the Euclidean distance between two points.

Many metrics can be defined in order to quantify the similarity/dissimilarity between two segmentations.

The Mean Absolute Distance (MAD)—defined in Equation (27)—quantifies the average error in the segmentation process. The Maximum Distance (MaxD)—defined in Equation (28)—measures the maximum difference between the two ROI boundaries. The Hausdorff Distance (HD) between the point sets A and T—defined in Equation (29)—measures the maximal distance from a point in the first set to a nearest point in the other one.

$$\mathrm{MAD} = \frac{1}{K} \sum_{i=1}^{K} d(a_i, T) \tag{27}$$

$$\mathrm{MaxD} = \max_{i \in 1,2,\ldots,K} \{d(a_i, T)\} \tag{28}$$

$$\mathrm{HD} = \max\{h(T, A), h(A, T)\}, \tag{29}$$

where $h(T, A) = \max_{t \in T}\{\min_{a \in A}\{d(t, a)\}\}$ is the so-called 'directed Hausdorff Distance'.

It is important to point out that all the measured distances are expressed in pixels: in this way, they result will be independent from the spatial resolution among different MRI datasets (i.e., pixel spacing).

5. Experimental Findings

5.1. Area-Based Metrics Segmentation Results

Area-based metrics obtained by each segmentation algorithm are shown in Tables 3 and 4—expressed as mean ± standard deviation. As easily appreciable, the results showed that the fuzzy framework offered by FCM and sFCM reflects the intrinsic uncertainty that characterizes medical images, allowing us to achieve better segmentation results compared to the hard clustering performed by k-means. Furthermore, it is worth to note that spatial constraints taken into account by sFCM contribute to reduce the standard deviation of the final result, thus ensuring higher reliability with respect to SMRG and k-means. On the other hand, the large value in standard deviation indicates that SMRG has a high variability of the results that affects its reliability. The boxplots in Figures 8 and 9 summarize the obtained results.

Regarding sensitivity, the results showed that the FCM-based approach offered better performance in terms of both mean value and standard deviation; on the contrary, the hard partition offered by k-means did not provide satisfying results. The specificity values showed that the clustering-based approaches obtained better performances compared to SMRG. In particular, sFCM and k-means achieved slightly better results than FCM.

Altogether, the results in Tables 3 and 4 showed that segmentation approaches based on soft-clustering techniques achieved better performance compared to SMRG and k-means. In fact, by explicitly exploiting the fuzziness, both FCM and sFCM better handled the intrinsic uncertainty and the natural variability of medical images. On the other hand, the crisp k-means allowed us to reach satisfying results in specificity values, but without granting good performance in sensitivity.

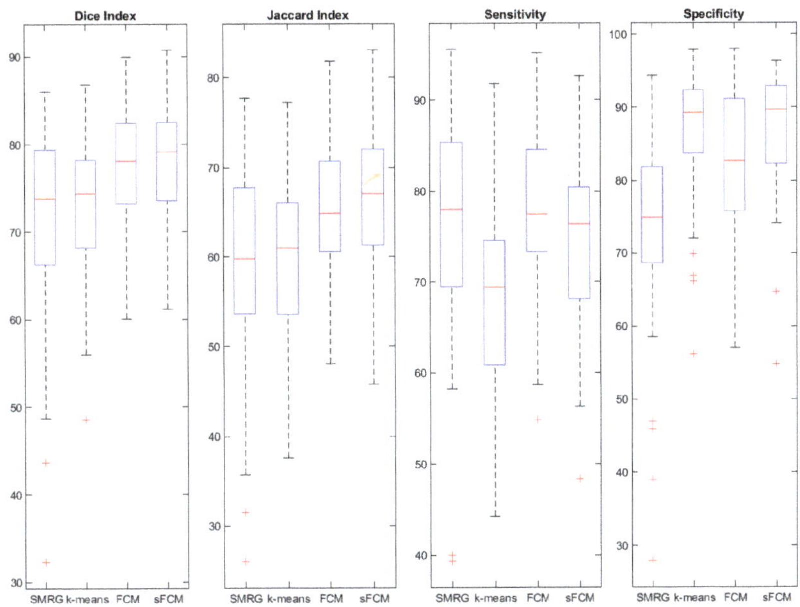

Figure 8. Boxplots of the area-based metrics achieved by the considered unsupervised segmentation approaches. From left to right: Dice Index, Jaccard Index, Sensitivity and Specificity. The lower and upper bounds of each box represent the first and the third quartiles of the metric distribution, respectively. The median is represented by a red line, while outliers are displayed as red crosses.

Table 3. Area-based metrics achieved by the considered unsupervised segmentation approaches: the results are expressed as average value ± standard deviation.

Method	DI	JI	Sensitivity	Specificity
SMRG	71.82 ± 10.83	58.98 ± 11.32	76.47 ± 12.41	73.65 ± 13.50
k-means	72.65 ± 8.26	59.87 ± 9.20	67.23 ± 10.78	86.97 ± 8.93
FCM	77.48 ± 6.77	65.22 ± 8.13	**77.84 ± 8.72**	81.94 ± 10.41
sFCM	**78.23 ± 6.50**	**65.90 ± 8.14**	74.69 ± 9.39	**87.10 ± 8.24**

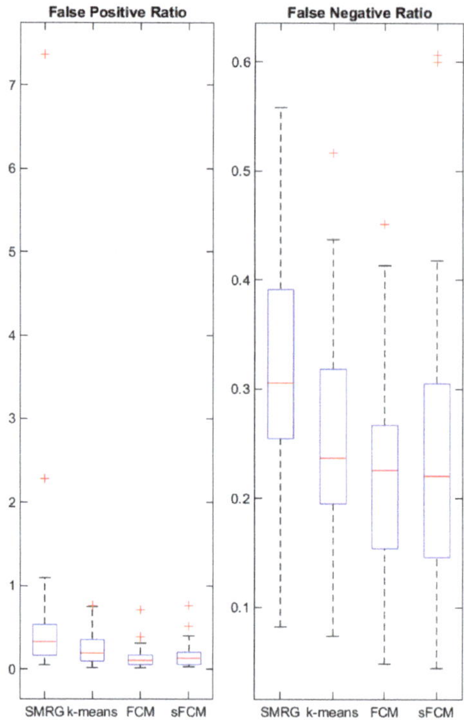

Figure 9. False Positive Ratio and False Negative Ratio boxplots. The lower and upper bounds of each box represent the first and the third quartiles of the metric distribution, respectively. The median is represented by a red line, while outliers are displayed as red crosses.

Table 4. False Positive and False Negative Ratios achieved by the considered unsupervised segmentation approaches: the results are expressed as average value ± standard deviation.

Method	False Positive Ratio	False Negative Ratio
SMRG	0.55 ± 1.05	0.33 ± 0.11
k-means	0.25 ± 0.20	0.25 ± 0.09
FCM	0.16 ± 0.13	0.23 ± 0.12
sFCM	**0.14 ± 0.12**	**0.22 ± 0.09**

Figure 10 shows some segmentation examples focusing on scenarios where the results are not particularly satisfactory in terms of FPs and FNs. In general, it is possible to note that SMRG and k-means have a greater tendency to leave out parts of the lesion (FN)—especially, when these are not uniform—and to include areas outside the lesion (FP), while the techniques based on clustering, and in particular the sFCM guarantees a better lesion detection.

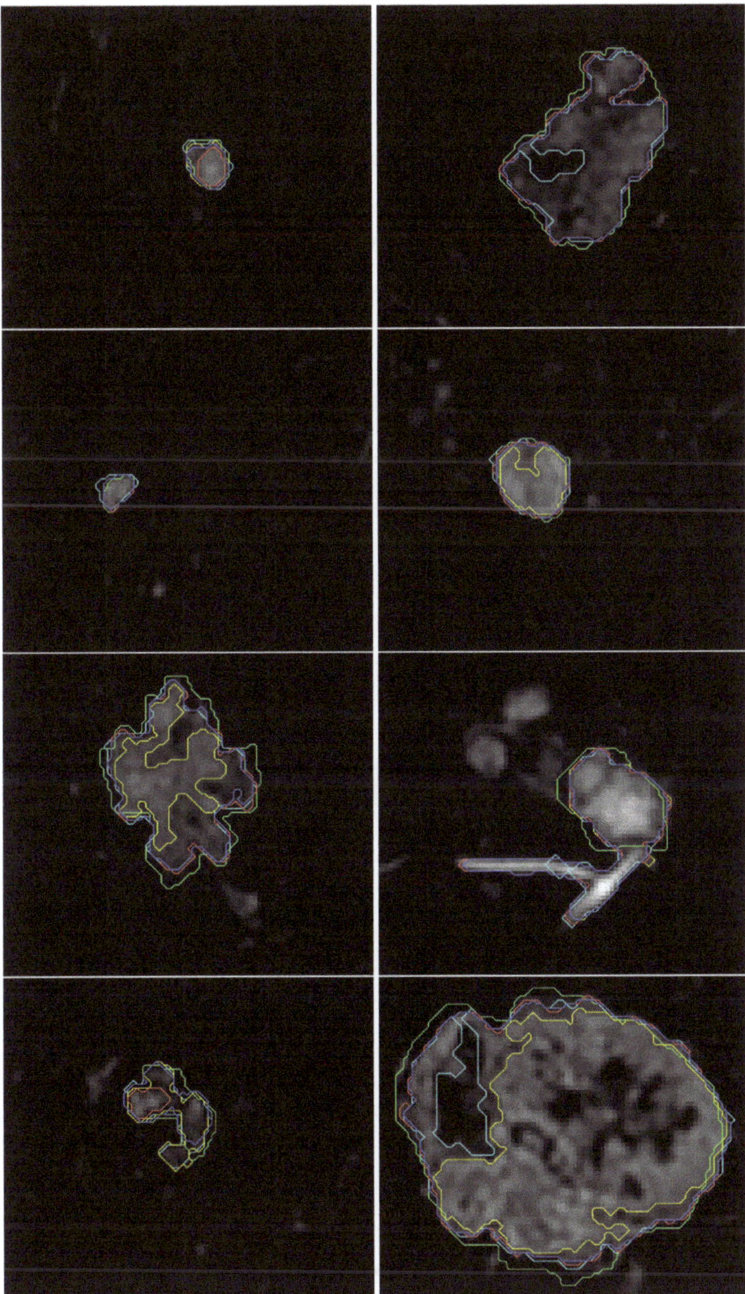

Figure 10. Examples of interesting segmentation results showing FPs and FNs. In particular, the examples compare results yielded by the four investigated unsupervised method—namely, SMRG (cyan), k-means (yellow), FCM (red), sFCM (blue)—against the ground-truth (green).

5.2. Spatial Distance-Based Metrics Segmentation Results

The need for using both area- and distance-based metrics comes out in considering that area-based metrics do not take in account pixels' spatial distribution. This leads to the need to quantify the distance between the boundaries computed by the semi-automatic methods and the ground-truth. The boxplots in Figure 11 summarize the obtained results.

In terms of HD, all the presented methods share similar characteristics with a mean value of ≈2.2 and a standard deviation of ≈0.43. This means that the boundaries of the semi-automatic masks are quite close to the manually traced ones.

The MAD metric is quite similar across the examined techniques, except for the SMRG, which shows the higher mean value and standard deviation. Clustering-based approaches, on the other hand, result more stable and precise, offering a better performance on the whole dataset.

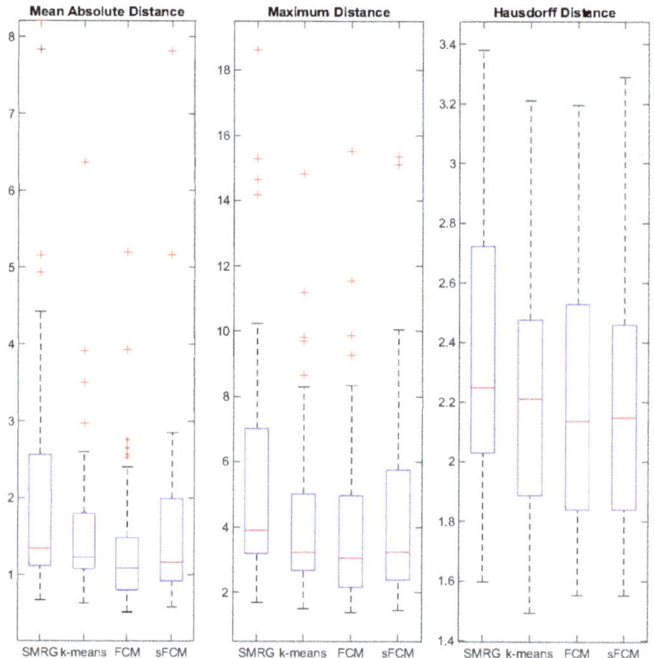

Figure 11. Boxplots of the spatial distance-based metrics achieved by the considered unsupervised segmentation approaches. From left to right: Mean Absolute Distance, Maximum Distance and Hausdorff Distance. The lower and upper bounds of each box represent the first and the third quartiles of the metric distribution, respectively. The median is represented by a red line, while outliers are displayed as red crosses.

Table 5 shows distance-based metrics obtained using the considered unsupervised segmentation approaches: lower distance values indicate better segmentation results. Observing the general trend, just a small deviation between the segmentations of the proposed methods and those of the experienced radiologist can be denoted. Furthermore, the achieved spatial distance-based indices are consistent with area-based metrics, confirming that clustering-based segmentation approaches allow us to reach better results with respect to SMRG.

Table 5. Spatial distance-based metrics achieved by the proposed segmentation approaches: the results are expressed as average value ± standard deviation.

Method	MAD	MaxD	HD
SMRG	2.00 ± 1.41	5.63 ± 3.81	2.32 ± 0.44
k-means	1.58 ± 0.99	4.22 ± 2.73	2.24 ± 0.42
FCM	1.58 ± 1.22	4.46 ± 3.12	**2.21 ± 0.43**
sFCM	**1.37 ± 0.90**	**4.04 ± 2.87**	2.21 ± 0.44

As shown at the top of Figure 12, FCM (with $p = 6.268 \times 10^{-6}$ and $p = 2.670 \times 10^{-7}$) and sFCM (with $p = 8.566 \times 10^{-5}$ and $p = 1.855 \times 10^{-7}$) clustering methods achieved significantly higher DI values compared to SMRG and k-means, respectively.

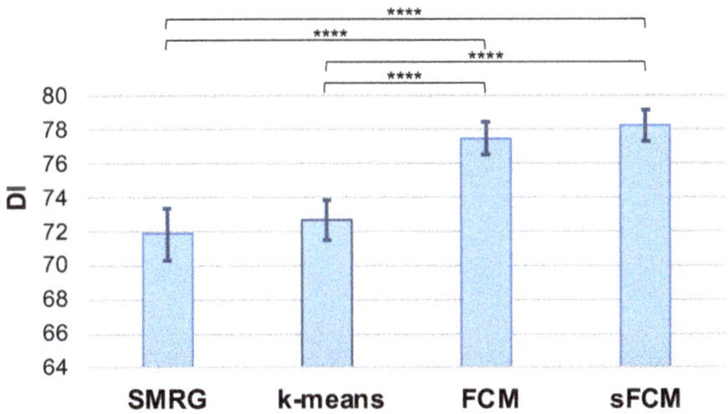

Figure 12. DI values achieved by the investigated traditional classic approaches. The bar graph and error bars denote the average value and the standard deviation DI values, respectively. The p-values, obtained from the statistical validation procedure, are shown at the top of the bars as brackets. The Wilcoxon rank-sum test for pairwise result comparison was used with the alternative hypothesis that the samples do not have equal medians of DI values. A significance level of $\alpha = 0.05$ with a correction using the Bonferroni–Holm method. Notation: **** $p < 0.0001$.

To statistically validate the obtained results, the two-sided Wilcoxon signed rank test [50] on paired DI results was performed with the null hypothesis that the samples come from continuous distributions with equal medians (considering a significance level of 0.05). Obtained p-values are shown in Table 6. With more details, this test on paired results was used to statistically compare the distributions of the DI values achieved by two competing methods and identify significant differences. The p-values were corrected by the Bonferroni–Holm method [51] for multiple comparisons.

The values reported in Tables 3–5 represent the average value ± standard deviation over all 50 breast masses. All the values obtained on each breast mass are reported in the Supplementary Materials.

Table 6. p-Values obtained from the statistical validation procedure. The Wilcoxon rank-sum test for pairwise result comparison was used with the alternative hypothesis that the samples do not have equal medians of DI. A significance level of $\alpha = 0.05$ with the Bonferroni-Holm correction for multiple comparisons was used. **Boldface** indicates that the null hypothesis can be rejected.

	SMRG	k-Means	FCM
k-means	0.973		
FCM	$\mathbf{6.268 \times 10^{-6}}$	$\mathbf{2.670 \times 10^{-7}}$	
sFCM	$\mathbf{8.566 \times 10^{-5}}$	$\mathbf{1.855 \times 10^{-7}}$	0.423

5.3. Processing Times

In order to evaluate the processing times, all segmentations were performed on the entire dataset, by calculating the average elapsed time and the corresponding standard deviation for each of the four investigated algorithms over all the analyzed images, obtaining the following values: SMRG: 2.78 ± 2.79 s; k-means: 1.35 ± 0.78 s; FCM: 1.65 ± 0.88 s; sFCM: 1.76 ± 0.83 s. These processing times were measured using the Matlab R2019b IDE (by means of the `tic` and `toc` stopwatch timer functions) running on a Windows 10 Pro general-purpose PC equipped with an Intel I7-3630QM@2.40 GHz CPU and 8 GB RAM.

As expected, SMRG had the longest processing time due to the two-stage approach and also considering the iterative processes employed during the Split-and-Merge and Region Growing executions. Interestingly, the introduction of the fuzzy logic is negligible compared to the crisp k-means, as well as the integration of the spatial constraints into the sFCM algorithm does not require a remarkable computational overhead in addition to the standard FCM clustering.

Overall, these results demonstrate the clinical feasibility of the investigated classic unsupervised methods also in terms of both computational resources and processing times, by considering that supervised CNN-based approaches for segmentation require a training phase and then an inference phase typically performed on Graphics Processing Units [18].

5.4. Difficult Cases

As previously highlighted, medical images are characterized by an intrinsic variability in which boundaries or anatomical details may be not well defined. Furthermore, noise corrupts digital images, thus affecting certain features within the original image. MRI suffers from various kinds of noise and artefacts because of the nature of the signal detection and spatial encoding [37,38]. For instance, hardware-induced errors are often caused by the complicated acquisition scheme depending on radiofrequency coils, while thermal noise can derive from transmission lines, receiver circuits and polarization magnetic field B_0 drift during the scan acquisitions. In addition, natural body motion (e.g., respiratory and cardiac motion) can degrade the image quality too. As a consequence, even after a proper pre-processing step, it is common to deal with low-contrast, noisy images. The proper lesion segmentation in this kind of images is not a trivial task and user interaction would be required to produce accurate results.

5.4.1. Case with Low Contrast-Enhanced Mass

In some cases, MR images are difficult to segment because the lesion itself does not appear brighter with respect to the muscle and adipose surrounding tissue. This kind of images exhibit a narrow histogram located typically toward the middle of the intensity scale, implying a washed-out grey look through the whole image. As a consequence, the meaningful partition of the original image results in a very difficult task leading to imprecise results. The opposite is true for the histogram of a high-contrast image, which covers a wide range of the intensity scale and has a pixel distribution not too far from uniform. The effect is an image that shows a great deal of gray-level detail and has a high dynamic range. As a matter of fact, a segmentation process on this kind of images will produce satisfying results with a high reproducibility. As easily appreciable in Figure 13, MR images offer a very difficult scenario where the lesion boundaries are not clearly distinguishable from the rest of the image. The manual segmentation (green contour) cannot properly segment the whole lesion because of the strong uncertainty due to the low contrast characterizing the whole image. On the other hand, the FCM segmentation (red contour) correctly identifies the lesion but, because of the low percentage of overlapping area with the manual mask, it does not ensure satisfying results.

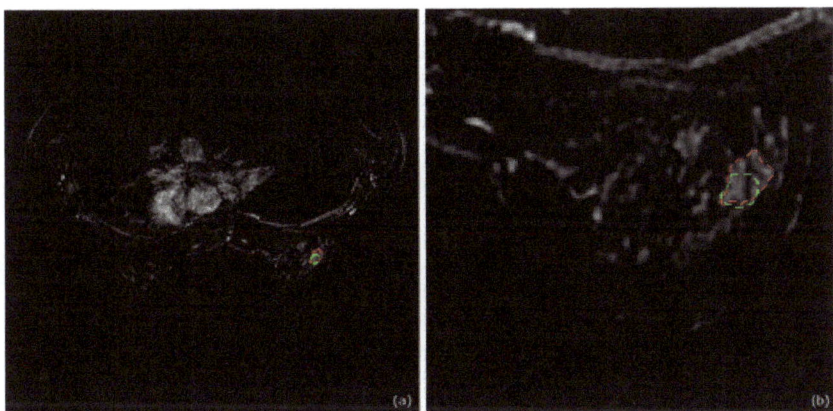

Figure 13. Case with a low-contrast lesion: (**a**) entire axial DCE-MRI slice; (**b**) crop of the slice shown in (**a**). Comparison between manual ground-truth (green) and automatic segmentation (red).

5.4.2. Case with Blurred Boundary Mass

From Figure 14, it is possible to appreciate how blur severely compromises the contrast of the original image, making lesion segmentation a very challenging task. As mentioned above, because of the strong uncertainty in boundaries delineation, the manual segmentation (green contour) identifies a very simple and smooth shape into which the lesion is included. Of course, this kind of strategy allows the identification of lesion's location, but also includes into the ROI a lot of false positives. On the other hand, the semi-automatic mask (red contour) identifies a more precise region avoiding the misclassification of FP pixels. Even in this case, because of the imperfect area, overlap area-based metrics will not yield a high score.

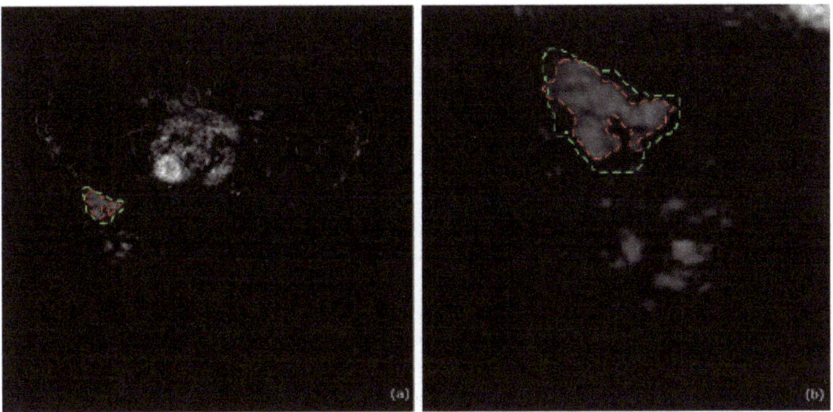

Figure 14. Case with a blurred boundary lesion: (**a**) entire axial DCE-MRI slice; (**b**) crop of the slice shown in (**a**). Comparison between manual ground-truth (green) and automated segmentation (red).

5.4.3. Case with Irregular Mass

Opposed to what is reported in the literature [39], breast lesions may sometimes exhibit irregular shapes and borders with internal divisions. In these cases, breast masses could cause difficulties during the manual segmentation process: in fact, as the lesion contour becomes more irregular, the manual tracing of the ROI becomes more challenging. As a consequence, the manual segmentation of breast lesions with unusual shapes does not always match the effective lesion contour. Figure 15 exhibits one of this cases with unusual elongated breast lesion. The manual segmentation (green contour) completely cut the left portion in the upper part of the lesion which is, instead, properly included in the semi-automatic boundary (red contour). As a consequence, computer-assisted segmentation process results in a more precise mask that properly reproduces the narrowed-shape of lesion.

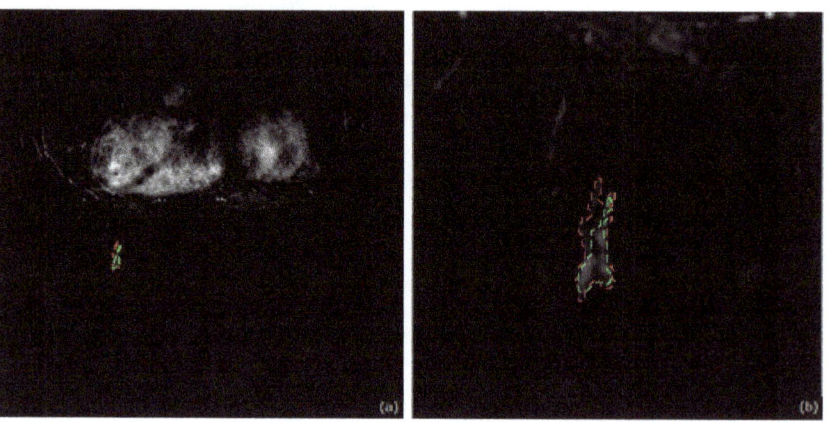

Figure 15. Case with an irregular lesion: (**a**) entire axial DCE-MRI slice; (**b**) crop of the slice shown in (**a**). Comparison between manual ground-truth (green) and automated segmentation (red).

6. Discussion and Conclusions

The main objective of this work was to offer a detailed analysis of well-established classical unsupervised segmentation techniques by carefully comparing them in a real clinical application. Breast cancer is the most common cause of cancer death in women worldwide [52,53] and the second most common cancer overall [54]. Fortunately, science evolution has led to the development of medical imaging techniques, which are used to detect abnormalities in breast parenchyma. Among imaging techniques, multiparametric MRI plays a crucial role and it is widely used in clinical applications, due to its high resolution images and the ability to precisely differentiate soft tissues.

As matter of fact, as a case study, the contrast-enhancing mass delineation in breast DCE-MRI was addressed by means of four popular unsupervised segmentation methods, namely: SMRG, k-means, FCM, and sFCM. Although they represent well-known approaches in the literature, they are still widely used in clinical tools. Starting from the basic versions of these approaches, during the initial analysis, we identified the shortcomings of each of them, developing and implementing improved versions, when possible.

Nowadays, deep learning techniques represent the state-of-the-art, allowing us to achieve high performance and accurate lesion segmentation for datasets of thousands of patients [14]. Deep learning approaches for image segmentation are generally supervised techniques that require a considerable computation times and a large amount of data for training [15]. In fact, these data must be representative of all the possible scenarios in which the deep neural network could operate, not always available in small- or medium-sized hospitals, and therefore not clinically feasible. Moreover, it should be noted that the authors of [18] showed that semi-automatic approached based on classic unsupervised techniques obtained results comparable or superior to the deep CNNs (namely, SegNet and U-Net). Therefore, this study was focused on classic pattern recognition approaches with the goal of providing an in-depth analysis.

It is important to point out that, even considering all the disadvantageous aspects related to manual segmentation, the contribution of an expert radiologist still remains essential at least to validate the results obtained by means a computer-assisted approach. In fact, clinicians rely on computational approaches with interpretable results [41]. From this point of view, the classical unsupervised approaches—such as those analyzed in this work—provide this important advantage. This aspect is even more critical in deep learning architectures, where CNNs are generally adopted as 'black-box', thus making it difficult to offer a physical interpretation to the features encoded in all the intermediate CNN layers [46].

The obtained experimental results, in terms of area- and distance-based metrics, encourage the use of unsupervised pattern recognition techniques in medical image segmentation. In particular, consistently with [55], clustering-based segmentation approaches achieved better performance compared to the SMRG, the only 'thresholding-based' approach considered. As a consequence, crisp segmentation techniques—such as k-means and SMRG—are not well-suited for medical images that are characterized by an uncertain/variability (sometimes related to the noise), yielding inaccurate boundaries and not well-defined details. Both FCM and sFCM clustering techniques—implementing fuzzy modeling that provides an intrinsic flexibility—significantly achieved the best performance. In fact, on area-based metrics, they obtained DI = 78.23% ± 6.50 (sFCM), JI = 65.90% ± 8.14 (sFCM), sensitivity = 77.84% ± 8.72 (FCM), and specificity = 87.10% ± 8.24 (sFCM), FPR = 0.14 ± 0.12 (sFCM), and FNR = 0.22 ± 0.09 (sFCM). On distance-based metrics, they obtained MAD = 1.37 ± 0.90 (sFCM), MaxD = 4.04 ± 2.87 (sFCM), and HD = 2.21 ± 0.43 (FCM).

A second segmentation of the same radiologist or the segmentation of a different radiologist would have certainly allowed us to quantify the inter-/intra-operator variability of the results. Nevertheless, as already observed in [56], the mean DI was 0.81 (range 0.19–0.96). The mean DI is higher for the 'easy tumors' compared to the 'challenging tumors' (0.83 *vs.* 0.75, respectively, $p < 0.001$). The mean DI for each observer combination separately, for all tumors, ranged between 0.78 and 0.83, where the segmentations of the breast radiologist and the medical student showed the highest overlap. These results confirm that the performance achieved by the best performing methods are in line with the inter-observer agreement, also in terms of metrics variability according to the lesion types.

As further developments, we plan to investigate innovative improvements to further improve the performance with fuzzy clustering, by using (*i*) more sophisticated membership functions, and (*ii*) more advanced pre- and post-processing steps. Moreover, investigating and comparing the latest machine learning techniques, such as Generative Adversarial Networks (GANs), for unsupervised detection and segmentation [57,58] would be relevant with a sufficient amount of data for training and test. Finally, the implementation of multiparametric or multimodal approaches [59], by using different types of co-registered medical images—i.e., Diffusion Weighted Imaging (DWI) and Positron Emission Tomography (PET)/MRI—probably would allow us to improve the detection performance [60,61].

Supplementary Materials: The following are available online at https://www.mdpi.com/article/10.3390/app12010162/s1: Table S1: Dice Index (DI) values obtained by the four investigated unsupervised methods on each of the 50 segmented breast masses on DCE-MRI. In the last row the mean value ± the standard deviation is reported; Table S2: Jaccard Index (JI) values obtained by the four investigated unsupervised methods on each of the 50 segmented breast masses on DCE-MRI. In the last row the mean value ± the standard deviation is reported; Table S3: Sensitivity values obtained by the four investigated unsupervised methods on each of the 50 segmented breast masses on DCE-MRI. In the last row the mean value ± the standard deviation is reported; Table S4: Specificity values obtained by the four investigated unsupervised methods on each of the 50 segmented breast masses on DCE-MRI. In the last row the mean value ± the standard deviation is reported; Table S5: False Positive Ratio (FPR) values obtained by the four investigated unsupervised methods on each of the 50 segmented breast masses on DCE-MRI. In the last row the mean value ± the standard deviation is reported; Table S6: False Negative Ratio (FNR) values obtained by the four investigated unsupervised methods on each of the 50 segmented breast masses on DCE-MRI. In the last row the mean value ± the standard deviation is reported; Table S7: Mean Absolute Distance (MAD) values obtained by the four investigated unsupervised methods on each of the 50 segmented breast masses on DCE-MRI. In the last row the mean value ± the standard deviation is reported; Table S8: Maximum

Distance (MaxD) values obtained by the four investigated unsupervised methods on each of the 50 segmented breast masses on DCE-MRI. In the last row the mean value ± the standard deviation is reported; Table S9: Hausdorff Distance (HD) values obtained by the four investigated unsupervised methods on each of the 50 segmented breast masses on DCE-MRI. In the last row the mean value ± the standard deviation is reported.

Author Contributions: Conceptualization, C.M. and A.R.; methodology, C.M. A.R. and L.R.; software, A.R.; validation, C.M. and L.R.; formal analysis, C.M. and L.R.; investigation, A.R.; resources, C.M., L.R. and I.D.; data curation, C.M. and I.D.; writing—original draft preparation, C.M., A.R. and L.R.; writing—review and editing, F.M., T.V.B., F.B. and G.R.; visualization, C.M. and L.R.; supervision, F.M., T.V.B., F.B. and G.R.; project administration, T.V.B., F.B. and G.R.; funding acquisition, L.R. and G.R. All authors have read and agreed to the published version of the manuscript.

Funding: This study has received funding by GeSeTon project, funded by Italian MISE Grant No. 489 of 21 February 2018. This study has also been partially supported by The Mark Foundation for Cancer Research and Cancer Research UK Cambridge Centre [C9685/A25177] and by the Royal Society for the International Exchanges 2020 Cost Share with the Italian CNR (project No. IEC/R2/202313). Additional support was also provided by the National Institute of Health Research (NIHR) Cambridge Biomedical Research Centre [BRC-1215-20014]. The views expressed are those of the authors and not necessarily those of the NHS, the NIHR, or the Department of Health and Social Care.

Institutional Review Board Statement: The study was conducted according to the guidelines of the Declaration of Helsinki, and approved by the Ethics Committee of "Azienda Ospedaliera Universitaria Policlinico P.Giaccone" of Palermo, Italy (protocol code n.1/2020-15/01/2020).

Informed Consent Statement: Retrospective data collection was approved by the Ethics Committee. The requirement for evidence of informed consent was waived because of the retrospective nature of our study.

Data Availability Statement: The data presented in this study may be available on reasonable request from the corresponding author. The data are not publicly available due to ethical and privacy restrictions.

Conflicts of Interest: The authors declare no conflict of interest.

References

1. Badr, E. Images in Space and Time: Real Big Data in Healthcare. *ACM Comput. Surv.* **2021**, *54*, 113. [CrossRef]
2. Duncan, J.S.; Ayache, N. Medical image analysis: Progress over two decades and the challenges ahead. *IEEE Trans. Pattern Anal. Mach. Intell.* **2000**, *22*, 85–106. [CrossRef]
3. Ghavami, N.; Hu, Y.; Gibson, E.; Bonmati, E.; Emberton, M.; Moore, C.M.; Barratt, D.C. Automatic segmentation of prostate MRI using convolutional neural networks: Investigating the impact of network architecture on the accuracy of volume measurement and MRI-ultrasound registration. *Med. Image Anal.* **2019**, *58*, 101558. [CrossRef]
4. Lee, B.; Yamanakkanavar, N.; Choi, J.Y. Automatic segmentation of brain MRI using a novel patch-wise U-net deep architecture. *PLoS ONE* **2020**, *15*, e0236493. [CrossRef]
5. Pasero, E.; Castagneri, C. Application of an automatic ulcer segmentation algorithm. In Proceedings of the International Forum on Research and Technologies for Society and Industry (RTSI), Modena, Italy, 11–13 September 2017; pp. 1–4. [CrossRef]
6. Yankeelov, T.E.; Mankoff, D.A.; Schwartz, L.H.; Lieberman, F.S.; Buatti, J.M.; Mountz, J.M.; Erickson, B.J.; Fennessy, F.M.; Huang, W.; Kalpathy-Cramer, J.; et al. Quantitative imaging in cancer clinical trials. *Clin. Cancer Res.* **2016**, *22*, 284–290. [CrossRef] [PubMed]
7. Aerts, H.J.; Velazquez, E.R.; Leijenaar, R.T.; Parmar, C.; Grossmann, P.; Carvalho, S.; Bussink, J.; Monshouwer, R.; Haibe-Kains, B.; Rietveld, D.; et al. Decoding tumour phenotype by noninvasive imaging using a quantitative radiomics approach. *Nat. Commun.* **2014**, *5*, 4006. [CrossRef] [PubMed]
8. Lambin, P.; Leijenaar, R.T.; Deist, T.M.; Peerlings, J.; de Jong, E.E.; van Timmeren, J.; Sanduleanu, S.; Larue, R.T.; Even, A.J.; Jochems, A.; et al. Radiomics: The bridge between medical imaging and personalized medicine. *Nat. Rev. Clin. Oncol.* **2017**, *14*, 749. [CrossRef]
9. Krupinski, E.A. Current perspectives in medical image perception. *Atten. Percept. Psychophys.* **2010**, *72*, 1205–1217. [CrossRef] [PubMed]
10. Rundo, L.; Pirrone, R.; Vitabile, S.; Sala, E.; Gambino, O. Recent advances of HCI in decision-making tasks for optimized clinical workflows and precision medicine. *J. Biomed. Inf.* **2020**, *108*, 103479. [CrossRef]
11. Lee, N.Y.; Lu, J.J. *Target Volume Delineation and Field Setup: A Practical Guide for Conformal and Intensity-Modulated Radiation Therapy*, 1st ed.; Springer: Berlin/Heidelberg, Germany, 2013. [CrossRef]

12. Hamamci, A.; Kucuk, N.; Karaman, K.; Engin, K.; Unal, G. Tumor-Cut: Segmentation of brain tumors on contrast enhanced MR images for radiosurgery applications. *IEEE Trans. Med. Imaging* **2012**, *31*, 790–804. [CrossRef]
13. Rundo, L.; Militello, C.; Russo, G.; Vitabile, S.; Gilardi, M.C.; Mauri, G. GTVcut for neuro-radiosurgery treatment planning: An MRI brain cancer seeded image segmentation method based on a cellular automata model. *Nat. Comput.* **2018**, *17*, 521–536. [CrossRef]
14. Litjens, G.; Kooi, T.; Bejnordi, B.E.; Setio, A.A.A.; Ciompi, F.; Ghafoorian, M.; Van Der Laak, J.A.; Van Ginneken, B.; Sánchez, C.I. A survey on deep learning in medical image analysis. *Med. Image Anal.* **2017**, *42*, 60–88. [CrossRef]
15. Tajbakhsh, N.; Jeyaseelan, L.; Li, Q.; Chiang, J.N.; Wu, Z.; Ding, X. Embracing imperfect datasets: A review of deep learning solutions for medical image segmentation. *Med. Image Anal.* **2020**, *63*, 101693. [CrossRef]
16. Tajbakhsh, N.; Shin, J.Y.; Gurudu, S.R.; Hurst, R.T.; Kendall, C.B.; Gotway, M.B.; Liang, J. Convolutional neural networks for medical image analysis: Full training or fine tuning? *IEEE Trans. Med. Imaging* **2016**, *35*, 1299–1312. [CrossRef] [PubMed]
17. Ravì, D.; Wong, C.; Deligianni, F.; Berthelot, M.; Andreu-Perez, J.; Lo, B.; Yang, G.Z. Deep learning for health informatics. *IEEE J. Biomed. Health Inform.* **2017**, *21*, 4–21. [CrossRef] [PubMed]
18. Militello, C.; Rundo, L.; Dimarco, M.; Orlando, A.; Conti, V.; Woitek, R.; D'Angelo, I.; Bartolotta, T.V.; Russo, G. Semi-automated and interactive segmentation of contrast-enhancing masses on breast DCE-MRI using spatial fuzzy clustering. *Biomed. Signal Process. Control.* **2022**, *71*, 103113. [CrossRef]
19. Acharya, R.; Wasserman, R.; Stevens, J.; Hinojosa, C. Biomedical imaging modalities: A tutorial. *Comput. Med. Imaging Graph.* **1995**, *19*, 3–25. [CrossRef]
20. Tirpude, N.N.; Welekar, R.R. Effect Of Global Thresholding On Tumor-Bearing Brain MRI Images. *Int. J. Eng. Comput. Sci.* **2013**, *2*, 728–731.
21. Militello, C.; Vitabile, S.; Rundo, L.; Russo, G.; Midiri, M.; Gilardi, M.C. A fully automatic 2D segmentation method for uterine fibroid in MRgFUS treatment evaluation. *Comput. Biol. Med.* **2015**, *62*, 277–292. [CrossRef] [PubMed]
22. Islam, M.R.; Imteaz, M.R.; Marium-E-Jannat. Detection and analysis of brain tumor from MRI by Integrated Thresholding and Morphological Process with Histogram based method. In Proceedings of the International Conference on Computer, Communication, Chemical, Material and Electronic Engineering (IC4ME2), Rajshahi, Bangladesh, 8–9 February 2018; pp. 1–5.
23. Rundo, L.; Militello, C.; Vitabile, S.; Casarino, C.; Russo, G.; Midiri, M.; Gilardi, M.C. Combining Split-and-Merge and Multi-Seed Region Growing Algorithms for Uterine Fibroid Segmentation in MRgFUS Treatments. *Med. Biol. Eng. Comput.* **2016**, *54*, 1071–1084. [CrossRef]
24. Horowitz, S.L.; Pavlidis, T. Picture Segmentation by a Tree Traversal Algorithm. *J. ACM* **1976**, *23*, 368–388. [CrossRef]
25. Manousakas, I.N.; Undrill, P.E.; Cameron, G.G.; Redpath, T.W. Split-and-Merge Segmentation of Magnetic Resonance Medical Images: Performance Evaluation and Extension to Three Dimensions. *Comput. Biomed. Res.* **1998**, *31*, 393–412. [CrossRef]
26. Saad, N.M.; Abu-Bakar, S.A.R.; Muda, S.; Mokji, M. Automated segmentation of brain lesion based on diffusion-weighted MRI using a split and merge approach. In Proceedings of the IEEE EMBS Conference on Biomedical Engineering and Sciences (IECBES), Kuala Lumpur, Malaysia, 30 November–2 December 2010; pp. 475–480.
27. Saad, N.M.; Abu-Bakar, S.A.R.; Muda, S.; Mokji, M.; Abdullah, A.R. Automated region growing for segmentation of brain lesion in diffusion-weighted MRI. In Proceedings of the International MultiConference of Engineers and Computer Scientists, IMECS 2012, Hong Kong, China, 14–16 March 2012; pp. 674–677.
28. Adams, R.; Bischof, L. Seeded region growing. *IEEE Trans. Pattern Anal.* **1994**, *16*, 641–647. [CrossRef]
29. Chang, Y.L.; Li, X. Adaptive image region-growing. *IEEE Trans. Image Process.* **1994**, *3*, 868–872. [CrossRef]
30. Otsu, N. A threshold selection method from gray-level histograms. *IEEE Trans. Syst. Man Cybern.* **1975**, *11*, 23–27. [CrossRef]
31. Joseph, R.P.; Senthil Singh, C.; Manikandan, M. Brain tumor MRI image segmentation and detection in image processing. *Int. J. Res. Eng. Technol.* **2014**, *3*, 1–5.
32. Kanungo, T.; Mount, D.M.; Netanyahu, N.S.; Piatko, C.D.; Silverman, R.; Wu, A.Y. An efficient k-means clustering algorithm: analysis and implementation. *IEEE Trans. Pattern Anal. Mach. Intell.* **2002**, *24*, 881–892. [CrossRef]
33. Bezdek, J.C. Objective function clustering. In *Pattern Recognition with Fuzzy Objective Function Algorithms*, 1st ed.; Springer: Secaucus, NJ, USA, 1981; pp. 43–93. [CrossRef]
34. Li, Y.l.; Shen, Y. An automatic fuzzy c-means algorithm for image segmentation. *Soft Comput.* **2010**, *14*, 123–128. [CrossRef]
35. Militello, C.; Rundo, L.; Vitabile, S.; Russo, G.; Pisciotta, P.; Marletta, F.; Ippolito, M.; D'Arrigo, C.; Midiri, M.; Gilardi, M. Gamma Knife treatment planning: MR brain tumor segmentation and volume measurement based on unsupervised Fuzzy C-Means clustering. *Int. J. Imaging Syst. Technol.* **2015**, *25*, 213–225. [CrossRef]
36. Rundo, L.; Militello, C.; Russo, G.; Garufi, A.; Vitabile, S.; Gilardi, M.C.; Mauri, G. Automated prostate gland segmentation based on an unsupervised fuzzy c-means clustering technique using multispectral T1w and T2w MR imaging. *Information* **2017**, *8*, 49. [CrossRef]
37. Caponetti, L.; Castellano, G.; Corsini, V. MR brain image segmentation: A framework to compare different clustering techniques. *Information* **2017**, *8*, 138. [CrossRef]
38. Rundo, L.; Militello, C.; Tangherloni, A.; Russo, G.; Vitabile, S.; Gilardi, M.C.; Mauri, G. NeXt for neuro-radiosurgery: A fully automatic approach for necrosis extraction in brain tumor MRI using an unsupervised machine learning technique. *Int. J. Imaging Syst. Technol.* **2018**, *28*, 21–37. [CrossRef]
39. Feder, J.M.; de Paredes, E.S.; Hogge, J.P.; Wilken, J.J. Unusual breast lesions: Radiologic-pathologic correlation. *Radiographics* **2019**, *19*, S11–S26. [CrossRef] [PubMed]

40. Chuang, K.S.; Tzeng, H.L.; Chen, S.; Wu, J.; Chen, T.J. Fuzzy c-means clustering with spatial information for image segmentation. *Comput. Med. Imaging Graph.* **2006**, *30*, 9–15. [CrossRef] [PubMed]
41. Rundo, L.; Beer, L.; Ursprung, S.; Martin-Gonzalez, P.; Markowetz, F.; Brenton, J.D.; Crispin-Ortuzar, M.; Sala, E.; Woitek, R. Tissue-specific and interpretable sub-segmentation of whole tumour burden on CT images by unsupervised fuzzy clustering. *Comput. Biol. Med.* **2020**, *120*, 103751. [CrossRef] [PubMed]
42. Patil, S. Preprocessing To Be Considered For MR and CT Images Containing Tumors. *IOSR J. Electr. Electron. Eng.* **2012**, *1*, 55–57. [CrossRef]
43. Vasuki, P.; Kanimozhi, J.; Devi, M.B. A survey on image preprocessing techniques for diverse fields of medical imagery. In Proceedings of the 2017 IEEE International Conference on Electrical, Instrumentation and Communication Engineering (ICEICE), Tamilnadu, India, 27–28 April 2017; pp. 1–6.
44. Behrenbruch, C.; Petroudi, S.; Bond, S.; Declerck, J.; Leong, F.; Brady, J. Image filtering techniques for medical image post-processing: An overview. *Br. J. Radiol.* **2004**, *77*, S126—S132. [CrossRef] [PubMed]
45. Seeram, E.; Seeram, D. Image Postprocessing in Digital Radiology—A Primer for Technologists. *J. Med. Imaging Radiat. Sci.* **2008**, *39*, 23–41. [CrossRef]
46. Castiglioni, I.; Rundo, L.; Codari, M.; Di Leo, G.; Salvatore, C.; Interlenghi, M.; et al. AI applications to medical images: From machine learning to deep learning. *Phys. Med.* **2021**, *83*, 9–24. [CrossRef]
47. Motwani, M.C.; Gadiya, M.C.; Motwani, R.C.; Harris, F.C. Survey of image denoising techniques. *Glob. Signal Process. Expo Conf. (GSPX)* **2004**, *27*, 27–30.
48. Marrone, S.; Piantadosi, G.; Fusco, R.; Petrillo, A.; Sansone, M.; Sansone, C. Breast segmentation using Fuzzy C-Means and anatomical priors in DCE-MRI. In Proceedings of the 23rd International Conference on Pattern Recognition (ICPR), Cancun, Mexico, 4–8 December 2016; pp. 1472–1477.
49. Piantadosi, G.; Fusco, R.; Petrillo, A.; Sansone, M.; Sansone, C. LBP-TOP for volume lesion classification in breast DCE-MRI. In *International Conference on Image Analysis and Processing*; Lecture Notes in Computer Science; Springer: Cham, Switzerland, 2015; pp. 647–657. [CrossRef]
50. Wilcoxon, F. Individual comparisons by ranking methods. *Biom. Bull.* **1980**, *1*, 196–202. [CrossRef]
51. Holm, S. A Simple Sequentially Rejective Multiple Test Procedure. *Scand. J. Stat.* **1979**, *6*, 65–70.
52. DeSantis, C.E.; Ma, J.; Gaudet, M.M.; Newman, L.A.; Miller, K.D.; Goding Sauer, A.; Jemal, A.; Siegel, R.L. Breast cancer statistics, 2019. *CA Cancer J. Clin.* **2019**, *69*, 438–451. [CrossRef]
53. Siegel, R.L.; Miller, K.D.; Fuchs, H.E.; Jemal, A. Cancer Statistics, 2021. *CA Cancer J. Clin.* **2021**, *71*, 7–33. [CrossRef] [PubMed]
54. International Agency for Research on Cancer. The Global Cancer Observatory. 2020. Available online: https://gco.iarc.fr/today/data/factsheets/cancers/20-Breast-fact-sheet.pdf (accessed on 28 July 2021).
55. Frackiewicz, M.; Koper, Z.; Palus, H.; Borys, D.; Psiuk-Maksymowicz, K. Breast lesion segmentation in DCE-MRI Imaging. In Proceedings of the 14th International Conference on Signal-Image Technology & Internet-Based Systems (SITIS), Las Palmas de Gran Canaria, Spain, 26–29 November 2018; pp. 308–313. [CrossRef]
56. Granzier, R.; Verbakel, N.; Ibrahim, A.; van Timmeren, J.; van Nijnatten, T.; Leijenaar, R.; Lobbes, M.; Smidt, M.; Woodruff, H. MRI-based radiomics in breast cancer: Feature robustness with respect to inter-observer segmentation variability. *Sci. Rep.* **2020**, *10*, 14163. [CrossRef] [PubMed]
57. Han, C.; Rundo, L.; Murao, K.; Noguchi, T.; Shimahara, Y.; Milacski, Z.Á.; Koshino, S.; Sala, E.; Nakayama, H.; Satoh, S. MADGAN: Unsupervised medical anomaly detection GAN using multiple adjacent brain MRI slice reconstruction. *BMC Bioinf.* **2021**, *22*, 31. [CrossRef]
58. Wu, X.; Bi, L.; Fulham, M.; Kim, J. Unsupervised Positron Emission Tomography Tumor Segmentation via GAN based Adversarial Auto-Encoder. In Proceedings of the 16th International Conference on Control, Automation, Robotics and Vision (ICARCV), Shenzhen, China, 13–15 December 2020; pp. 448–453. [CrossRef]
59. Rundo, L.; Stefano, A.; Militello, C.; Russo, G.; Sabini, M.G.; D'Arrigo, C.; Marletta, F.; Ippolito, M.; Mauri, G.; Vitabile, S.; et al. A fully automatic approach for multimodal PET and MR image segmentation in Gamma Knife treatment planning. *Comput. Methods Programs Biomed.* **2017**, *144*, 77–96. [CrossRef]
60. Woitek, R.; McLean, M.A.; Gill, A.B.; Grist, J.T.; Provenzano, E.; Patterson, A.J.; Ursprung, S.; Torheim, T.; Zaccagna, F.; Locke, M.; et al. Hyperpolarized 13C MRI of tumor metabolism demonstrates early metabolic response to neoadjuvant chemotherapy in breast cancer. *Radiol. Imaging Cancer* **2020**, *2*, e200017. [CrossRef]
61. Carmona-Bozo, J.C.; Manavaki, R.; Woitek, R.; Torheim, T.; Baxter, G.C.; Caracò, C.; Provenzano, E.; Graves, M.J.; Fryer, T.D.; Patterson, A.J.; et al. Hypoxia and perfusion in breast cancer: Simultaneous assessment using PET/MR imaging. *Eur. Radiol.* **2021**, *31*, 333–344. [CrossRef]

Article

Development of Detection and Volumetric Methods for the Triceps of the Lower Leg Using Magnetic Resonance Images with Deep Learning

Yusuke Asami [1], Takaaki Yoshimura [2,3], Keisuke Manabe [1], Tomonari Yamada [1] and Hiroyuki Sugimori [4,5,*]

1 Graduate School of Health Sciences, Hokkaido University, Sapporo 060-0812, Japan; yusuke12@eis.hokudai.ac.jp (Y.A.); ksk0843@eis.hokudai.ac.jp (K.M.); tomonarihandball@eis.hokudai.ac.jp (T.Y.)
2 Department of Health Sciences and Technology, Faculty of Health Sciences, Hokkaido University, Sapporo 060-0812, Japan; takaaki.ysm@med.hokudai.ac.jp
3 Department of Medical Physics, Hokkaido University Hospital, Sapporo 060-8648, Japan
4 Department of Biomedical Science and Engineering, Faculty of Health Sciences, Hokkaido University, Sapporo 060-0812, Japan
5 Clinical AI Human Resources Development Program, Faculty of Medicine, Hokkaido University, Sapporo 060-8648, Japan
* Correspondence: sugimori@hs.hokudai.ac.jp; Tel.: +81-11-706-3410

Abstract: Purpose: A deep learning technique was used to analyze the triceps surae muscle. The devised interpolation method was used to determine muscle's volume and verify the usefulness of the method. Materials and Methods: Thirty-eight T1-weighted cross-sectional magnetic resonance images of the triceps of the lower leg were divided into three classes, i.e., gastrocnemius lateralis (GL), gastrocnemius medialis (GM), and soleus (SOL), and the regions of interest (ROIs) were manually defined. The supervised images were classified as per each patient. A total of 1199 images were prepared. Six different datasets separated patient-wise were prepared for K-fold cross-validation. A network model of the DeepLabv3+ was used for training. The images generated by the created model were divided as per each patient and classified into each muscle types. The model performance and the interpolation method were evaluated by calculating the Dice similarity coefficient (DSC) and error rates of the volume of the predicted and interpolated images, respectively. Results: The mean DSCs for the predicted images were >0.81 for GM and SOL and 0.71 for GL. The mean error rates for volume were approximately 11% for GL, SOL, and total error and 23% for GL. DSCs in the interpolated images were >0.8 for all muscles. The mean error rates of volume were <10% for GL, SOL, and total error and 18% for GM. There was no significant difference between the volumes obtained from the supervised images and interpolated images. Conclusions: Using the semantic segmentation of the deep learning technique, the triceps were detected with high accuracy and the interpolation method used in this study to find the volume was useful.

Keywords: deep learning; semantic segmentation; triceps surae muscle

Citation: Asami, Y.; Yoshimura, T.; Manabe, K.; Yamada, T.; Sugimori, H. Development of Detection and Volumetric Methods for the Triceps of the Lower Leg Using Magnetic Resonance Images with Deep Learning. *Appl. Sci.* **2021**, *11*, 12006. https://doi.org/10.3390/app112412006

Academic Editors: Leonardo Rundo, Carmelo Militello and Andrea Tangherloni

Received: 9 November 2021
Accepted: 14 December 2021
Published: 16 December 2021

1. Introduction

Deep learning technology has been widely used in recent years for automatic driving, drones, weather forecasting, and games [1–4]. The deep learning techniques include classification [5], object detection [6], and semantic segmentation. Semantic segmentation has been used to visualize the three-dimensional (3D) anatomical structures of multiple organs in 3D computed tomography images [7], to facilitate the quantitative coronary angiography-based diagnosis of major vessels in X-ray coronary angiography [8], and to quantitate whole breast image analysis in diffusion-weighted images [9]. An applied method using Conditional Generative Adversarial Networks (cGANs) has been reported [10] to be highly accurate for in segmenting peri-knee tissue using magnetic resonance imaging (MRI).

These techniques are used in various body parts and modalities. This study focused on semantic segmentation of the triceps surae muscle, which consists of the gastrocnemius and soleus (SOL) muscles. The gastrocnemius is a biceps muscle that comprises the gastrocnemius lateralis (GL) and gastrocnemius medialis (GM). The triceps is an important muscle because it makes activities, such as walking and standing, possible by its repeated contraction and relaxation. Triceps muscle weakness increases the risk of falling and prevents these essential activities [11]. The muscle volume is commonly used as a measure of the muscle size, and it is frequently measured in studies because it correlates with various functional parameters [12]. In a study by Thom et al., the volume of the triceps femoris muscle was calculated to assess the loss of muscle mass with aging [13]. The volume was calculated by the product of the cross-sectional area and slice thickness obtained by manual segmentation using an imaging software; however, this procedure takes a long time [14]. Friedberger et al. [15] showed that it is now possible to perform semi-automatic volume calculations using the Random Forest classifier for hand muscle segmentation. However, this method is still semi-automatic, and even though it reduces the effort of manual segmentation, it still requires modifications and is significantly time-consuming. In addition, since magnetic resonance imaging of the lower extremities is often performed with thicker slices and more spacing between images, volume calculations using only the acquired images are likely to differ significantly from the actual volume. Therefore, it was believed that by slice interpolating the MRI images and calculating the volume using the interpolated images, this difference from the actual image could be reduced.

Many studies have been conducted on the segmentation of quadriceps [16–21], there have been no previous studies focusing on the segmentation of triceps with deep learning. In a study conducted by Essafi et al. [22], the authors performed the segmentation of the medial gastrocnemius, one of the triceps muscles, without deep learning and reported that the average Dice similarity coefficient (DSC) in 25 subjects was 0.55. In the studies of thigh muscle segmentation using deep learning [16–19], the higher the DSC > 0.9, the higher the accuracy of muscle detection. Furthermore, in the studies of segmentation of thigh muscles without deep learning [20,21], there are examples of detection with relatively high accuracy, although not as high compared with studies that have used deep learning. Based on these results, that the triceps could be detected with high accuracy using the deep learning method.

In this study, deep learning was used to detect the triceps surae muscle. Then, the devised interpolation method was used to determine the muscle volume and verify the usefulness of the method.

The manuscript is structured as follows. Section 2 outlines the preprocessing of the dataset and the methodology and evaluations for transfer learning and the image interpolation method for semantic segmentation. Section 3 describes the created models and the evaluation of the image interpolation method. In Section 4, the results obtained in this study are compared with those of other papers, and the limitations of this study are discussed. Finally, the conclusions and future directions of this study are presented in Section 5.

2. Materials and Methods

2.1. Subjects

The 38 consecutive patients who underwent a noncontrast-enhanced lower leg MRI examination were retrospectively evaluated. The Digital Imaging and Communications in Medicine (DICOM) images were subsequently converted to anonymized DICOM files from the image server. These images have been approved by the Ethics Committee of the Hokkaido University Hospital. The details information of the obtained images is shown in Table 1.

Table 1. The details information of the obtained images (N = 38).

	Mean ± SD [Min–Max]
Number of slices	17.9 ± 4.9 [3–29]
Field of view [mm]	342.1 × 295.0 [160 × 160–500 × 425]
Acquisition matrix size [pixel]	446.0 × 402.4 [320 × 224–672 × 672]
Pixel size [mm]	0.764 × 0.764 [0.559 × 0.559–0.928 × 0.928]
Slice thickness [mm]	5.8 ± 0.5 [4–6]
Slice gap [mm]	13.6 ± 4.7 [4.8–22.8]
Length of acquisition [mm]	231.0 ± 76.2 [42–342]

2.2. Preprocessing

For supervised image creation, the DICOM images were converted to 8-bit using the MATLAB (The MathWorks, Inc., Natick, MA, USA) "mat2gray" function because the bit depth of DICOM images is 16-bit. Those with only one leg captured were resized to 256 × 256 portable network graphics (PNG) images, whereas those with both legs captured were trimmed to the appropriate size for each leg and resized to 256 × 256. The left leg was reversed and oriented in the same way as the right leg (Figure 1). The coordinates of the trimmed area were also recorded to estimate the pixel size and the number of pixels in a trimmed image for volume calculation. Images of 67 leg muscles were obtained from 38 patients. The trimmed images were used for training and evaluation.

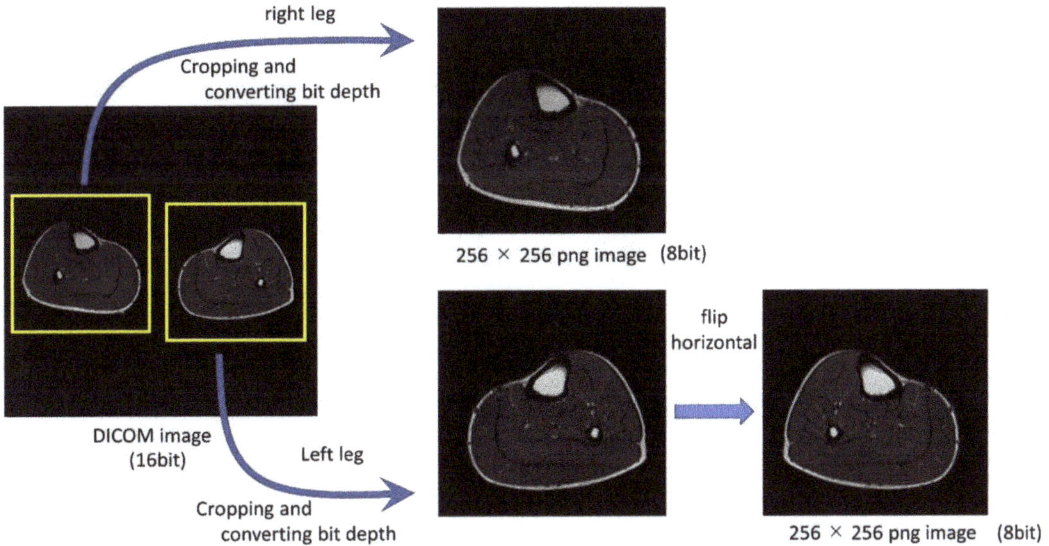

Figure 1. The conversion process from DICOM to PNG image.

2.3. Dataset

These images were saved as PNG images by dividing the triceps into three classes (GM, GL, and SOL) with an in-house MATLAB software; the regions of interest (ROIs) were also manually defined (Figure 2). The ROIs were manually defined by the author alone and were then verified by two radiological technologists with 5 (T.Y.) and 20 years of experience (H.S.), respectively. A total of 1199 images were prepared from 38 patients, and the Train:Test was set to 10:2. The 38 patients were classified into six groups, and six datasets were created so that one could be used as a test (Figure 2). Six different datasets with 1199 images were prepared for K-fold cross-validation so that the number of tests would be approximately 200. Data augmentation for improving the training [6] was performed on the training images, and they were rotated by 5° from −45 to 45°, increasing the number of images by 19 times.

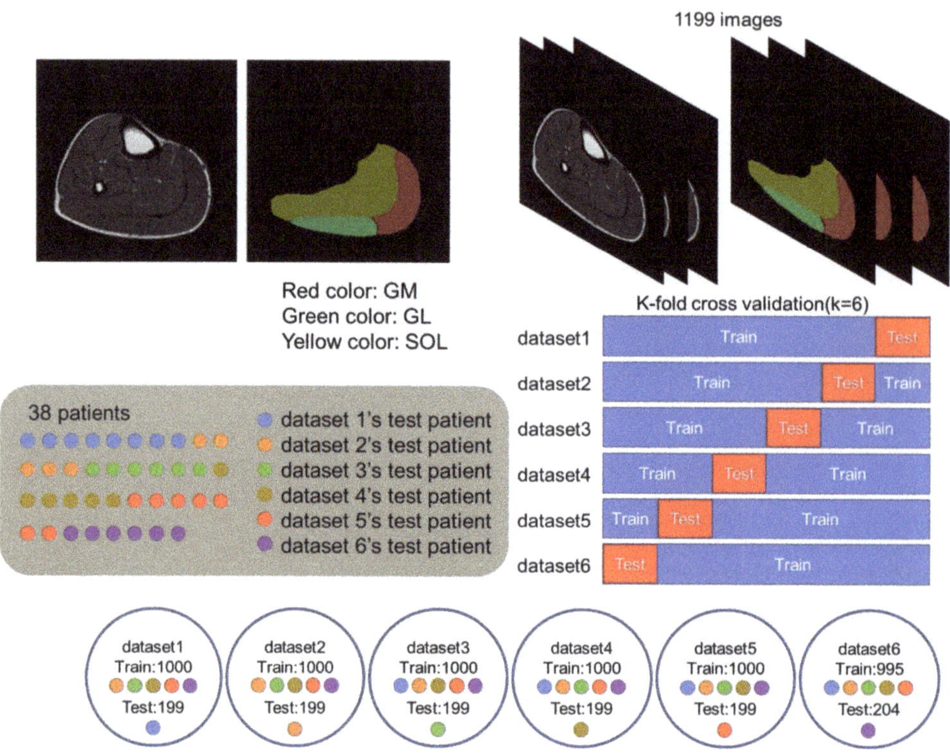

Figure 2. Six divided datasets and 6-fold cross-validation.

2.4. Training for Creating Models

The software was developed with MATLAB software, and a computer with NVIDIA GeForce GTX 1080Ti 12GB (NVIDIA Corporation, Santa Clara, CA, USA) was used. DeepLabv3+ was used for the architecture for implementation of image segmentation models. The model was inputted with 2D images taken by MRI. The following training parameters were used: the batch size for the number of training samples was 32, the number of epochs was 100, and the initial learning rate was 0.0001. Using this network model, six different datasets were trained to create six models.

2.5. Interpolation

The supervised images and the images generated using the devised model were divided in each patient and classified into three muscle types (GM, GL, and SOL). Only the images with muscles present in the supervised images were extracted and interpolated by excluding even-numbered images from the predicted images in the same position (Figures 3 and 4). The interpolation method is described in Figure 5. Interpolation was performed in the slice direction from two images of the muscle. The contours of the two muscles were represented by points. The centroid of the larger muscle was calculated and connected to the centroid by a point on the larger muscle. The distance between the connecting line and all points on the smaller muscle was calculated, and the point with the smallest distance was adopted. Moreover, the midpoint was calculated by connecting the adopted point to a point on the larger muscle. The same process was performed for all points of the larger muscle, and by connecting all the calculated midpoints, the interpolated image of the muscle was completed.

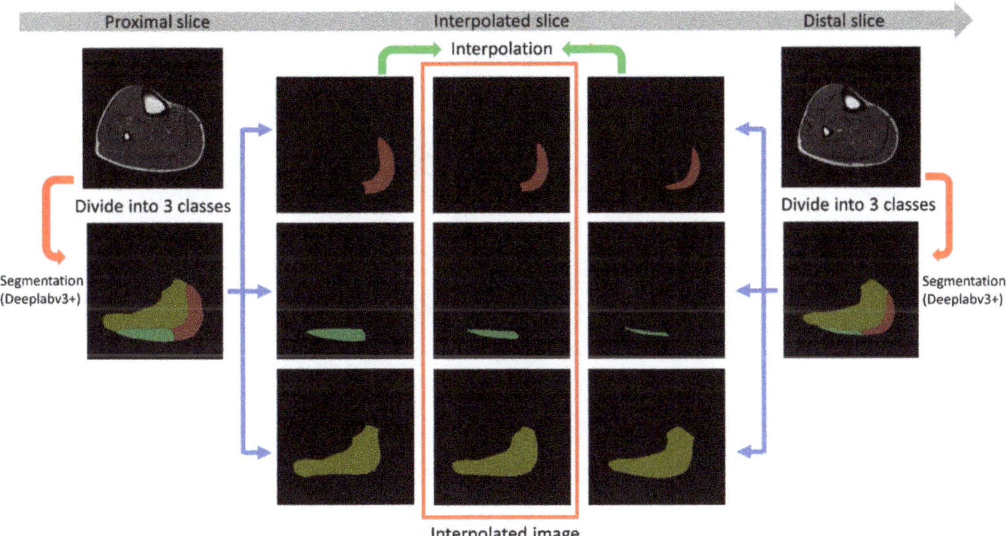

Figure 3. Classifying muscles into three for interpolation.

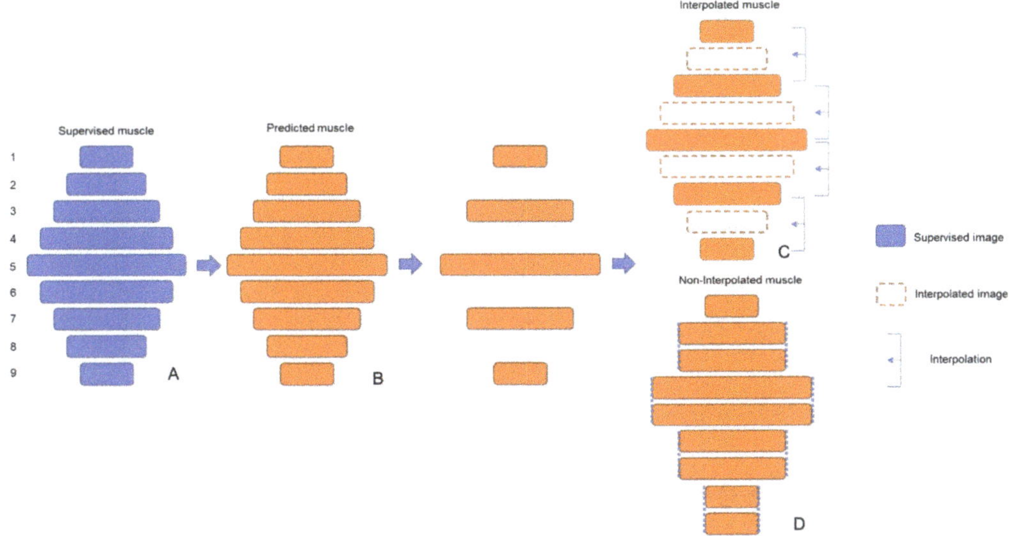

Figure 4. Creating the interpolated images from the predicted images based on the presence or absence of muscles in the supervised images. (**A**) Muscles in supervised images; (**B**) Predictive muscles using segmentation model; (**C**) Interpolated muscles using the interpolation method by thinning out the even number of predicted images; (**D**) Muscles with even-numbered predicted images thinned out and slice thickness doubled without using the interpolation method; 1–9; The number of the slice where the muscle is located.

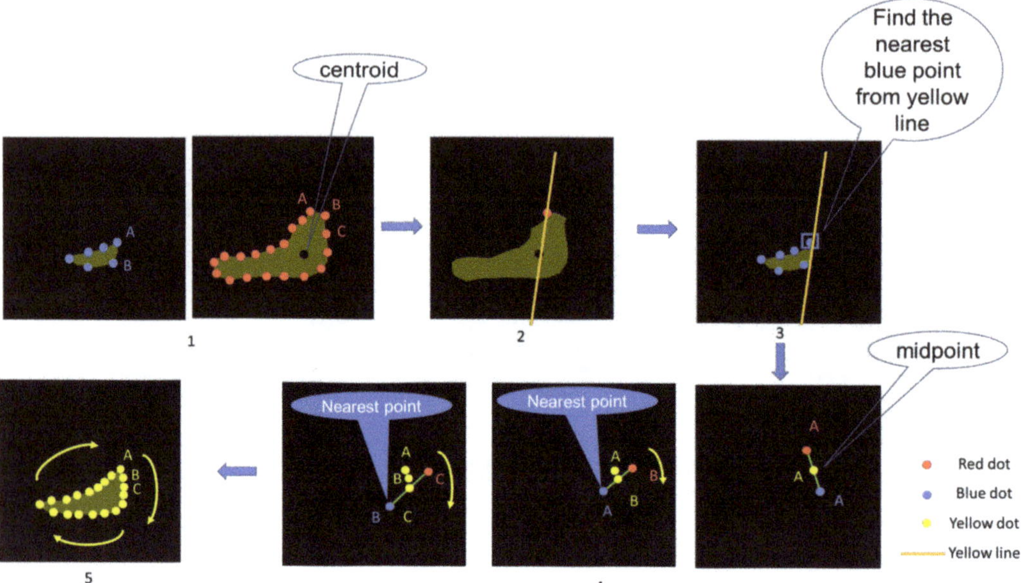

Figure 5. Procedure of interpolation: (1) The outline with dots and the calculated centroid of the muscle with the most dots were expressed; (2) a line was drawn connecting the red dot and the centroid; (3) measured the distance between all the blue points and the yellow line, found the nearest blue point from yellow line; (4) the midpoint of the red dot and the selected blue dot was marked; (5) a different red dot was selected and steps 2 through 4 were repeated for as many red dots as there are red dots, connecting all the yellow dots.

2.6. Indicators Used for Evaluation

2.6.1. DSC

When the supervised images were set as A and the predicted images were set as B, the following formula was used to calculate DSC. This value is an index to check the agreement between the images, and the closer the value is to 1, the better the agreement is.

$$DSC = \frac{2 \times |A \cap B|}{|A| + |B|} \tag{1}$$

2.6.2. Calculation of the Volume and Error Rate

The spacing between the slices and pixels were obtained from the DICOM data. The number of pixels used in the labeled muscle was determined, and the volume was calculated from the product of these values. The error rate was calculated to compare the volume of the supervised images with that of the predicted images. The error rate was calculated using the following formula.

$$\left| \frac{Volume\ of\ predicted\ images\ -\ Volume\ of\ supervised\ images}{Volume\ of\ supervised\ images} \right| \times 100\ (\%) \tag{2}$$

2.7. Evaluation of the Created Models

DSCs were calculated for each muscle using the abovementioned formula, and the averages were calculated for each dataset to show the overall average. All images were used for this evaluation. The volumes were calculated for each patient and muscle, and the average error rates of the datasets were calculated using the mentioned formula. For this evaluation, only the images with muscle present in both the supervised and predicted images were used to calculate the error rate.

2.8. Evaluation of the Interpolation Method

Only the interpolated images were used to obtain DSCs, and the average DSC for each muscle was calculated. DSCs were calculated from the interpolated image and the supervised image of the same number (Figure 4C). The volumes were calculated from the interpolated images, and the error rates were calculated for comparison with the volumes of the supervised images. To confirm that the interpolation method is useful in terms of reducing volumetric errors, the error rate was calculated by comparing the volume of the supervised image and the volume of the image with the spacing between its slices doubled without interpolation while excluding even numbered images (Figure 4D). All patients with all three muscles and all predicted images that could be interpolated were used for these evaluations. Statistical tests were performed to compare the interpolated and supervised volumes. First, the Shapiro–Wilk test was applied to evaluate normality; if normality was accepted, the *t*-test was used; otherwise, the Wilcoxon signed-ranked test, a nonparametric test, was used. The differences in the statistical analyses were considered statistically significant when p was <0.05.

3. Results

3.1. Evaluation of the Created Models

The DSC values per dataset are shown in Table 2. The following table shows the DSC values of all 67 leg muscles. GM and SOL were detected with relatively high accuracy. However, the DSCs of GL were lower than those of the others. The following are some representative examples of segmentation successes and failures (Figures 6 and 7)

Table 2. Average DSC values per dataset for the predicted images.

Position	Dataset1	Dataset2	Dataset3	Dataset4	Dataset5	Dataset6	Mean ± SD
GM	0.897	0.754	0.876	0.830	0.776	0.745	0.813 ± 0.064
GL	0.804	0.591	0.765	0.773	0.639	0.704	0.713 ± 0.084
SOL	0.889	0.820	0.791	0.823	0.840	0.844	0.835 ± 0.033

GM: gastrocnemius medialis, GL: gastrocnemius lateralis, SOL: soleus, SD: standard deviation.

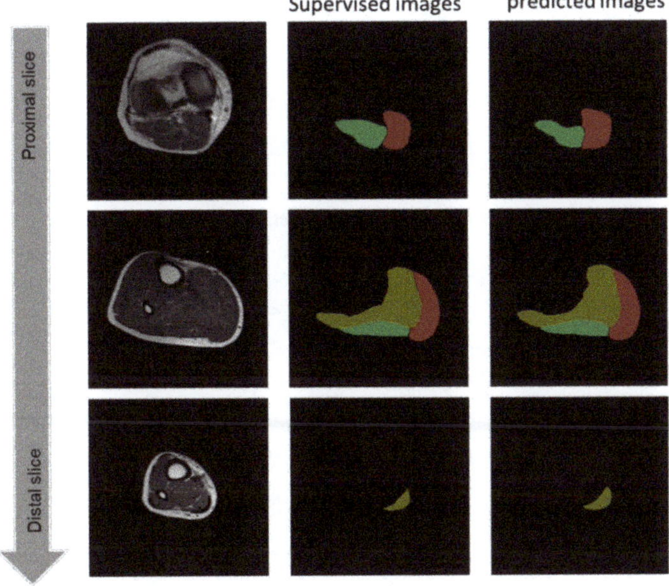

Figure 6. Representative examples of accurately predicted images.

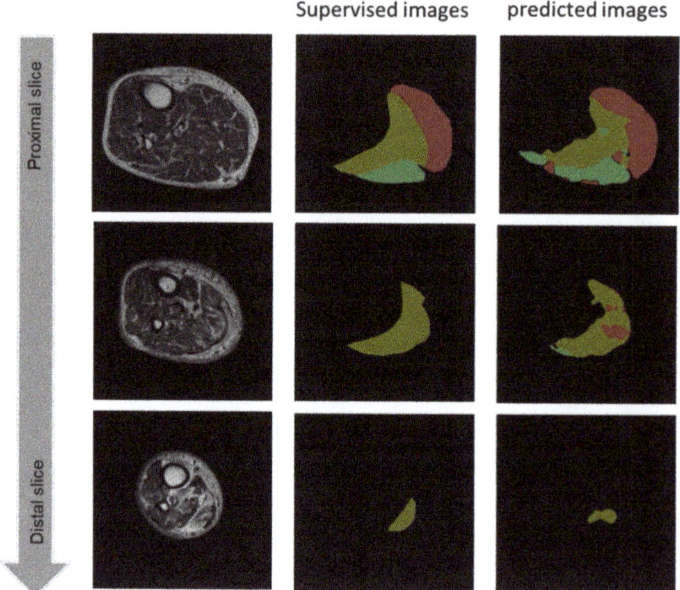

Figure 7. Representative examples of partially incorrectly predicted images.

The error rates of volume per dataset are shown in Table 3. The following table shows the error rates of the volumes of all 67 legs that were present in the supervised and predicted images. The total error represents the error rate of all three muscle volumes. The error rate of MG, SOL, and sum was 10% but that of the GL volume was large.

Table 3. The error rate of volume per dataset for the predicted images.

Position	Dataset1	Dataset2	Dataset3	Dataset4	Dataset5	Dataset6	Mean ± SD
GM	5.68	11.84	8.96	10.28	8.80	24.48	11.67 ± 6.60
GL	27.46	23.03	18.74	27.14	32.71	12.36	23.57 ± 7.22
SOL	11.71	17.22	14.60	6.87	5.92	9.24	10.93 ± 4.43
total error	11.07	13.38	11.87	7.27	7.84	12.43	10.64 ± 2.51

[%], GM: gastrocnemius medialis, GL: gastrocnemius lateralis, SOL: soleus, SD: standard deviation.

3.2. Evaluation of the Interpolation Method

The DSC values of the interpolated images are shown in Table 4. The following table shows the DSC of 60 GMs, 54 GLs, and 66 SOLs in the interpolated images. All muscles had mean DSC larger than 0.8.

Table 4. The DSC values of the interpolated images.

Position	DSC
GM	0.877 ± 0.134
GL	0.809 ± 0.170
SOL	0.867 ± 0.078

GM: gastrocnemius medialis, GL: gastrocnemius lateralis, SOL: soleus, DSC: Dice similarity coefficient.

The error rates of volume for the interpolated images are shown in Table 5. The following table shows the error rate between the volume of the supervised images (Figure 4A) and the volume calculated from the 52 legs to which all the three muscles could be interpolated (Figure 4C). For GM, SOL, and the total error, the percentage was <10%. In fact, the percentage of the total error was as low as 7.5%. The usefulness of this interpolation method in reducing volume calculation errors was evaluated by comparing the error rate between the volume of the supervised image (Figure 4A) and the volume of the non-interpolated image (Figure 4D). In the comparison between the supervised and non-interpolated muscles, all muscles were >12%. Both volumes were calculated for each leg and were then statistically tested. p value results showed that there were no significant differences between the volumes of the supervised and interpolated muscles. the same sequence for the supervised and non-interpolated muscle were performed, and the p value results also showed that there were no significant differences.

Table 5. The error rate of volume for the interpolated images.

Comparison Target	Position	Error Rate [%]	p Value
Supervised muscle and interpolated muscle	GM	9.41 ± 7.65	0.5052
	GL	17.89 ± 6.37	0.3098
	SOL	9.43 ± 5.07	0.7280
	total error	7.69 ± 3.49	-
Supervised muscle and non-interpolated muscle	GM	20.28 ± 9.68	0.1499
	GL	33.93 ± 7.54	0.2539
	SOL	12.70 ± 5.79	0.3442
	total error	14.97 ± 3.98	-

GM: gastrocnemius medialis; GL: gastrocnemius lateralis; SOL: soleus. Supervised muscle: Figure 4A; interpolated muscle: Figure 4C; non-interpolated muscle: Figure 4D.

4. Discussion

With regards to the evaluation of the created models, DSCs were used to compare the agreement of the images and evaluate whether the detection was correct in the two dimensions. Based on the results of the evaluation of the created models, MG and SOL were detected with higher accuracy than GL, with DSCs exceeding 0.8. As for MG and SOL, the DSC of segmentation in the thigh muscle, which exceeded 0.8 in many previous studies, were found to be lower than the respective values in DSCs, but the values themselves cannot be compared because the target areas were different. However, this study could detect MGs with a higher accuracy than the reported segmentation of MGs without using deep learning [22]. For GM and SOL, there were many large muscles present, but there were some false positives, which were detected with a DSC of >0.8 on average. As for the DSCs of GL, GL was shorter in the craniocaudal direction than the other two muscles, and the number of slices in which GL exists is small. The slices at the edge of the GLs were so small that it was difficult for the human eye to recognize them, and the ratio of the number of small GLs slices was higher than that of other muscles. In small muscles where the other muscles and surrounding fat tissues are mistakenly detected as GL, false positives were more common and this was believed to be the cause of the decline in DSCs. Figure 7 is a representative example of partially incorrectly predicted images. Not only are there false detections of the SOL as the GL and GM, but there are also examples of inaccurate location. As per Ding's study [17], when false positives, which were not present in the surveillance images, appeared in the prediction images, DSC was zero and affected the average DSC. In addition, some images used in the test were blurry, and false positives were particularly common in these. The triceps surae is a long muscle that exists in a craniocaudal direction, and the shape of the muscle changes, particularly at the level of the knee and calf. In this study, the number of slices was uneven in each position; therefore, the number of slices at the knee level was small, which affected DSCs. As there were no reports on the segmentation of the triceps surae, the results were compared with those of the quadriceps. In a study by Kemnitz et al. [16], using T1-weighted 250 images of thighs, including the quadriceps, hamstrings, sutures, adductor muscles, and others, were prepared and trained with U-Net, which reported a muscle-only DSC of >0.9. However, as this was only a result for a specific anatomical location, it may not be suitable for the volume calculation and interpolation that were performed in this study. In a report by Ding et al. [17], 46 thigh

4968-slice Fat-water decomposition MRI images were trained with U-Net. The mean DSC of the quadriceps, hamstrings, sutures, and thin muscles was 0.89, and the mean percent difference in volume was 7.57%. A similar study was conducted to determine DSC and error rates for different image types and regions used; the results showed a similar trend to the present study, indicating the validity of the present results. With regards to volumetry without interpolation, the agreement was represented in three dimensions by calculating the error rates of the volume.

Based on the average error rates of volumes per dataset, the error rate of GM, SOL, and sum was 10%. The error rate in the GL volume was large, probably due to the presence of many false positives as in DSC. In addition, the predicted images were overestimated in all muscles, which might be due to the false positives because they were surrounded by other muscles and fat. Of the images used for training, the images with all three muscles present were the most common; however, there were few images at the level of the knee where only the gastrocnemius muscle was present, and the quadriceps muscles were mistakenly detected as gastrocnemius. In addition, images with contrast and roughness were also detected that made it difficult to see the boundaries between the muscles, and false positives were frequent in these images.

The interpolation method was evaluated in-plane by calculating the DSC of the supervised and interpolated images. All muscles had mean DSC larger than 0.8. The evaluation of Table 4 was DSC evaluated only with interpolated images; hence, it can be said that the interpolation was performed with high accuracy. In this interpolation method, the centroid of the larger muscle contour was discovered, and the centroid was connected to the dots that make up the contour; the dot with the shortest distance between the line and that of the smaller muscle contour were adopted. If a dot in the false positive area was adopted, the DSC was lowered. Even if they were not considered in this interpolation, all of them were >0.8, so the correct dot with a higher probability among the corresponding dots in the small contour was selected. In a study by Yap Abdullah et al. [23], the results of an interpolation to measure the intracranial volume were comparable to those of manual segmentation, and its use led to the development of a software for rapid measurement. In a study by Nordez et al. [24], to calculate the volume of the quadriceps muscle in MRI images, four different methods were used to calculate the error rate; it was reported that the interpolation method was the best. These results using the interpolation method showed the same trend as these studies.

The interpolation method was evaluated along the slice direction by calculating and comparing the error rates of muscle volumes, including the supervised and interpolated images. The volume contained interpolated images generated from the two images but with low error rates. Therefore, the results in Table 5 demonstrate that the interpolation method is effective in reducing the volumetric error. The percentages of GM, SOL, and total errors were >10%; for the total error, the percentage was as low as 7.5% because if there were no false positives in the images before and after the interpolated images, there were no false positives in the interpolated images. This was believed to lead to a decreased error rate because of the possibility of reduced false positives. Even if false positives existed, it seemed that the probability of adopting dots that constitute false positives was low. There was no significant difference between the volumes of the supervised and interpolated images, indicating that this method can be used without any problems in calculating the volume. From the results in Tables 4 and 5, interpolation was performed by excluding the even numbered images from the subsequent evaluation. As these findings underline that the interpolation was accomplished with high accuracy, while observing at the same time the effect of volume reduction, the error with the actual volume could be further reduced by interpolating the actual captured images that were found.

This study has some limitations. First, this study was unable to interpolate all images used in the test. Because muscles were present in the supervised images and the even number of those present in the predicted images were excluded, if the muscles were small in the supervised images, they could not be detected and the images could not be complemented. Therefore, the solution is to improve the detection capability by training more images or to prepare a test image where the number of interpolatable images exists. Second, the error rate of the volume calculated in this study could not be calculated in the legs without gastrocnemius or SOL muscles. Therefore, to get a more correct error rate, it would be necessary to have all the three muscle types in the legs used for training or testing. Third, most studies on quadriceps segmentation used U-Net [16–18]. In these studies [25,26], network models were compared for segmentation, and deeplabv3puls was shown to have the highest detection ability. Therefore, DeepLabv3+ was used for segmentation instead of U-Net because Deeplabv3+, which has a higher detection capability, was more likely to be able to detect muscles with higher accuracy than U-net. Using this algorithm, the triceps could be detected immediately and determine its volume. However, the recent advances in the channel and spatial attention mechanisms [27–29] have definitely led to the evolution of semantic segmentation using U-Net. Therefore, a further study on CNNs in the state of the art is needed. Fourth, the hyperparameters were not adjusted because this study did not aim to tune the hyperparameters as in the study by Chieh et al. [30]. Setting and training the optimal hyperparameters may lead to more accurate detection with Bayesian Optimization. Fifth, the sample size of data is small. Not only is it difficult to guarantee generalization performance with small sample size, but it is also possible to improve accuracy by increasing sample size. Additionally, in this study, training and test images were divided 10:2 for six-fold cross validation, but five- or ten-fold is general [31,32]. This value was obtained by dividing the data by person, and it is thought that cross-validation can be performed with general values by increasing the sample size. Sixth, there are no studies of triceps surae muscle segmentation using Deep Learning. Therefore, this study was compared with a study of quadriceps segmentation [15–20] and a study of triceps surae muscle segmentation that did not use deep learning [21]. The results showed that the DSC of the quadriceps was higher than the DSC of this study, which suggests that this study has room for DSC improvement. In addition, it was more accurate than the reported segmentation of the triceps surae muscle. This interpolation method can be used to calculate their volume without any problems found. Although muscle perimeter length and cross-sectional area, which are correlated with muscle volume [33,34], are sometimes used as indices, it is more accurate to calculate volume from images of the entire muscle [35]. However, it is more accurate to calculate the volume from the whole muscle image. Therefore, this study, which can calculate the actual volume by immediately calculating the volume and interpolating between slices, suggests the possibility of advancing research in the field of physical therapy. The technique of detecting the triceps and calculating their volume can be used to other parts and organs, and it was assumed that the interpolation method can be used to determine the volume more accurately. The interpolation technique may also be useful in smoothing out the images captured with a thicker slice for multiplanar reconstruction processing and 3D visualization.

5. Conclusions

This deep learning segmentation technique could detect the triceps with relatively high DSCs. This eliminates the time needed to manually identify the muscles, and thus enables us to immediately calculate the respective volume. The ability to immediately calculate the volume of the triceps muscle will provide a distinct contribution to research studies in the field of physical therapy, including the ability to immediately perform the quantitative evaluations of rehabilitation devices in leg muscles. The interpolation method used in this study to determine the volume was useful. Using the interpolation method allowed us to find that the difference between the volume calculated by the interpolation method and the actual measurement could be smaller than the volume calculated solely from the captured image. In the future, this interpolation method can be used to calculate the volume of other muscles and organs as well, thereby making it possible to calculate the respective volume in a more accurate manner.

Author Contributions: Y.A. contributed to the data analysis, algorithm construction, and the writing and editing of the manuscript. T.Y. (Takaaki Yoshimura), K.M. and T.Y. (Tomonari Yamada) reviewed and edited the manuscript. H.S. proposed the idea and contributed to the data acquisition, performed supervision, project administration, and reviewed and edited the paper. All authors have read and agreed to the published version of the manuscript.

Funding: This research received no external funding.

Institutional Review Board Statement: The study was conducted according to the guidelines of the Declaration of Helsinki and approved by the Hokkaido University Hospital.

Informed Consent Statement: Informed consent was waived because of the retrospective nature of the study and the analysis used anonymous clinical data.

Data Availability Statement: The created models in this study are available on request from the corresponding author. However, the image datasets presented in this study are not publicly available due to ethical reasons, e.g., because they contain sensitive information that could compromise the privacy of the participants. The source code of this study is available at https://github.com/MIA-laboratory/TricepsMRIsegmentation (accessed on 7 December 2021).

Acknowledgments: The authors would like to thank the laboratory members of the Medical Image Analysis laboratory and Yoshimura's laboratory for their help. The authors would also like to thank Masayuki Kugimoto for providing us with useful advice and information.

Conflicts of Interest: The authors declare that no conflict of interest exist.

References

1. Son, S.; Jeong, Y.; Lee, B. An audification and visualization system (AVS) of an autonomous vehicle for blind and deaf people based on deep learning. *Sensors* **2019**, *19*, 5053. [CrossRef]
2. Chen, Y.; Aggarwal, P.; Choi, J.; Jay, C.C. A deep learning approach to drone monitoring. In Proceedings of the 2017 Asia-Pacific Signal and Information Processing Association Annual Summit and Conference (APSIPA ASC), Kuala Lumpur, Malaysia, 12–15 December 2017; pp. 686–691. [CrossRef]
3. Ghaderi, A.; Sanandaji, B.M.; Ghaderi, F. Deep Forecast: Deep Learning-based Spatio-Temporal Forecasting. In Proceedings of the International Conference on Machine Learning, Time Series Workshop, Sydney, NSW, Australia, 6–11 August 2017.
4. Silver, D.; Schrittwieser, J.; Simonyan, K.; Antonoglou, I.; Huang, A.; Guez, A.; Hubert, T.; Baker, L.; Lai, M.; Bolton, A.; et al. Mastering the game of Go without human knowledge. *Nature* **2017**, *550*, 354–359. [CrossRef]
5. Sugimori, H.; Hamaguchi, H.; Fujiwara, T.; Ishizaka, K. Classification of type of brain magnetic resonance images with deep learning technique. *Magn. Reson. Imaging* **2021**, *77*, 180–185. [CrossRef]
6. Sugimori, H.; Kawakami, M. Automatic detection of a standard line for brain magnetic resonance imaging using deep learning. *Appl. Sci.* **2019**, *9*, 3849. [CrossRef]
7. Zhou, X. Automatic Segmentation of Multiple Organs on 3D CT Images by Using Deep Learning Approaches. *Adv. Exp. Med. Biol.* **2020**, *1213*, 135–147. [CrossRef] [PubMed]
8. Yang, S.; Kweon, J.; Roh, J.H.; Lee, J.H.; Kang, H.; Park, L.J.; Kim, D.J.; Yang, H.; Hur, J.; Kang, D.Y.; et al. Deep learning segmentation of major vessels in X-ray coronary angiography. *Sci. Rep.* **2019**, *9*, 16897. [CrossRef]
9. Zhang, L.; Mohamed, A.A.; Chai, R.; Guo, Y.; Zheng, B.; Wu, S. Automated deep learning method for whole-breast segmentation in diffusion-weighted breast MRI. *J. Magn. Reson. Imaging* **2020**, *51*, 635–643. [CrossRef] [PubMed]
10. Kessler, D.A.; MacKay, J.W.; Crowe, V.A.; Henson, F.M.D.; Graves, M.J.; Gilbert, F.J.; Kaggie, J.D. The optimisation of deep neural networks for segmenting multiple knee joint tissues from MRIs. *Comput. Med. Imaging Graph.* **2020**, *86*, 101793. [CrossRef] [PubMed]
11. Cattagni, T.; Scaglioni, G.; Laroche, D.; Gremeaux, V.; Martin, A. The involvement of ankle muscles in maintaining balance in the upright posture is higher in elderly fallers. *Exp. Gerontol.* **2016**, *77*, 38–45. [CrossRef]
12. Belavý, D.L.; Miokovic, T.; Rittweger, J.; Felsenberg, D. Estimation of changes in volume of individual lower-limb muscles using magnetic resonance imaging (during bed-rest). *Physiol. Meas.* **2011**, *32*, 35–50. [CrossRef]
13. Tortorella, C.; Simone, O.; Piazzolla, G.; Stella, I.; Cappiello, V.; Antonaci, S. Role of phosphoinositide 3-kinase and extracellular signal-regulated kinase pathways in granulocyte macrophage-colony-stimulating factor failure to delay fas-induced neutrophil apoptosis in elderly humans. *J. Gerontol. Ser. A* **2006**, *61*, 1111–1118. [CrossRef]
14. Karamanidis, K.; Epro, G.; König, M.; Mersmann, F.; Arampatzis, A. Simplified Triceps Surae Muscle Volume Assessment in Older Adults. *Front. Physiol.* **2019**, *10*, 1299. [CrossRef]
15. Friedberger, A.; Figueiredo, C.; Bäuerle, T.; Schett, G.; Engelke, K. A new method for quantitative assessment of hand muscle volume and fat in magnetic resonance images. *BMC Rheumatol.* **2020**, *4*, 72. [CrossRef] [PubMed]

16. Kemnitz, J.; Baumgartner, C.F.; Eckstein, F.; Chaudhari, A.; Ruhdorfer, A.; Wirth, W.; Eder, S.K.; Konukoglu, E. Clinical evaluation of fully automated thigh muscle and adipose tissue segmentation using a U-Net deep learning architecture in context of osteoarthritic knee pain. *Magn. Reson. Mater. Phys. Biol. Med.* **2020**, *33*, 483–493. [CrossRef] [PubMed]
17. Ding, J.; Cao, P.; Chang, H.C.; Gao, Y.; Chan, S.H.S.; Vardhanabhuti, V. Deep learning-based thigh muscle segmentation for reproducible fat fraction quantification using fat–water decomposition MRI. *Insights Imaging* **2020**, *11*, 128. [CrossRef]
18. Gadermayr, M.; Li, K.; Müller, M.; Truhn, D.; Krämer, N.; Merhof, D.; Gess, B. Domain-specific data augmentation for segmenting MR images of fatty infiltrated human thighs with neural networks. *J. Magn. Reson. Imaging* **2019**, *49*, 1676–1683. [CrossRef] [PubMed]
19. Ghosh, S.; Ray, N.; Boulanger, P. A Structured Deep-Learning Based Approach for the Automated Segmentation of Human Leg Muscle from 3D MRI. In Proceedings of the 2017 14th Conference on Computer and Robot Vision (CRV), Edmonton, AB, Canada, 16–19 May 2017; pp. 117–123. [CrossRef]
20. Andrews, S.; Hamarneh, G. The Generalized Log-Ratio Transformation: Learning Shape and Adjacency Priors for Simultaneous Thigh Muscle Segmentation. *IEEE Trans. Med. Imaging* **2015**, *34*, 1773–1787. [CrossRef] [PubMed]
21. Baudin, P.-Y.; Azzabou, N.; Carlier, P.G.; Paragios, N. Automatic Skeletal Muscle Segmentation through Random Walks and Graph-Based Seed Placement. In Proceedings of the 2012 9th IEEE International Symposium on Biomedical Imaging (ISBI), Barcelona, Spain, 2–5 May 2012; pp. 1036–1039. [CrossRef]
22. Essafi, S.; Langs, G.; Deux, J.F.; Rahmouni, A.; Bassez, G.; Paragios, N. Wavelet-driven knowledge-based MRI calf muscle segmentation. In Proceedings of the 2009 IEEE International Symposium on Biomedical Imaging: From Nano to Macro, Boston, MA, USA, 28 June–1 July 2009; pp. 225–228. [CrossRef]
23. Abdullah, J.Y.; Rajion, Z.A.; Martin, A.G.; Jaafar, A.; Ghani, A.R.I.; Abdullah, J.M. Shape-based interpolation method in measuring intracranial volume for pre- and post-operative decompressive craniectomy using open source software. *Neurocirugia* **2019**, *30*, 115–123. [CrossRef]
24. Nordez, A.; Jolivet, E.; Südhoff, I.; Bonneau, D.; De Guise, J.A.; Skalli, W. Comparison of methods to assess quadriceps muscle volume using magnetic resonance imaging. *J. Magn. Reson. Imaging* **2009**, *30*, 1116–1123. [CrossRef]
25. Ahmed, I.; Ahmad, M.; Khan, F.A.; Asif, M. Comparison of deep-learning-based segmentation models: Using top view person images. *IEEE Access* **2020**, *8*, 136361–136373. [CrossRef]
26. Khan, Z.; Yahya, N.; Alsaih, K.; Ali, S.S.A.; Meriaudeau, F. Evaluation of deep neural networks for semantic segmentation of prostate in T2W MRI. *Sensors* **2020**, *20*, 3183. [CrossRef]
27. Rundo, L.; Han, C.; Nagano, Y.; Zhang, J.; Hataya, R.; Militello, C.; Tangherloni, A.; Nobile, M.S.; Ferretti, C.; Besozzi, D.; et al. USE-Net: Incorporating Squeeze-and-Excitation blocks into U-Net for prostate zonal segmentation of multi-institutional MRI datasets. *Neurocomputing* **2019**, *365*, 31–43. [CrossRef]
28. Schlemper, J.; Oktay, O.; Schaap, M.; Heinrich, M.; Kainz, B.; Glocker, B.; Rueckert, D. Attention gated networks: Learning to leverage salient regions in medical images. *Med. Image Anal.* **2019**, *53*, 197–207. [CrossRef]
29. Yeung, M.; Sala, E.; Schönlieb, C.-B.; Rundo, L. Focus U-Net: A novel dual attention-gated CNN for polyp segmentation during colonoscopy. *Comput. Biol. Med.* **2021**, *137*, 104815. [CrossRef] [PubMed]
30. Chen, L.C.; Zhu, Y.; Papandreou, G.; Schroff, F.; Adam, H. Encoder-decoder with atrous separable convolution for semantic image segmentation. In *Lecture Notes in Computer Science*; Springer: Berlin/Heidelberg, Germany, 2018; pp. 833–851. [CrossRef]
31. Do, D.T.; Le, T.Q.T.; Le, N.Q.K. Using deep neural networks and biological subwords to detect protein S-sulfenylation sites. *Brief. Bioinform.* **2021**, *22*, bbaa128. [CrossRef]
32. Le, N.Q.K.; Huynh, T.T. Identifying SNAREs by Incorporating Deep Learning Architecture and Amino Acid Embedding Representation. *Front. Physiol.* **2019**, *10*, 1501. [CrossRef] [PubMed]
33. Miyachi, R.; Yamazaki, T.; Ohno, N.; Miyati, T. Relationship between muscle cross-sectional area by mri and muscle thickness by ultrasonography of the triceps surae in the sitting position. *Healthcare* **2020**, *8*, 166. [CrossRef] [PubMed]
34. Henninger, H.B.; Christensen, G.V.; Taylor, C.E.; Kawakami, J.; Hillyard, B.S.; Tashjian, R.Z.; Chalmers, P.N. The Muscle Cross-sectional Area on MRI of the Shoulder Can Predict Muscle Volume: An MRI Study in Cadavers. *Clin. Orthop. Relat. Res.* **2020**, *478*, 871–883. [CrossRef] [PubMed]
35. Akagi, R.; Takai, Y.; Ohta, M.; Kanehisa, H.; Kawakami, Y.; Fukunaga, T. Muscle volume compared to cross-sectional area is more appropriate for evaluating muscle strength in young and elderly individuals. *Age Ageing* **2009**, *38*, 564–569. [CrossRef]

MDPI

Article

Automated Breast Lesion Detection and Characterization with the Wavelia Microwave Breast Imaging System: Methodological Proof-of-Concept on First-in-Human Patient Data

Angie Fasoula [1,*], Luc Duchesne [1], Julio Daniel Gil Cano [1], Brian M. Moloney [2,3], Sami M. Abd Elwahab [4] and Michael J. Kerin [3,4]

1 Medical Imaging Department, MVG Industries, 91140 Villejust, France; luc.duchesne@mvg-world.com (L.D.); julio_daniel.gil_cano@mvg-world.com (J.D.G.C.)
2 Department of Radiology, Galway University Hospital, Saolta University Healthcare Group, H91 YR71 Galway, Ireland; brianmoloney1@hotmail.com
3 Discipline of Surgery, Lambe Institute for Translational Research, School of Medicine, National University of Ireland Galway, H91 TK33 Galway, Ireland; michael.kerin@nuigalway.ie
4 Department of Surgery, Galway University Hospital, Saolta University Healthcare Group, H91 YR71 Galway, Ireland; sami.elwahab@hse.ie
* Correspondence: angie.fasoula@mvg-world.com

Abstract: Microwave Breast Imaging (MBI) is an emerging non-ionizing imaging modality, with the potential to support breast diagnosis and management. Wavelia is an MBI system prototype, of 1st generation, which has recently completed a First-In-Human (FiH) clinical investigation on a 25-symptomatic patient cohort, to explore the capacity of the technology to detect and characterize malignant (invasive carcinoma) and benign (fibroadenoma, cyst) breast disease. Two recent publications presented promising results demonstrated by the device in this FiH study in detecting and localizing, as well as delineating size and malignancy risk, of malignant and benign palpable breast lesions. In this paper, the methodology that has been employed in the Wavelia semi-automated Quantitative Imaging Function (QIF), to support breast lesion detection and characterization in the FiH clinical investigation of the device, is presented and the critical design parameters are highlighted.

Keywords: breast cancer detection; microwave breast imaging; computer-aided diagnosis (CAD); first-in-human (FiH) study

Citation: Fasoula, A.; Duchesne, L.; Gil Cano, J.D.; Moloney, B.M.; Abd Elwahab, S.M.; Kerin, M.J. Automated Breast Lesion Detection and Characterization with the Wavelia Microwave Breast Imaging System: Methodological Proof-of-Concept on First-in-Human Patient Data. *Appl. Sci.* **2021**, *11*, 9998. https://doi.org/10.3390/app11219998

Academic Editors: Leonardo Rundo, Carmelo Militello and Andrea Tangherloni

Received: 28 September 2021
Accepted: 19 October 2021
Published: 26 October 2021

1. Introduction

Microwave Breast Imaging (MBI) uses the scattering wave, or reflected wave, that arises from the contrast in dielectric properties between the various breast tissues, in the microwave frequency range [1]. The increased volume of water within the denser breast tissues is responsible for the detectable electromagnetic scattering associated with microwave imaging. The increase in sodium and water, particularly in-bound water within the tumor cells, is expected to lead to even greater conductivity and permittivity of the tumorous tissues [2,3]. Due to the dielectric contrast, back-scattered radar signals are physically generated, when the breast is illuminated with low-power electromagnetic waves in the microwave frequency range.

MBI has been investigated as a novel modality for the detection of breast disease, offering a non-ionizing, non-compressive approach [4–6] and as a potential diagnostic management strategy in the monitoring of neoadjuvant chemotherapy [7]. To date, a total of at least 10 MBI system prototypes have been employed in human subject tests, to investigate the clinical utility of MBI [8–11]. Despite encouraging clinical results being reported, several recurrent limitations, as outlined in [12], remain unresolved across most studies and justify further clinical research with alternative MBI systems, such as Wavelia.

Wavelia^{TM} is an MBI system prototype, of 1st generation, which demonstrated the ability to detect dielectric contrast between tumor phantoms and synthetic fibroglandular tissue in preclinical studies [13] and has recently completed a First-In-Human (FiH) clinical investigation on a 25-symptomatic patient cohort, hosted in NUIG Clinical Research Facility Galway, Ireland. In this study (ClinicalTrials.gov NCT03475992), the Wavelia MBI system was evaluated in the clinical setting for the first time, using mammography as the reference conventional imaging modality and post-surgery histology data to assess the size of the cancers. Ultrasound, MRI and core biopsy data were also collected as a reference and were available as part of the patient's standard of care. In this FiH study, Wavelia demonstrated the capacity to detect and approximate underlying breast abnormalities to the appropriate location, in patients with palpable biopsy-confirmed invasive carcinomas and benign breast lesions, such as cysts and fibroadenomas [12]. The device also demonstrated promising results in delineating the size and malignancy risk of the detected breast lesions [14].

The methodology that was employed in the Wavelia semi-automated Quantitative Imaging Function (QIF) during this FiH study, to support morphological breast lesion detection based on persistence, lesion sizing and lesion characterization in a low-dimensional feature space, spanning shape and texture-based features, is presented in this article.

2. Materials and Methods

The Wavelia MBI Quantitative Imaging Function (QIF) was initially conceived using experimental MBI datasets from anthropomorphic breast phantoms [13,15] and was further developed and configured following training on the available FiH patient datasets [12,16]. The MBI parametric radar image formation and clinical feature extraction are performed offline at this stage of development of Wavelia.

2.1. Wavelia MBI: Parametric Radar Image Formation

The Wavelia MBI system operates using 18 antennae arranged in a circle in a horizontal plane outside a cylinder. With the patient lying in the prone position, one breast is submerged at a time into the cylinder, which is filled with a creamy transition liquid. The liquid has dielectric properties similar to the ones of the human skin within the microwave frequency spectrum, thus favoring the penetration of the electromagnetic waves in the breast. The device illuminates the breast using low-power electromagnetic waves in the frequency range [0.5–4] GHz. The probe array moves vertically below the examination table and illuminates the breast at regular intervals of 5 mm. Coronal sections of the breast, of a given thickness (10 mm) are generated using the MBI data at each vertical scan position of the probe array. Partially overlapping consecutive coronal breast sections, formed per azimuthal sector of illumination based on multi-static radar detection technology, are integrated to form a 3D MBI image of the dielectrically contrasted interior breast tissues.

As specified in prior publications on Wavelia MBI [13,15], the multi-static radar imaging algorithm, which is employed for MBI image formation, is the Time-Reversal Multiple SIgnal Classification (TR-MUSIC) algorithm, which was originally conceived for the detection of obscured radar targets in heavily cluttered environments [17]. The intensity of the TR-MUSIC images gets maximized in the imaging pixels where the MBI sensor array illumination vector is more orthogonal to the noise subspace; thus, the image intensity is indicative of the probability for the presence of a dielectrically contrasted scatterer on each pixel of the image. The noise subspace is estimated at each frequency, by means of decomposition and analysis of the Multi-Static Frequency Response Matrix (MFRM) of the imaging array. The illumination vector of the imaging array, at each pixel p of the imaging scene and each frequency f, is defined as:

$$\mathbf{G}_{\mathrm{sect}}(\mathrm{p},\mathrm{f}) = \left[\ \mathbf{g}_0\left(\mathrm{p}_{\mathrm{TRx}_{\mathrm{sect},1}},\mathrm{p},\mathrm{f}\right)\quad \mathbf{g}_0\left(\mathrm{p}_{\mathrm{TRx}_{\mathrm{sect},2}},\mathrm{p},\mathrm{f}\right)\quad \cdots\quad \mathbf{g}_0\left(\mathrm{p}_{\mathrm{TRx}_{\mathrm{sect},\mathrm{N_s}}},\mathrm{p},\mathrm{f}\right)\ \right]^{\mathrm{T}} \tag{1}$$

with:

$$\mathbf{g}_0\left(\mathrm{p}_{\mathrm{TRx}_{\mathrm{sect},i}},\mathrm{p},\mathrm{f}\right) = \mathrm{j}\cdot\mathbf{H}_0^{(1)}\left(\frac{2\pi\mathrm{f}}{\mathrm{c}_0}\cdot\left(\sqrt{\mathbf{e}_{\mathbf{r,trans}}(\mathrm{f})}\cdot\hat{\mathrm{d}}_{\mathrm{OutOfBreast},i,p}+\sqrt{\hat{\mathbf{e}}_{\mathbf{r,InBreast}}(\mathrm{f})}\cdot\hat{\mathrm{d}}_{\mathrm{InBreast},i,p}\right)\right) \tag{2}$$

the assumed underlying ElectroMagnetic (EM) wave propagation model for the antenna element at position $p_{TRx_{sect,i}}$, $\mathbf{H}_0^{(1)}$ the Hankel function of 1st kind and 0th order, c_0 the speed of light in vacuum, $\mathbf{e}_{\mathbf{r,trans}}(f)$ the known permittivity of the transition liquid at the frequency f, $\hat{d}_{OutOfBreast,i,p}$ an estimate of the distance travelled by the EM wave in the transition liquid up to reaching the imaging pixel p, $\hat{d}_{InBreast,i,p}$ an estimate of the distance travelled by the EM wave within the breast up to reaching the pixel p, and:

$$\hat{\mathbf{e}}_{\mathbf{r,InBreast}}(f) = \left(pc_{fib} \cdot \hat{\mathbf{e}}_{\mathbf{r,fibroglandular}}(f) + (1 - pc_{fib}) \cdot \hat{\mathbf{e}}_{\mathbf{r,adipose}}(f)\right) \cdot 10^{-2} \tag{3}$$

the average permittivity of the background healthy tissues of the breast, defined as a weighted average (weighting by pc_fib) of the adipose tissue and fibro-glandular tissue "mean" dielectric properties, as derived by Sugitani et al. [18].

The breast external envelope is reconstructed at first using the Wavelia MBI scan data. The geometry is exploited to split the imaging scene in "Out of breast" and "In breast" segments and further estimate $\hat{d}_{OutOfBreast,i,p}$ and $\hat{d}_{InBreast,i,p}$ for each transceiver i and each pixel p in the imaging scene.

The critical elements of the tailored implementation of TR-MUSIC in the Wavelia QIF are summarized below and in the flowchart of Figure 1.

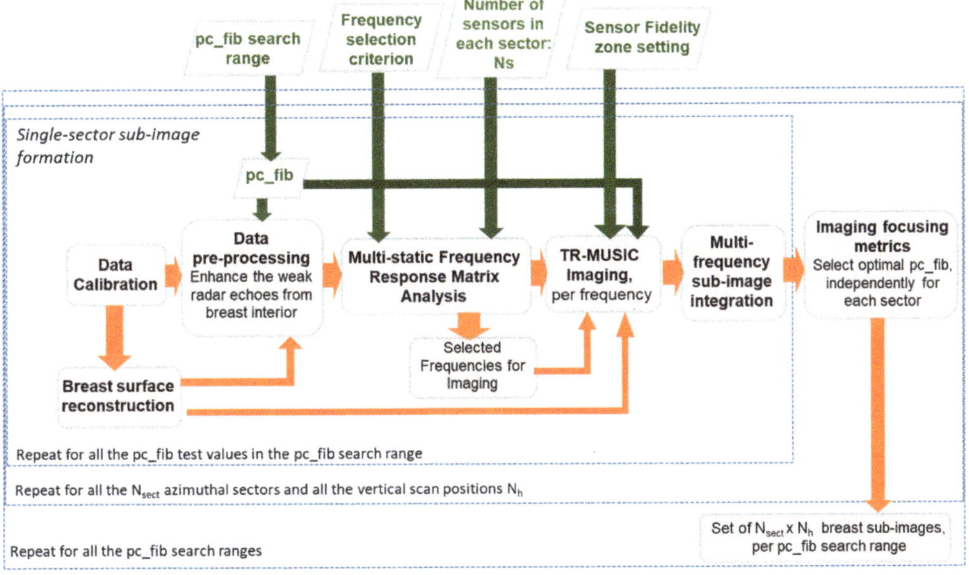

Figure 1. Wavelia MBI radar image formation: the critical parameters.

- Sectorization of the imaging scene: Considering the high level of heterogeneity of the breast tissues, but also the potentially irregularly shaped breast tumours, MBI image formation is performed using sectorized subsets of the circular sensor network of the Wavelia MBI system, at each vertical scan position. Both the physical size of the sensor sub-array (sector) and the number Ns of sensors used for the elementary sub-image formation is critical to the achievable performance of the MBI system, in terms of unambiguous detection and valid characterization of breast lesions. Ns = 6 was fixed in the Wavelia QIF for the FiH clinical investigation.
- Sensor fidelity zone setting: Bounding the portion of the imaging scene, being efficiently illuminated by each sensor sub-array. In the current implementation of the

Wavelia QIF, a spatial filter activates the contribution of a given antenna to a given pixel p in the imaging scene, only if the Euclidean distance between the pixel and the phase centre of the antenna is inferior to a pre-set value d_{max}.

- Automated, data-driven, frequency selection for imaging: The TR-MUSIC imaging is performed per frequency. In the Wavelia QIF, a limited number of frequency points is automatically selected for imaging in each azimuthal sector of illumination, based on the information content of the MFRM at each frequency. The employed frequency selection criterion was previously defined in [15].
- pc_fib parameter setting in multiple search ranges: A large variability exists in the dielectric properties of each breast tissue type over the population, as demonstrated by multiple studies involving ex-vivo dielectric measurements of a large sample of excised breast tissues [18–21]. Considering that the full dielectric map of each breast cannot become practically available, data-driven techniques are employed in the Wavelia QIF to deduce the unknown dielectric properties of the healthy breast tissue in each breast, by assessing the pc_fib parameter. The pc_fib parameter, which is involved in the formulation of the illumination vector of the MBI sensor array, is physically associated with the percentage of fibro-glandular tissue along the propagation path within the breast, from a given transmitting antenna to the interrogated imaging pixel and back to a given receiving antenna, as defined in Equations (1)–(3). The Wavelia QIF generates a set of parametric MBI radar images under various assumptions on pc_fib. The generated set of parametric images is further evaluated in terms of focusing, using the image curvature [22,23] as a focusing quality measure. To better handle the heterogeneity of the breast and potentially better reveal the non-uniform angular response of the breast lesions to MBI, the pc_fib parameter setting is performed independently in each azimuthal imaging sector, while employing multiple search ranges. In the Wavelia QIF, X1 wide and X2 narrow pc_fib parameter search ranges are systematically employed for image formation, thus a total number of X = (X1 + X2) MBI images are formed per patient's breast.

The wide pc_fib search ranges result in 3D MBI images including the most complete representations of the detected breast lesion shape. The narrow pc_fib search ranges are expected to lead to partial representations of the detectable breast lesions. The X = 5 pc_fib search ranges, which were systematically employed during the FiH clinical investigation of Wavelia, are listed below:

- Wide pc_fib search range #1 (W1): pc_fib ϵ [10 20 30 40 50 60]%
- Wide pc_fib search range #2 (W2): pc_fib ϵ [20 30 40 50]%
- Narrow pc_fib search range #1 (n1): pc_fib ϵ [10 20]%
- Narrow pc_fib search range #2 (n2): pc_fib ϵ [30 40]%
- Narrow pc_fib search range #3 (n3): pc_fib ϵ [50 60]%

As explained in the next subsection, the persistent presence of a Region-Of-Interest (ROI) in the set of X MBI parametric radar images of a given breast is further exploited, to support the association of automatically extracted ROIs with breast lesions and validate their reporting for clinical analysis.

2.2. Morphological MBI Image Post-Processing: Breast Lesion Detection Based on Persistence

Automated breast lesion detection is performed in the Wavelia QIF by means of morphological post-processing of the set of parametric radar images, which are formed with the employment of X = 5 search ranges for the pc_fib parameter. Automated segmentation of ROIs and association, or not, to a breast lesion is based on morphological properties (solidity and volume) of the ROI and its persistence on the set of parametric radar images, which is evaluated by means of spatial clustering. The persistent visibility of a ROI over multiple pc_fib search ranges is indicative of the association of the ROI with a physical object (breast lesion) in the MBI image. On the other hand, the presence of a ROI in the minority of the pc_fib search ranges under test is indicative of it being associated with an imaging artefact. This setting has been inspired by the "breast mass" definition for mammography, as the space-occupying a 3D lesion seen in two different projections [24]. To the authors' knowledge, no such breast lesion detection method, based on persistence in a set of parametric images carrying redundant information content, has ever been integrated into any of the state-of-the-art MBI systems before. A second novel element of the proposed method is the coupling of morphological properties (solidity) with the notion of persistence to validate a ROI detection.

The automated breast lesion detection method, as designed and integrated into the Wavelia QIF for the FiH clinical investigation of device prototype #1, is outlined below and in the block diagram in Figure 2.

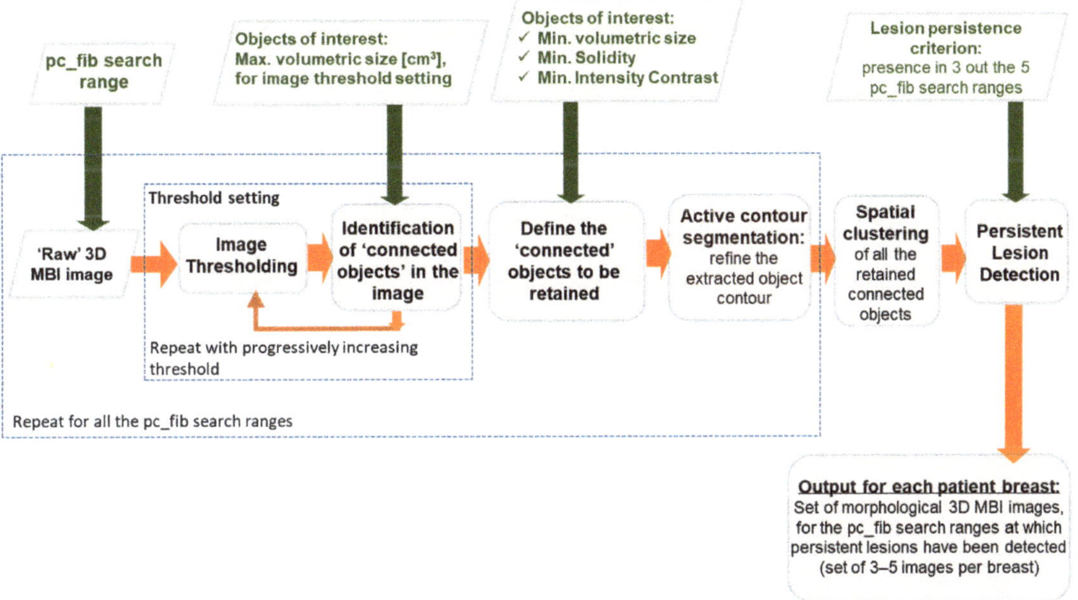

Figure 2. Wavelia MBI: morphological image post-processing for breast lesion detection based on persistence.

1. Iterative Image threshold setting: The following operation is first-of-all performed to set a threshold for the raw 3D MBI image, with no a priori available on how normal breast tissue is represented in this type of image:
   a. Progressive increase in the image threshold, starting from the null threshold,
   b. At each iteration, identification of the "connected" objects in the thresholded image,
   c. Threshold setting based on the maximal accepted volumetric size of "connected" objects, potentially defining a breast mass in the image (default value 3 cm^3, in this implementation).
2. Semi-automated ROI extraction, based on morphological properties: The connected objects to be retained as ROIs, are defined based on a set of user-defined characteristics, including:
   a. Volumetric size: all the small objects, of volume inferior to 1 cm^3, are removed from the FiH clinical data analysis, considering the status of the Wavelia system prototype #1 in terms of minimum size of detectable lesions.
   b. Solidity: this structural feature measures the density (or convexity) of an object. A measure of solidity can be obtained as a ratio of the volume of the object to the volume of a convex hull of the object. A value of 100% indicates a solid object,

and a value less than 100% indicates an object having an irregular boundary or containing holes. All connected objects with solidity >30% have been ultimately retained, for the data analysis of the FiH clinical investigation of Wavelia.

c. Intensity Contrast: In case of ambiguity (i.e., multiple connected objects to be retained in a single 3D MBI image, based on the volume and solidity criteria), each connected object is retained as ROI, only if it is associated with the maximal intensity in the image and is minimally contrasted (by at least 5%) against the intensity of all the "competing" connected objects in the image.

3. Refinement of the ROI segmentation: An Active Contour segmentation module (Chan-Vese algorithm for segmentation without edges [25]) has been configured and employed to refine the contour of the extracted ROIs, in order to enable a more valid characterization of the lesions in terms of shape and texture, as defined in the following sub-section. This module may have a critical impact, especially in the case of small lesions, which are low contrasted against the background healthy breast tissue.
4. Spatial clustering of the ROIs which have been extracted on the set of X MBI images, formed with varying pc_fib search ranges, is performed. In the current implementation of the Wavelia QIF, the Euclidean distance between the centroids of two ROIs is required to be shorter than d_R = 1 cm for the ROIs to be associated with the same lesion in the breast.
5. Persistence over varying pc-fib search range: Five pc-fib search ranges (2 wide and 3 narrow) are systematically used during the FiH clinical investigation of Wavelia to generate parametric 3D MBI images of each patient's breast, as earlier stated. Detection of a breast lesion in a minimum number of parametric images (3 out of the 5 pc_fib search ranges) is required for a ROI to be considered persistent and validated.

2.3. *Combination of 3D Shape Descriptors and Texture Features for Breast Lesion Characterization with Microwave Breast Imaging (MBI)*

Apart from using reflected microwave energy to reconstruct images of the breast, additional information on the size, shape, and surface texture can be extracted and potentially exploited for discrimination between benign and malignant breast lesions using microwaves [1]. Malignant tumors usually present the following characteristics: irregular and asymmetric shapes, blurred boundaries (lack of sharpness), rough and complex surfaces with spicules or micro-lobules, non-uniform permittivity, and irregular tissue density. Conversely, benign tumors tend to have the following characteristics: well-circumscribed contours, compactness, and a smooth surface. Previous research works on breast lesion characterization/classification with MBI [26–30] considered principally the MBI received signals as input to a classifier, with or without dimensionality reduction. These state-of-the-art research works [26–30] have been based on simulated datasets and/or simplified experimental setups; no evaluation of such methods on patient clinical datasets has been published to date. Among the state-of-the-art MBI prototypes which have been tested on clinical datasets, two of them published studies on breast lesion classification with MBI. Early concept work on the exploitation of the pattern of the frequency-domain Radio-Frequency (RF) responses of the ROIs representing the breast lesions in the MBI image was published in [31] for the MARIA M5 [4] MBI system. For the MammoWave MBI system, machine learning methods were employed with raw received signals in the frequency domain to classify them as healthy or non-healthy responses [32].

In the Wavelia QIF a module is integrated for the characterization of the ROIs which have been prior detected and validated based on morphological properties and their persistence, and thus associated with breast lesions. This module includes the following operations:

- Breast Lesion sizing: by means of fitting an ellipsoid to the ROI associated with the persistent lesion detection, in the 3D MBI images that have been generated by applying either of the two wide pc_fib search ranges. The greatest linear dimension of the lesion is defined as the length of the longest axis of the fitted ellipsoid. This definition is compatible with the conventional method that is applied for sizing breast abnormalities based on 2D mammography and ultrasound images [24,33]. During the FiH clinical investigation, the Wavelia MBI system showed promise for measuring lesion size with a more favorable linear trend between MBI and post-surgery histological lesion size, compared to the results obtained for conventional imaging [14]. Two challenging patient cases in terms of breast lesion sizing are indicatively discussed in Results Section 3.2, to better highlight the status of the MBI lesion sizing method, as integrated into the current version of the Wavelia QIF.

- Malignant-to-benign Breast Lesion labelling is performed in the Wavelia QIF using a combination of 3 features, extracted from morphologically validated ROIs in the MBI images. The selected features include: a shape descriptor [34], a Gray-Level Co-occurrence Matrix (GLCM) texture feature [35] and a Neighborhood Gray Tone Difference Matrix (NGTDM) texture feature [36]. The 3 features which were selected in the Wavelia QIF implementation for the FiH study of the investigational device are more specifically the following:
    - Shape descriptors—Solidity: This feature measures the density, or the convexity, of an object. It is computed as the ratio of the volume of the object to the volume of the convex hull of the object, as illustrated in Figure 3. Breast lesion scoring, in terms of risk for malignancy, is routinely based on visual inspection and evaluation of the shape and margins of the imaged breast lesion, as per BIRADS [24,33]. Shape descriptors have been earlier considered for breast lesion classification with mammography [34,37] and ultrasound [38].
    - GLCM texture—Correlation: The GLCM texture features measure the spatial relationship between pixels per specific directions, thus highlighting the properties of uniformity, homogeneity, randomness, and linear dependency of the image [35]. More specifically, the "correlation" feature varies between 0 (uncorrelated) and 1 (perfectly correlated), showing the linear dependency of gray level values to their respective voxels, as graphically illustrated in Figure 3.
    - NGTDM texture—Busyness: The NGTDM texture features measure the spatial relationship among three or more pixels neighborhood, closely approaching the human perception of the image [36], as graphically illustrated in Figure 3 More specifically, for the "busyness" feature, a high value indicates a "busy" image, with rapid changes of intensity between pixels and its neighborhood.

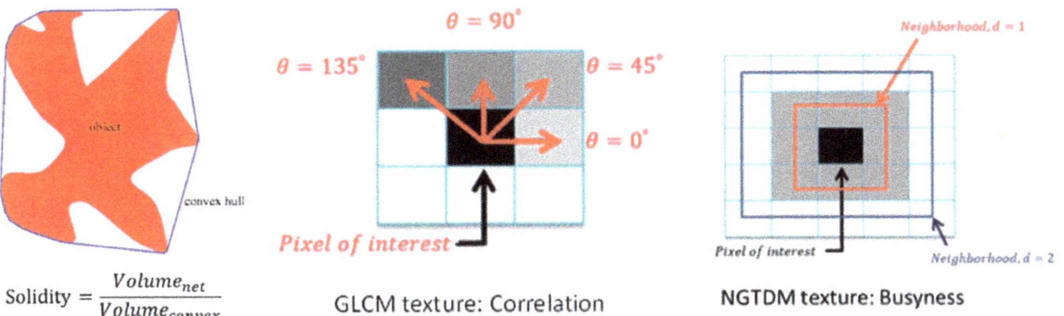

Figure 3. The 3-d feature vector employed in Wavelia for malignant-to-benign breast lesion discrimination.

Texture-based features have been earlier considered in Radiomics Research for cancerous lesions identification on CT, PET and MRI images [39–41]. Breast lesion classification, employing texture features on multi-parametric breast MRI images has also been introduced in the state-of-the-art [42].

In contrast to the Radiomics Research studies, which suggest the employment of high-dimensional feature vectors (typical size > 30) [39–41], appropriate feature selection has been considered in the Wavelia QIF, to achieve malignant-to-benign lesion separability in a feature space of low dimensionality. To the authors' knowledge, no shape-based or texture-based feature extraction from Microwave Breast Images (MBI) has ever been considered in the past.

The 3-dimensional (3-d) lesion feature vector data [Solidity; Correlation; Busyness] is exploited in a malignant-to-benign breast lesion classification framework in the Wavelia QIF. A 2-class discrimination problem is defined, with: (i) Class #1: Malignant breast lesions, and (ii) Class #2: Benign breast lesions. Two classifiers have been trained in this 3-d feature space. The two classifiers, i.e., a Naïve Bayesian (NB) classifier and a Quadratic Discriminant Analysis (QDA) classifier, were selected such that their decision hypersurface partitions the 3-d feature space in two disjoint and continuous manifolds (malignant lesions subspace vs. benign lesions subspace).

In the Wavelia FiH clinical investigation [12], female patients were recruited from the symptomatic unit to one of three groups: Biopsy-proven breast cancers (Group-1), unaspirated cysts (Group-2) and biopsy-proven benign breast lesions (Group-3). For the training of the 2 classifiers:

- The Group-1 patient datasets were labelled as Class #1.
- The Group-2 and Group-3 patient datasets were labelled as Class #2.

A total of 25 patients underwent MBI in this FiH study. Of these, 24 were included in the final data analysis (11 Group-1, 8 Group-2 and 5 Group-3 patients). The patient who was excluded from the final analysis was a patient who presented with a palpable lump which was determined to be normal breast tissue, and who also had small, scattered, cysts appearing in a different breast quadrant.

Given the small total number of analyzed patients, the number of training data samples which was extracted from each patient dataset equals the number of pc_fib search ranges for which the detection of each breast lesion was morphologically validated based on persistence. This implies that each detected breast lesion was represented by 3–5 points in the 3-d feature space, as depicted in Figure 4c. The confusion matrix and classification loss were estimated for the two trained classifiers by means of 10-fold cross-validation (i.e., 10 partitions of the full dataset in disjoint training and test datasets) to evaluate the potential for discrimination between malignant and benign breast lesions with Wavelia MBI. The confusion matrices and the decision surfaces are shown for the two classifiers in Figure 4a–c. This proof-of-concept FiH patient dataset suggested the good potential for discrimination between malignant and benign lesions in the defined 3-d feature space. The two classifiers demonstrated very comparable performance and associated classification loss 11.5–12.5%, as depicted in Figure 4.

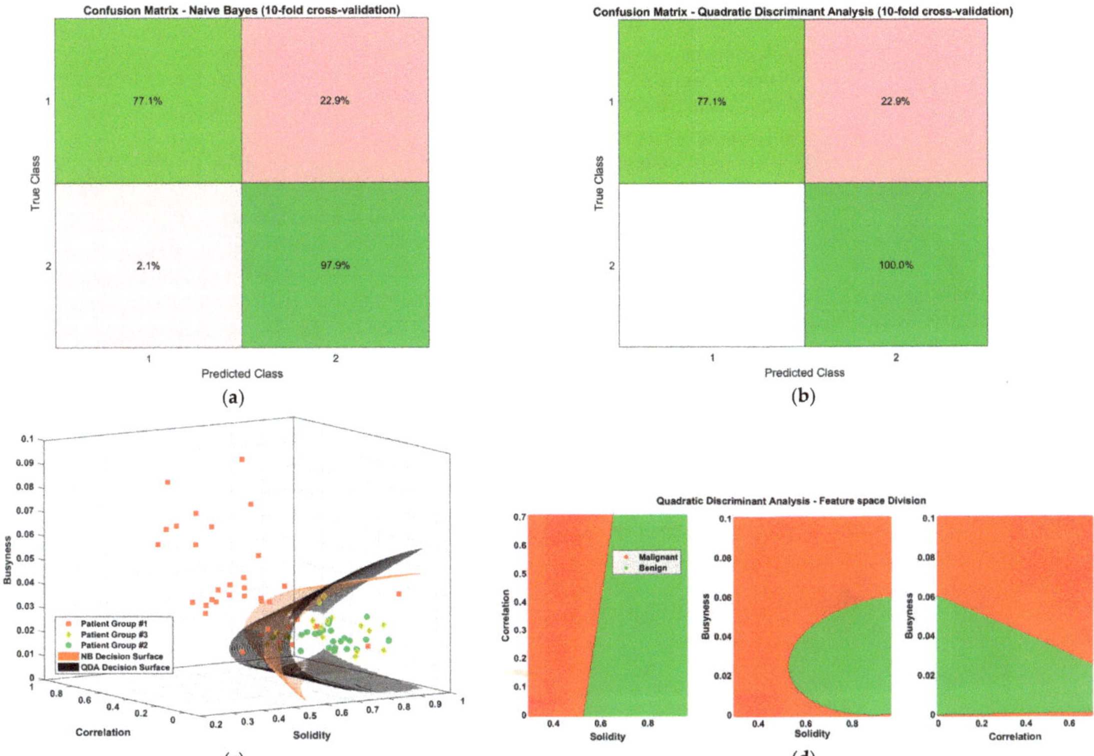

Figure 4. Average confusion matrices for the 2 trained classifiers, estimated with 10-fold cross-validation, (**a**) NB classifier, (**b**) QDA classifier, (**c**) the training dataset and the decision hypersurfaces of the 2 classifiers, (**d**) Partitioning of the 3-d feature space illustrated with 3 cuts for the QDA classifier.

3. Results

3.1. Semi-Automated Breast Lesion Detection Based on Persistence

The Wavelia MBI algorithm for morphological breast lesion detection based on persistence has been specified in the previous section. The lesion persistence is assessed over a set of MBI images that were generated under varying assumptions on the dielectric properties of the healthy tissue of the breast (varying pc_fib parameter search ranges). Lesions that are morphologically detected in at least 3 out of the 5 pc_fib search ranges under evaluation are considered persistent and validated. The principle of the breast lesion detection method is illustrated in Figures 5–8 on two indicative patient test cases.

- Patient 032: Group-1: 54-years old patient with an Invasive Lobular Carcinoma (ILC) of size 30 mm (MRI data) at the 12 o'clock position of the Right Breast. Breast density: BIRADS Category c, Volumetric Breast Density (VBD) = 13.3%.
- Patient 031: Group 3: 38-years old patient with a Fibroadenoma of size 19 mm (Ultrasound) in the lateral Left Breast. Breast density: BIRADS Category c, VBD = 10.8%.

The achieved persistence level of each breast lesion on MBI may vary depending on the histological type of the lesion and the density of the breast. It is interesting to note that the 30 mm ILC of Patient 032, which was not clearly visible on both mammogram and ultrasound, was persistent at 60% (i.e., 3 out of the 5 raw MBI images formed with the employment of distinct pc_fib search ranges) with MBI. More than a single dominant ROIs were visible on the raw MBI images. In Figure 6, the ROI which was extracted and validated in terms of morphological properties, sufficient intensity contrast against the other competing ROIs in each image and persistence over varying pc_fib search ranges is presented encircled in the 3 out of the 5 raw MBI images in which it was detectable. As illustrated in Figures 7 and 8, the fibroadenoma of Patient 031 was persistent at 100% (i.e., 5 out of the 5 raw MBI images formed with the employment of distinct pc_fib search ranges) and was predominantly visible with MBI. Both patients had comparably dense breasts (P032: VBD = 13.3%, P031, VBD = 10.8%); however, the difference in terms of consistency of the two lesions may have been the principal reason for the distinct level of persistence of the response of the two lesions to MBI. The MBI scan datasets for the two patients have been processed using the same configuration of the Wavelia QIF.

In future upgraded implementations of the Wavelia QIF, both the persistence level over the varying assumption of the dielectric properties of the healthy tissue in the breast (varying pc_fib search ranges), but also the presence of a single dominant ROI or various competing ROIs in the image, may serve to define a confidence level for each MBI lesion detection, to better support the diagnosis.

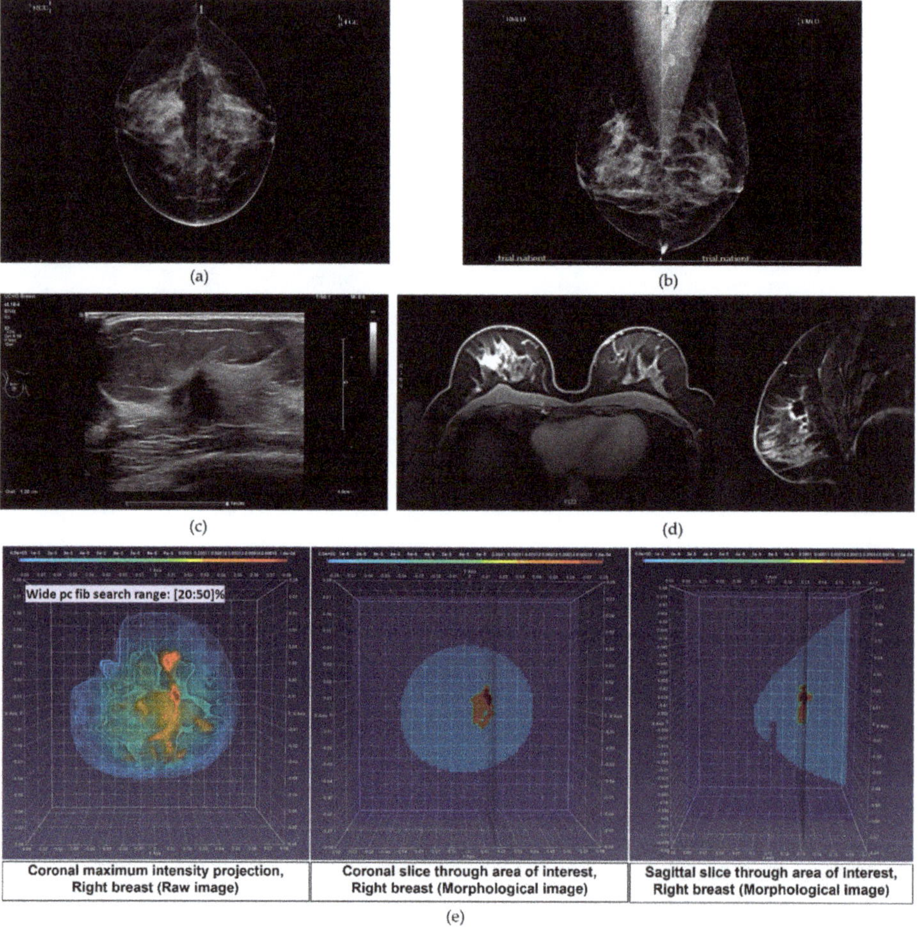

Figure 5. Patient 032, ILC in the Right Breast: (**a**) Bilateral mammogram Cranio-Caudal (CC) view, (**b**) Bilateral mammogram Medio-Lateral Oblique (MLO) view, (**c**) Ultrasound scan, Right Breast, (**d**) MRI scan, bilateral axial image and sagittal image of the Right breast (**e**) MBI test results, Right Breast.

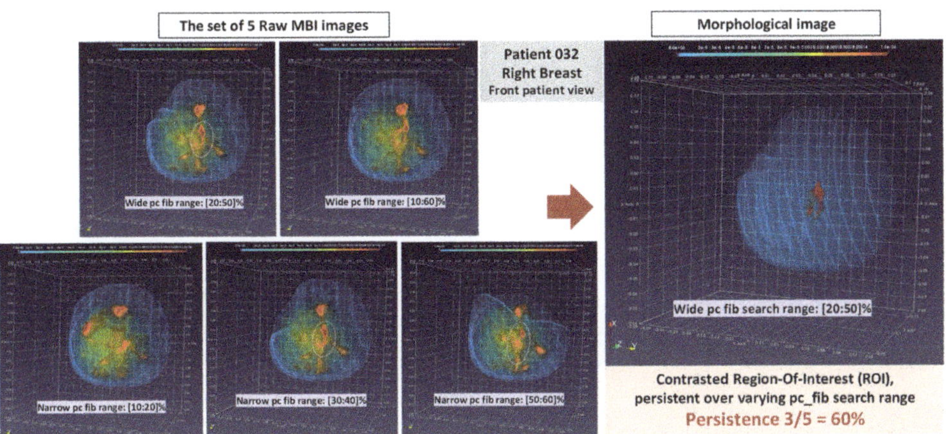

Figure 6. Patient 032, Right breast: ILC morphological detection based on persistence, with Wavelia #1 MBI.

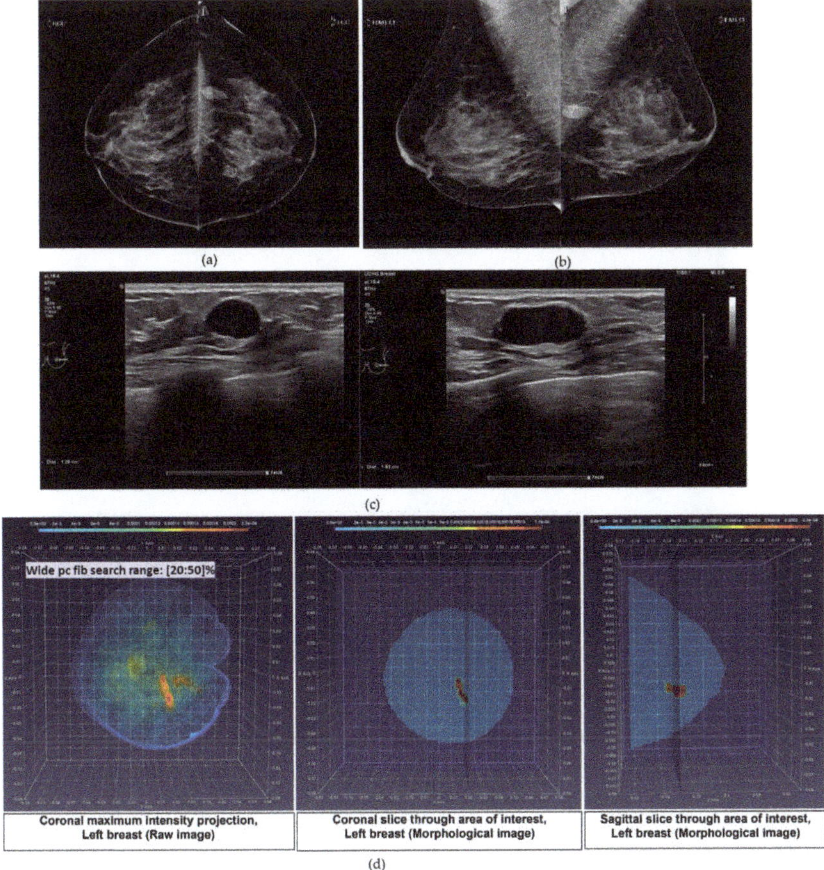

Figure 7. Patient 031, Fibroadenoma in the Left Breast: (**a**) Bilateral mammogram Cranio-Caudal (CC) view, (**b**) Bilateral mammogram Medio-Lateral Oblique (MLO) view, (**c**) Ultrasound scan, Left Breast, (**d**) MBI test results, Left Breast.

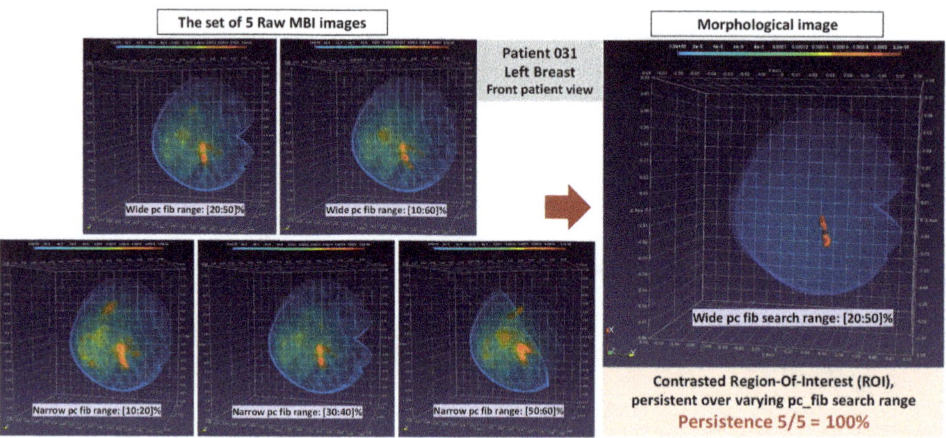

Figure 8. Patient 031, Left Breast: Fibroadenoma morphological detection based on persistence, with Wavelia MBI.

3.2. Breast Lesion Sizing: Correlation with Conventional Imaging and Post-Surgery Histology

For two of the cancer patients, post-surgery histological analysis of the excised tumor demonstrated total tumor sizes which were much larger than the invasive tumor size. This was the case with Patient 002 and Patient 029, as reported in Table 1. The conventional imaging data (Mammography, Ultrasound) and the MBI imaging test results are depicted in Figure 9 for the case of Patient 002 and in Figure 10 for Patient 029.

It is interesting to observe in Table 1 and in Figure 9 that for the Patient 002 case the MBI lesion size estimate varies considerably depending on the pc_fib search range. Maximal linear dimension [34–51] mm, overestimated against the conventional imaging but better fitting to the total tumor size as confirmed with post-surgery histological analysis of the excised tumor, was retrieved with MBI for this lesion. For a subset of 3 out of the five 5 pc_fib search ranges being systematically evaluated in the Wavelia QIF, the irregularly shaped finding of the MBI system extended over a large volume (maximal linear dimension = 51 mm), including the core of the invasive tumor, as identified at triple assessment. By comparison with the patient's mammograms, it was deemed reasonable to consider that the Wavelia MBI system detected either the total tumor, or the invasive tumor and a concentration of fibro-glandular tissue adjacent to it. For the second subset of pc_fib search ranges, the size of the MBI detection was smaller, and its location seemed to correlate closely with the invasive tumor site. Due to the uncertainty on the orientation and the deformability of the patient's breast during the MBI scan, inaccuracies in the 3D reconstruction and localization of the tumor may arise when compared to conventional imaging data. This difficulty is not considered to be MBI-specific though. The registration of multi-modality imaging data of any kind, in the case of soft and deformable organs, like the breast, is a challenging task due to variations in the natural suspended position of the breast in the upright, supine and prone position.

Table 1. Breast lesion size: maximal linear dimension [mm].

Patient ID	Post-Surgery Histology		MBI			Conventional Imaging		
	Invasive Tumor	Total Tumor	W1	W2	Max.	Mammography	Ultrasound	Max.
P002	20	40	51	34	51	25	15	25
P029	22	35	24.5	19.8	24.5	20	37	37

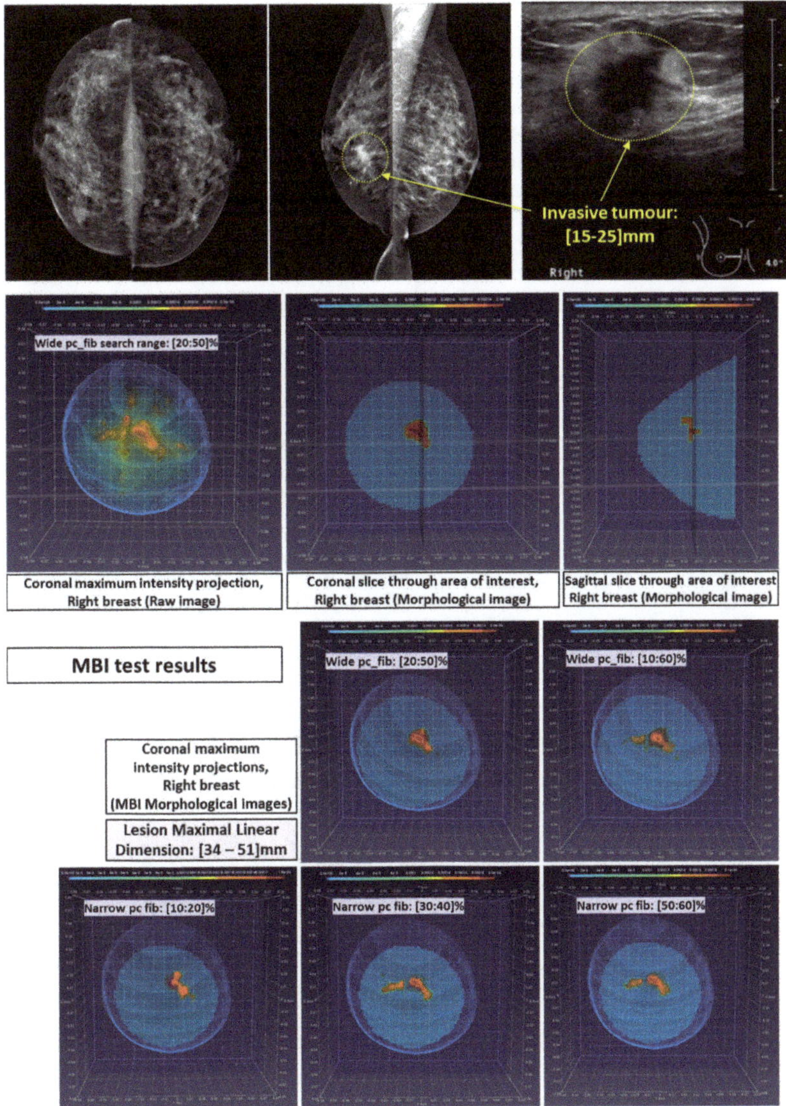

Figure 9. Patient 002, Invasive Ductal Carcinoma (IDC) at the 3 o'clock position of the Right breast (Breast density: BIRADS Category c, VBD = 8.5%): Bilateral Mammogram, Ultrasound and MBI test results of the Right Breast.

Patient 029 was a patient with very dense breasts (VBD = 15.4%). It is clear in Figure 10 that the delineation of the margins of the tumor was not evident on the mammogram. In the radiology report, the presence of a 20 mm spiculated mass in the lower outer quadrant of the Left breast and calcifications extending anteriorly and medially from the mass and measuring up to (42 mm) × (47 mm) was reported. Ultrasound scan of the Left breast highlighted a 23 mm irregular hypoechoic mass in the lower outer quadrant (concurring with the invasive tumor), and a smaller node of 7 mm with indeterminate appearance, immediately superior to the mass. The total inclusive diameter of both lesions was reported to be 37 mm in the craniocaudal direction, concurring with the total tumor size. MBI highlighted the presence of two persistent ROIs. The two ROIs are clearly visible in the raw MBI image shown in Figure 10. The ROI which was morphologically validated was located in the lower outer quadrant of the breast, had a maximal linear dimension of 24.5 mm and was associated with the invasive tumor in this analysis. The second ROI of volumetric size > 3 cm^3 and rather low solidity ($\approx$0.4) was present in the upper breast and could be, interestingly, associated with the extended zone of calcifications, as reported on the patient's mammogram. In the current version of the Wavelia QIF, the ROI definition is based on the notion of pixel connectivity, thus "discontinuous" constellations such as the one highlighted on ultrasound for this patient case and concurring with the total tumor size (as confirmed with post-surgery histology) could not be revealed. This patient case represents a limitation, which may be addressed in subsequent versions of the Wavelia semi-automated lesion sizing method.

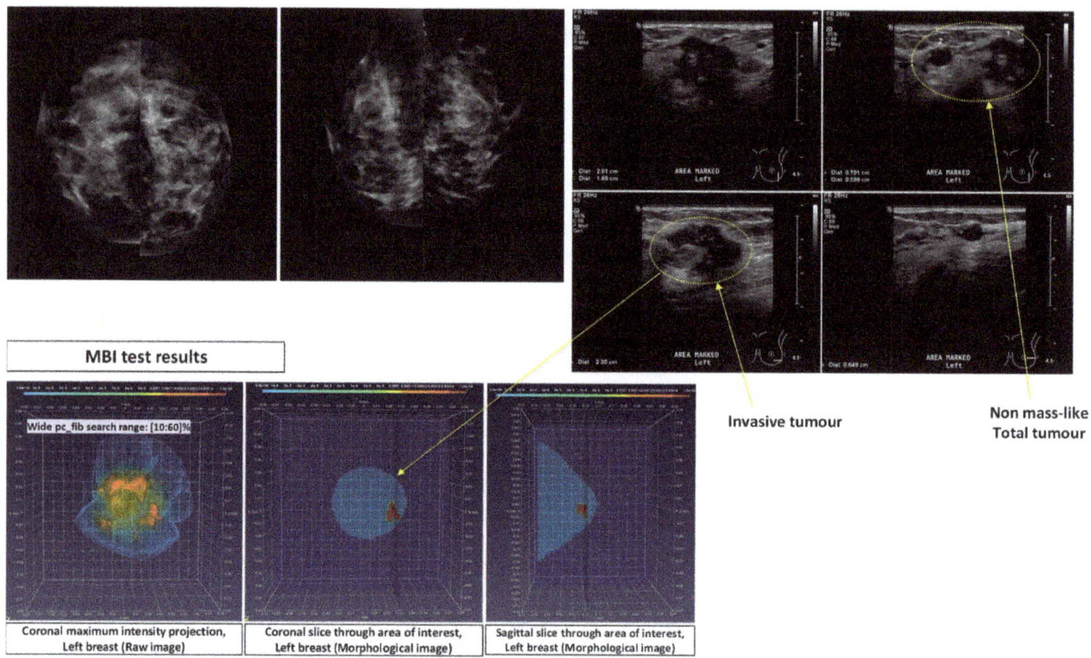

Figure 10. Patient 029, IDC in the Left breast (Breast density: BIRADS Category c, VBD = 15.4%): Bilateral Mammogram, Ultrasound Scan and MBI test results of the Left Breast.

3.3. *Discrimination between Malignant and Benign Breast Lesions in a 3-d Feature Space (Shape-Based and Texture-Based Feature Employment)*

While three narrow pc_fib search ranges are systematically used, together with the two wide pc_fib search ranges, to analyze the persistence of radar echoes for lesion detection, for the characterization of the detected lesions, i.e., sizing, shape and texture analysis, the wide pc_fib search ranges are mostly adequate to be employed, as they are expected to be associated with the most complete representations of the lesions in the available set of MBI images. In the course of the FiH clinical investigation of Wavelia, mapping of the wide pc_fib search range detections in the 3-d feature space was performed and the posterior probability for each detection to be associated to "Class #1 = Malignant lesion" (i.e., probability of malignancy) was computed, as predicted by the trained QDA classifier.

If a breast lesion was detected and validated in both wide pc_fib search range MBI images, two probabilities of malignancy were reported for the breast lesion, as depicted in

Figure 11. The maximal probability of malignancy was ultimately considered to represent a unique MBI classification score for the lesion in the data analysis. Patient 027 was the only Group-1 patient (IDC) for whom the probability of malignancy was inferior to 50%, for both wide pc_fib search ranges. Patient 029 (IDC) was an ambiguous case, with a probability of malignancy 16.5% and 55.7% for the 2 wide pc_fib search ranges, correspondingly. As depicted in Figure 10, this was a patient with very dense breasts, thus rendering the ROI delineation sensitive to the specific parameterization of the MBI morphological detector (Wavelia QIF), in its current version. The probability of malignancy was superior to 95% for all the other Group-1 patients (invasive carcinomas), inferior to 38% for all the Group-3 lesion detections (biopsied benign lesions) and inferior to 14% for all the Group-2 lesion detections (cysts).

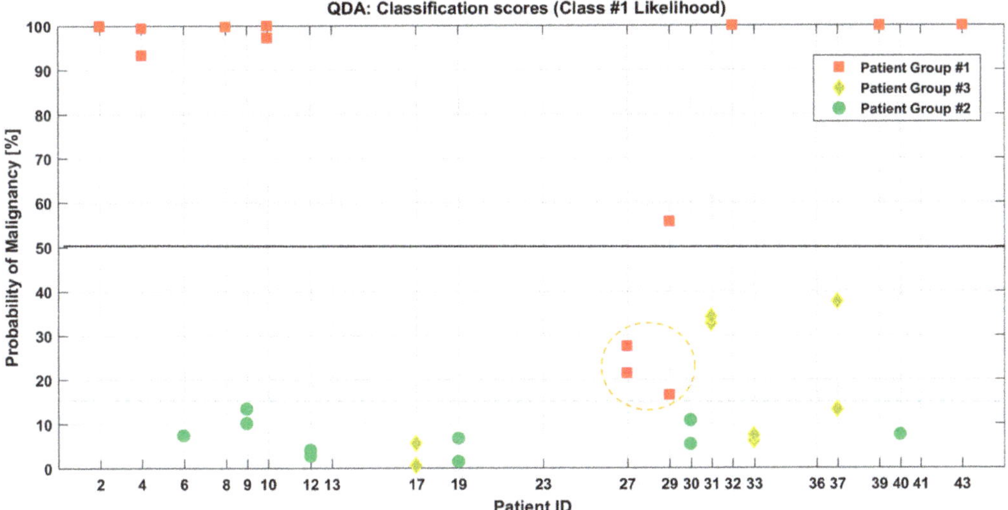

Figure 11. Posterior probability of malignancy (classification score) per patient case in the Wavelia FiH study.

In Figure 12, four patient cases are used to illustrate the impact of the three selected features (solidity, correlation, busyness) on the MBI lesion classification score. The MBI morphological images (i.e., ROIs detected and validated based on morphological properties and persistence) are superimposed with the outer surface of the breast, as reconstructed using the auxiliary Wavelia Optical Breast Contour Detection (OBCD) subsystem, in Figure 12a for the four patient test cases. The Wavelia OBCD subsystem which is employed to reconstruct the external surface of the breast with high resolution, based on optical data, was earlier introduced in [12,13,16]. This superposition serves to better highlight the location of the MBI breast lesion detection with reference to the nipple of the breast, which is visible in the OBCD reconstruction. Ultrasound images of the four patient test cases are included in Figure 12b, for a straightforward comparison with the MBI findings, both in terms of the morphology of each lesion and its localization in the breast. Mapping of the four breast lesions (1 IDC, 1 ILC, 1 fibroadenoma and 1 cyst) on the 3-d feature space of Wavelia MBI, together with the QDA decision surface, are shown in Figure 12c. The probability of malignancy, as predicted by the trained QDA classifier, is also annotated for each of the four lesions. The values of the three features and the associated probabilities of malignancy are reported in Table 2. The morphological detection (W1 or W2 pc_fib search range) which was associated with the highest probability of malignancy, has been used to represent each patient test case in Figure 12, and in Table 2.

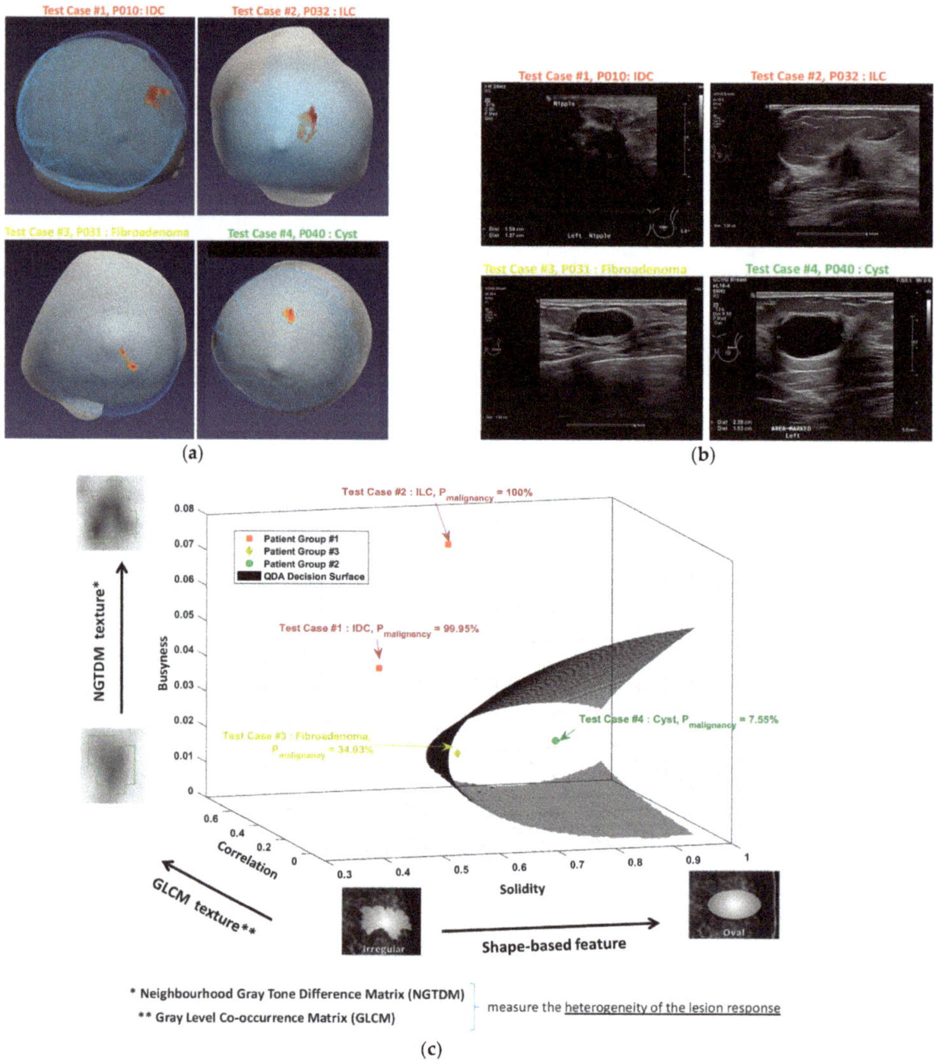

Figure 12. Breast Lesion Characterization in the Wavelia 3-d feature space—Four illustrative patient test cases: (**a**) morphologically validated persistent ROI detections with Wavelia MBI, (**b**) Ultrasound images, (**c**) mapping in the Wavelia 3-d feature space, probability of malignancy and QDA decision surface.

Table 2. Wavelia MBI lesion feature values and derived probability of malignancy.

Patient ID	Lesion Histological Type	Features			Probability of Malignancy [%]
		Solidity	Correlation	Busyness	
P010	Invasive Ductal Carcinoma	0.529	0.496	0.041	99.95%
P032	Invasive Lobular Carcinoma	0.661	0.556	0.074	100%
P031	Fibroadenoma	0.643	0.403	0.018	34.03%
P040	Simple Cyst	0.816	0.422	0.02	7.55%

It is interesting to observe the following:

- the clear differentiability of the simple cyst (Patient 040) in terms of higher solidity,
- the fibroadenoma and the two cancerous lesions had similar solidity levels, however clear distinction was achieved between the benign and the two malignant lesions in terms of texture features,
- both the correlation and the busyness feature values were slightly increased in the case of the malignant lesions, with the increase being more notable on the busyness feature,
- substantially increased busyness value was associated with more heterogeneous lesion patterns, which may be interpretable as being indicative of distributed non-mass like ILC's, such as in the Patient 032 case.

This illustration highlights the physical reasoning behind the selection of the three specific features, for benign-to-malignant MBI lesion classification, based on shape and texture in the Wavelia QIF. It also serves to justify the achievable level of separability with this small proof-of-concept FiH dataset, while working with continuous subspace manifolds and very simple classifier models. The potential generalization of the above findings is intended to be confirmed with future clinical investigations, involving larger and more diverse patient datasets. Expansion of the feature space to include additional dimensions (features), supporting the generalization of the above findings on larger patient cohorts, will be also evaluated further during the development of Wavelia.

4. Discussion and Conclusions

In this article, the methodology that was employed in the Wavelia semi-automated Quantitative Imaging Function (QIF) during the FiH clinical investigation of the device, to support morphological breast lesion detection based on persistence, lesion sizing and lesion characterization in a low-dimensional feature space, spanning shape and texture-based features, has been outlined and the critical design parameters highlighted.

Semi-automated breast lesion detection using morphological post-processing of a set of parametric radar images, which are formed with Wavelia MBI under varying assumptions on the dielectric properties of the healthy background tissue of the breast was introduced in the Wavelia QIF. Automated segmentation of ROIs and association, or not, to a breast lesion is based on morphological properties (solidity, volume) of the ROI and its persistence on the set of parametric images, as evaluated based on spatial clustering. The novelty of the proposed method lies in the exploitation of the notion of persistence and its combination with the solidity feature to support validation of ROI detection in MBI images.

A methodology for malignant-to-benign breast lesion discrimination, based on mapping in a low-dimensional feature space, which spans both shape-based features (solidity) and texture-based features (correlation, busyness), and training of a Naïve Bayes (NB) and Quadratic Discriminant Analysis (QDA) classifier was also introduced in the Wavelia QIF. An interesting level of separability between malignant and benign breast lesions was achieved, with a classification loss of 11.5% estimated with 10-fold cross-validation for the trained QDA classifier. This is a result pending to be verified, reproduced, and validated with larger datasets in future clinical investigations of Wavelia. While extensive research work is already published on Radiomics applied to the well-established breast imaging modalities, to our knowledge, it is the first time that shape descriptors and texture-based features are computed for ROIs extracted from MBI images, to support breast lesion characterization and malignant/benign lesion labelling.

In this FiH study, which was conducted in 25 patients, the Wavelia #1 prototype system demonstrated the preliminary potential to detect and discriminate between malignant and benign palpable breast lumps, the imaging procedure had no safety issues and patients reported a favorable experience of the MBI scan. Although the number of subjects included in the FiH study was small and was not intended to permit a clinically meaningful statistical analysis, the promising findings from this study provided initial data to support the valid clinical association of the technology and warranted the preparation of further clinical investigations, with an upgraded prototype version of the Wavelia system (Wavelia #2) and its semi-automated QIF, to progressively address the identified technological challenges. Larger and more diverse patient datasets are needed to validate these findings and delineate the cases where the Wavelia MBI modality may offer a beneficial adjunct to current diagnostic protocols.

For the first conception of the Wavelia QIF and its feasibility testing for the first time in the clinical setting, simplifications were imposed to the FiH data processing for the analysis to become more straightforward. The two most important limitations of this analysis, which are intended to be loosened in future clinical investigations for the Wavelia QIF to become sufficiently realistic and clinically meaningful, are discussed in the two following paragraphs.

Palpable breast lesions larger than 1 cm^3 were only considered in the current implementation of the Wavelia QIF. All the objects with a volumetric size inferior to 1 cm^3 were excluded and not morphologically validated with the Wavelia QIF. This setting was fixed in accordance with the expected minimum size of the detectable lesion with the first prototype of Wavelia and in order to avoid extreme degradation of the overall specificity of the system.

Detection of a single abnormality zone, which remained persistently contrasted against the surrounding breast tissues, was targeted with the applied MBI imaging algorithm and morphological lesion detector, at this preliminary stage of development, and in the context of the FiH feasibility study of Wavelia. Patients with bilateral breast disease were excluded and the focus was on the detection of a single (the largest) cyst, in the case of patients with multiple cysts in their breasts. This constraint is planned to be loosened in future clinical investigations, while evolving towards more realistic and generalized subsequent phases of the clinical evaluation of the Wavelia QIF.

The MBI scans were performed at least two weeks following the biopsy, in the case of Group-1 (invasive carcinoma) and Group-3 (benign biopsied lesion) patients. The two-week time lapse was considered sufficient to allow healing of the biopsy site in the breast. It is noteworthy though that, in most of these patients, a metallic biopsy clip was placed in their breast, as standard-of-care practice, to mark the tumor site. The size of the biopsy clip was small (~3 mm) compared to the targeted tumor sizes in this study (all palpable lumps), therefore its impact on the MBI images and the associated breast lesion detectability and characterization results were not considered to be significant. However, as the impact of the presence of a biopsy clip has not been characterized so far, future clinical investigations enabling the MBI examination to be performed prior to biopsy will be needed to investigate the impact of the biopsy clip. In this FiH study, there were only two Group-1 patients (P008, P043—both ILC's) and one Group-3 patient (P017—fibroadenoma) with no biopsy clip placed in their breast prior to the MBI examination and positive MBI findings nevertheless.

Quantitative evaluation, by means of a computable confidence level, is also planned to be implemented in subsequent versions of the Wavelia QIF, such that the imaging system performance can be assessed both in terms of lesion detectability rate and detection confidence level, in the case of various lesion types (solid/liquid, mass-like/non-mass-like, malignant/benign) and different breast density levels. Other factors, such as the breast size, the location of the lesion in the breast (superficial/deep, distance to the chest wall), the size of the lesion, the patient's age and breast deformability in the scanner, will also be investigated in terms of achievable lesion detectability rate and detection confidence level, as well as malignant-to-benign lesion separability. The assessment of these factors will be only feasible at a pivotal clinical investigation stage, and after sufficient stabilization of the Wavelia MBI scanning system and the associated QIF.

5. Patents

Two patents have been filed resulting from the work reported in this manuscript. The first patent covers the morphological breast lesion detection method, based on persistence, including sectorization of the imaging scene and MBI image reconstruction using multiple pc_fib search ranges. The second patent covers the malignant-to-benign breast lesion discrimination method, based on mapping in a low-dimensional feature space employing both the solidity feature and texture-based features, applied to ROIs extracted from MBI images.

Author Contributions: Conceptualization, A.F., L.D., B.M.M. and M.J.K.; methodology, A.F., L.D. and J.D.G.C.; software, A.F. and J.D.G.C.; validation, A.F., L.D., J.D.G.C.; formal analysis, A.F., B.M.M., L.D., J.D.G.C., S.M.A.E. and M.J.K.; investigation, B.M.M., S.M.A.E. and M.J.K.; resources, L.D., A.F., and M.J.K.; data curation, A.F., B.M.M., J.D.G.C.; writing—original draft preparation, A.F.; writing—review and editing, L.D., J.D.G.C., B.M.M. and M.J.K.; visualization, A.F. and J.D.G.C.; supervision, L.D., A.F. and M.J.K.; project administration, L.D. and A.F.; funding acquisition, L.D. and A.F. All authors have read and agreed to the published version of the manuscript.

Funding: This research received no external funding.

Institutional Review Board Statement: The study was conducted according to the guidelines of the Declaration of Helsinki and approved by the Institutional Review Board (or Ethics Committee) of Galway University Hospital.

Informed Consent Statement: Informed consent was obtained from all subjects involved in the study.

Conflicts of Interest: Authors L. Duchesne, A. Fasoula and J.D. Gil Cano are employed by MVG Industries, the company that has funded this study and is currently conducting clinical investigations of Wavelia, and have a financial interest in the outcome of those clinical investigations. Authors B. M. Moloney, S. M. Abd Elwahab and M.J. Kerin were investigators for the First-in-Human (FiH) clinical investigation of Wavelia and were funded by their institution. The authors declare no conflict of interest.

References

1. Conceição, R.; Mohr, J.; O'Halloran, M. *An Introduction to Microwave Imaging for Breast Cancer Detection*; Conceição, R.C., Mohr, J.J., O'Halloran, M., Eds.; Biological and Medical Physics, Biomedical Engineering; Springer International Publishing: Cham, Switerland, 2016; ISBN 978-3-319-27865-0.
2. Gabriel, C. Compilation of the Dielectric Properties of Body Tissues at RF and Microwave Frequencies. *Environ. Health* **1996**. Report N.AL/OE-TR-1996-0037. Available online: http://www.brooks.af.mil/HSC/AL/OE/OER/Title/Title.html (accessed on 18 October 2021).
3. Campbell, A.M.; Land, D.V. Dielectric properties of female human breast tissue measured in vitro at 3.2 GHz. *Phys. Med. Biol.* **1992**, *37*, 193–210. [CrossRef] [PubMed]
4. Shere, M.; Lyburn, I.; Sidebottom, R.; Massey, H.; Gillett, C.; Jones, L. MARIA® M5: A multicentre clinical study to evaluate the ability of the Micrima radio-wave radar breast imaging system (MARIA®) to detect lesions in the symptomatic breast. *Eur. J. Radiol.* **2019**, *116*, 61–67. [CrossRef]
5. Sani, L.; Ghavami, N.; Vispa, A.; Paoli, M.; Raspa, G.; Ghavami, M.; Sacchetti, F.; Vannini, E.; Ercolani, S.; Saracini, A.; et al. Novel microwave apparatus for breast lesions detection: Preliminary clinical results. *Biomed. Signal Process. Control* **2019**, *52*, 257–263. [CrossRef]
6. Janjic, A.; Cayoren, M.; Akduman, I.; Yilmaz, T.; Onemli, E.; Bugdayci, O.; Aribal, M.E. SAFE: A Novel Microwave Imaging System Design for Breast Cancer Screening and Early Detection—Clinical Evaluation. *Diagnostics* **2021**, *11*, 533. [CrossRef]
7. Meaney, P.M.; Kaufman, P.A.; Muffly, L.S.; Click, M.; Poplack, S.P.; Wells, W.A.; Schwartz, G.N.; di Florio-Alexander, R.M.; Tosteson, T.D.; Li, Z.; et al. Microwave imaging for neoadjuvant chemotherapy monitoring: Initial clinical experience. *Breast Cancer Res.* **2013**, *15*, R35. [CrossRef]
8. Benny, R.; Anjit, T.A.; Mythili, P. An overview of microwave imaging for breast tumor detection. *Prog. Electromagn. Res. B* **2020**, *87*, 61–91. [CrossRef]
9. Moloney, B.M.; O'Loughlin, D.; Abd Elwahab, S.; Kerin, M.J. Breast Cancer Detection—A Synopsis of Conventional Modalities and the Potential Role of Microwave Imaging. *Diagnostics* **2020**, *10*, 103. [CrossRef]
10. O'Loughlin, D.; O'Halloran, M.; Moloney, B.M.; Glavin, M.; Jones, E.; Elahi, M.A. Microwave Breast Imaging: Clinical Advances and Remaining Challenges. *IEEE Trans. Biomed. Eng.* **2018**, *65*, 2580–2590. [CrossRef]
11. Kwon, S.; Lee, S. Recent Advances in Microwave Imaging for Breast Cancer Detection. *Int. J. Biomed. Imaging* **2016**, *2016*, 5054912. [CrossRef] [PubMed]
12. Moloney, B.M.; McAnena, P.F.; Abd Elwahab, S.M.; Fasoula, A.; Duchesne, L.; Gil Cano, J.D.; Glynn, C.; O'Connell, A.; Ennis, R.; Lowery, A.J.; et al. Microwave Imaging in Breast Cancer—Results from the First-In-Human Clinical Investigation of the Wavelia System. *Acad. Radiol.* **2021**. [CrossRef] [PubMed]
13. Fasoula, A.; Duchesne, L.; Gil Cano, J.; Lawrence, P.; Robin, G.; Bernard, J.-G. On-Site Validation of a Microwave Breast Imaging System, before First Patient Study. *Diagnostics* **2018**, *8*, 53. [CrossRef] [PubMed]

14. Moloney, B.M.; McAnena, P.F.; Elwahab, S.M.; Fasoula, A.; Duchesne, L.; Gil Cano, J.D.; Glynn, C.; O'Connell, A.; Ennis, R.; Lowery, A.J.; et al. The Wavelia Microwave Breast Imaging system–tumour discriminating features and their clinical usefulness. *Br. J. Radiol.* **2021**. epub ahead of print. [CrossRef]
15. Fasoula, A.; Moloney, B.M.; Duchesne, L.; Cano, J.D.G.; Oliveira, B.L.; Bernard, J.; Kerin, M.J. Super-resolution radar imaging for breast cancer detection with microwaves: The integrated information selection criteria. In Proceedings of the 41st Annual International Conference of the IEEE Engineering in Medicine & Biology Society (EMBC), Berlin, Germany, 23–27 July 2019.
16. Fasoula, A.; Duchesne, L.; Moloney, B.M.; Gil Cano, J.D.; Chenot, C.; Oliveira, B.L.; Bernard, J.-G.; Abd Elwahab, S.M.; Kerin, M.J. Pilot patient study with the Wavelia Microwave Breast Imaging system for breast cancer detection: Clinical feasibility and identified technical challenges. In Proceedings of the 2020 14th European Conference on Antennas and Propagation (EuCAP), Copenhagen, Denmark, 15–20 March 2020.
17. Devaney, A. Super-resolution processing of multi-static data using time reversal and MUSIC. *J. Acoust. Soc. Am.* **2000**. Available online: https://ece.northeastern.edu/fac-ece/devaney/preprints/paper02n_00.pdf (accessed on 18 October 2021).
18. Sugitani, T.; Kubota, S.; Kuroki, S.; Sogo, K.; Arihiro, K.; Okada, M.; Kadoya, T.; Hide, M.; Oda, M.; Kikkawa, T. Complex permittivities of breast tumor tissues obtained from cancer surgeries. *Appl. Phys. Lett.* **2014**, *104*, 253702. [CrossRef]
19. Lazebnik, M.; Popovic, D.; McCartney, L.; Watkins, C.B.; Lindstrom, M.J.; Harter, J.; Sewall, S.; Ogilvie, T.; Magliocco, A.; Breslin, T.M.; et al. A large-scale study of the ultrawideband microwave dielectric properties of normal, benign and malignant breast tissues obtained from cancer surgeries. *Phys. Med. Biol.* **2007**, *52*, 6093–6115. [CrossRef] [PubMed]
20. Martellosio, A.; Bellomi, M.; Pasian, M.; Bozzi, M.; Perregrini, L.; Mazzanti, A.; Svelto, F.; Summers, P.E.; Renne, G.; Preda, L. Dielectric Properties Characterization From 0.5 to 50 GHz of Breast Cancer Tissues. *IEEE Trans. Microw. Theory Tech.* **2017**, *65*, 998–1011. [CrossRef]
21. Summers, P.E.; Vingiani, A.; Di Pietro, S.; Martellosio, A.; Espin-Lopez, P.F.; Di Meo, S.; Pasian, M.; Ghitti, M.; Mangiacotti, M.; Sacchi, R.; et al. Towards mm-wave spectroscopy for dielectric characterization of breast surgical margins. *The Breast* **2019**, *45*, 64–69. [CrossRef]
22. Pertuz, S.; Puig, D.; Garcia, M.A. Analysis of focus measure operators for shape-from-focus. *Pattern Recognit.* **2013**, *46*, 1415–1432. [CrossRef]
23. O'loughlin, D.; Krewer, F.; Glavin, M.; Jones, E.; O'halloran, M. Focal quality metrics for the objective evaluation of confocal microwave images. *Int. J. Microw. Wirel. Technol.* **2017**, *9*, 1365–1372. [CrossRef]
24. Sickles, E.A.; D'Orsi, C.J.; Bassett, L.W.; Appleton, C.M.; Berg, W.A.; Burnside, E.S. Acr Bi-Rads® Mammography. In *ACR BI-RADS® Atlas, Breast Imaging Reporting and Data System*; American College of Radiology: Reston, VA, USA, 2013; Volume 5.
25. Chan, T.F.; Vese, L.A. Active contours without edges. *IEEE Trans. Image Process.* **2001**, *10*, 266–277. [CrossRef]
26. Chen, Y.; Gunawan, E.; Low, K.S.; Wang, S.C.; Soh, C.B.; Putti, T.C. Effect of Lesion Morphology on Microwave Signature in 2-D Ultra-Wideband Breast Imaging. *IEEE Trans. Biomed. Eng.* **2008**, *55*, 2011–2021. [CrossRef] [PubMed]
27. Davis, S.K.; Van Veen, B.D.; Hagness, S.C.; Kelcz, F. Breast Tumor Characterization Based on Ultrawideband Microwave Backscatter. *IEEE Trans. Biomed. Eng.* **2008**, *55*, 237–246. [CrossRef]
28. Gerazov, B.; Conceicao, R.C. Deep learning for tumour classification in homogeneous breast tissue in medical microwave imaging. In Proceedings of the IEEE EUROCON 2017-17th International Conference on Smart Technologies, Ohrid, Macedonia, 6–8 July 2017; pp. 564–569.
29. Oliveira, B.; Godinho, D.; O'Halloran, M.; Glavin, M.; Jones, E.; Conceição, R. Diagnosing Breast Cancer with Microwave Technology: Remaining challenges and potential solutions with machine learning. *Diagnostics* **2018**, *8*, 36. [CrossRef] [PubMed]
30. Conceição, R.C.; Medeiros, H.; Godinho, D.M.; O'Halloran, M.; Rodriguez-Herrera, D.; Flores-Tapia, D.; Pistorius, S. Classification of breast tumor models with a prototype microwave imaging system. *Med. Phys.* **2020**, *47*, 1860–1870. [CrossRef]
31. Doshi, T.; Lyburn, I.; Sidebottom, R.; Gibbins, D. Radio-wave imaging: Frequency response as an aid to lesion characterization. Early concept work. In Proceedings of the Symposium Mammographicum, Liverpool, UK, 8–10 July 2018.
32. Rana, S.P.; Dey, M.; Tiberi, G.; Sani, L.; Vispa, A.; Raspa, G.; Duranti, M.; Ghavami, M.; Dudley, S. Machine Learning Approaches for Automated Lesion Detection in Microwave Breast Imaging Clinical Data. *Sci. Rep.* **2019**, *9*, 10510. [CrossRef]
33. Mendelson, E.B.; Böhm-Vélez, M.; Berg, W.A. ACR BI-RADS® Ultrasound. In *ACR BI-RADS® Atlas, Breast Imaging Reporting and Data System*; American College of Radiology: Reston, VA, USA, 2013.
34. De Brito Silva, T.F.; de Paiva, A.C.; Silva, A.C.; Braz Júnior, G.; de Almeida, J.D.S. Classification of breast masses in mammograms using geometric and topological feature maps and shape distribution. *Res. Biomed. Eng.* **2020**, *36*, 225–235. [CrossRef]
35. Haralick, R.M.; Shanmugam, K.; Dinstein, I. Textural Features for Image Classification. *IEEE Trans. Syst. Man. Cybern.* **1973**, *SMC-3*, 610–621. [CrossRef]
36. Amadasun, M.; King, R. Textural features corresponding to textural properties. *IEEE Trans. Syst. Man. Cybern.* **1989**, *19*, 1264–1274. [CrossRef]
37. Safdarian, N.; Hedyezadeh, M. Detection and Classification of Breast Cancer in Mammography Images Using Pattern Recognition Methods. *Multidiscip. Cancer Investig.* **2019**, *3*, 13–24. [CrossRef]
38. Sadad, T.; Hussain, A.; Munir, A.; Habib, M.; Ali Khan, S.; Hussain, S.; Yang, S.; Alawairdhi, M. Identification of Breast Malignancy by Marker-Controlled Watershed Transformation and Hybrid Feature Set for Healthcare. *Appl. Sci.* **2020**, *10*, 1900. [CrossRef]
39. Vallières, M.; Freeman, C.R.; Skamene, S.R.; El Naqa, I. A radiomics model from joint FDG-PET and MRI texture features for the prediction of lung metastases in soft-tissue sarcomas of the extremities. *Phys. Med. Biol.* **2015**, *60*, 5471–5496. [CrossRef] [PubMed]

40. Parekh, V.; Jacobs, M.A. Radiomics: A new application from established techniques. *Expert Rev. Precis. Med. Drug Dev.* **2016**, *1*, 207–226. [CrossRef] [PubMed]
41. Coroller, T.P.; Grossmann, P.; Hou, Y.; Rios Velazquez, E.; Leijenaar, R.T.H.; Hermann, G.; Lambin, P.; Haibe-Kains, B.; Mak, R.H.; Aerts, H.J.W.L. CT-based radiomic signature predicts distant metastasis in lung adenocarcinoma. *Radiother. Oncol.* **2015**, *114*, 345–350. [CrossRef] [PubMed]
42. Parekh, V.S.; Jacobs, M.A. Multiparametric radiomics methods for breast cancer tissue characterization using radiological imaging. *Breast Cancer Res. Treat.* **2020**, *180*, 407–421. [CrossRef]

applied sciences

MDPI

Article

Artificial Neural Network-Derived Cerebral Metabolic Rate of Oxygen for Differentiating Glioblastoma and Brain Metastasis in MRI: A Feasibility Study

Hakim Baazaoui [1], Simon Hubertus [1], Máté E. Maros [2,3], Sherif A. Mohamed [4], Alex Förster [2], Lothar R. Schad [1] and Holger Wenz [2,*]

1 Computer Assisted Clinical Medicine, Medical Faculty Mannheim, Heidelberg University, 68167 Mannheim, Germany; hb12@medma.ad.uni-heidelberg.de (H.B.); simon.hubertus@medma.uni-heidelberg.de (S.H.); lothar.schad@medma.uni-heidelberg.de (L.R.S.)

2 Department of Neuroradiology, Medical Faculty Mannheim, Heidelberg University, 68167 Mannheim, Germany; mate.maros@umm.de (M.E.M.); alex.foerster@umm.de (A.F.)

3 Department of Biomedical Informatics at the Center for Preventive Medicine and Digital Health, Medical Faculty Mannheim, Heidelberg University, 68167 Mannheim, Germany

4 Department of Diagnostic and Interventional Radiology, Heidelberg University Hospital, 69120 Heidelberg, Germany; sherif.mohamed@med.uni-heidelberg.de

* Correspondence: holgerwenz.hw@gmail.com

Featured Application: MR-derived cerebral metabolic rate of oxygen in contrast-enhancing and peritumoral non-enhancing regions, as calculated by an artificial neural network, allows for robust differentiation of glioblastoma and brain metastasis.

Abstract: Glioblastoma may appear similar to cerebral metastasis on conventional MRI in some cases, but their therapies differ significantly. This prospective feasibility study was aimed at differentiating them by applying the quantitative susceptibility mapping and quantitative blood-oxygen-level-dependent (QSM + qBOLD) model to these entities for the first time. We prospectively included 15 untreated patients with glioblastoma ($n = 7$, median age: 68 years, range: 54–84 years) or brain metastasis ($n = 8$, median age 66 years, range: 50–78 years) who underwent preoperative MRI including multi-gradient echo and arterial spin labeling sequences. Oxygen extraction fraction (OEF), cerebral blood flow (CBF) and cerebral metabolic rate of oxygen ($CMRO_2$) were calculated in the contrast-enhancing tumor (CET) and peritumoral non-enhancing T2 hyperintense region (NET2), using an artificial neural network. We demonstrated that OEF in CET was significantly lower ($p = 0.03$) for glioblastomas than metastases, all features were significantly higher ($p = 0.01$) in CET than in NET2 for metastasis patients only, and the ratios of CET/NET2 for CBF ($p = 0.04$) and $CMRO_2$ ($p = 0.01$) were significantly higher in metastasis patients than in glioblastoma patients. Discriminative power of a support-vector machine classifier was highest with a combination of two features, yielding an area under the receiver operating characteristic curve of 0.94 with 93% diagnostic accuracy. QSM + qBOLD allows for robust differentiation of glioblastoma and cerebral metastasis while yielding insights into tumor oxygenation.

Keywords: brain metastasis; glioblastoma; machine learning; oxygenation; tumor infiltration

Citation: Baazaoui, H.; Hubertus, S.; Maros, M.E.; Mohamed, S.A.; Förster, A.; Schad, L.R.; Wenz, H. Artificial Neural Network-Derived Cerebral Metabolic Rate of Oxygen for Differentiating Glioblastoma and Brain Metastasis in MRI: A Feasibility Study. *Appl. Sci.* **2021**, *11*, 9928. https://doi.org/10.3390/app11219928

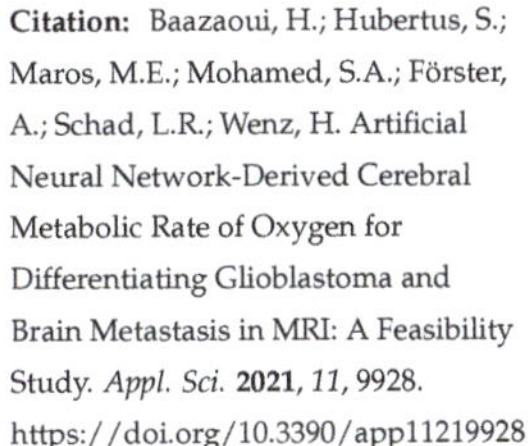

Academic Editor: Fabio La Foresta

Received: 26 August 2021
Accepted: 22 October 2021
Published: 24 October 2021

1. Introduction

Glioblastoma (GBM) and cerebral metastasis (cMET) are the most common brain tumors in adult patients [1]. Reliably differentiating GBM and cMET based on their conventional magnetic resonance imaging (MRI) characteristics has proven difficult [2,3], as both tumor types can show necrotic centers, contrast-enhancing peripheral areas and peritumoral edema (Figure 1) [4]. However, studies employing advanced MR-imaging techniques focusing on the tumor microenvironment and hypoxia-induced changes in the microvasculature found that an elevated cerebral blood flow (CBF) and proxies for increased metabolic activity including a higher resulting cerebral metabolic rate of oxygen ($CMRO_2$) were associated with high-grade gliomas [5,6]. Therefore, we performed a prospective feasibility study to differentiate between GBM and cMET by using an artificial neural network (ANN) approach for non-invasive estimation of $CMRO_2$, combining quantitative susceptibility mapping (QSM) and the quantitative blood-oxygenation-level-dependent effect (qBOLD). To the best of the authors' knowledge, the concept of using the combined QSM + qBOLD approach for estimation of $CMRO_2$ is new for differentiating these two entities.

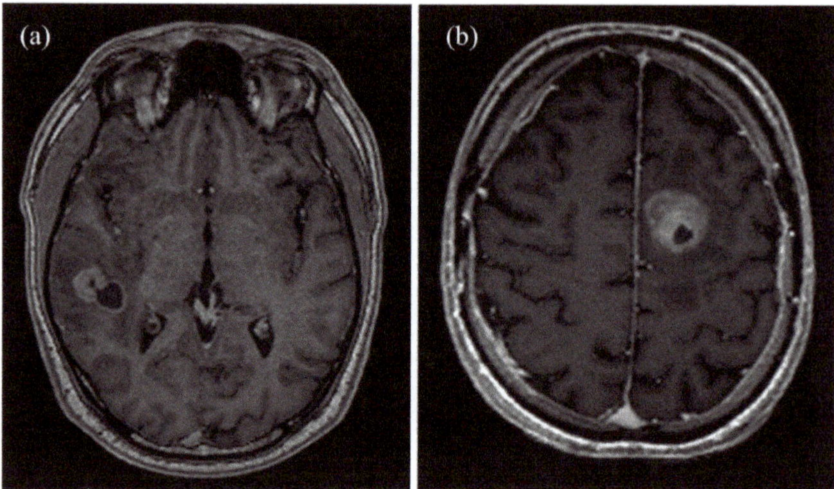

Figure 1. Axial contrast-enhanced magnetization prepared rapid gradient echo (MP-RAGE) brain images comparing (**a**) a right temporal metastasis from known esophageal cancer in a 50-year-old male patient and (**b**) a glioblastoma in the frontal lobe of the left hemisphere of a 71-year-old male patient. Both tumors show cystic elements and peripheral contrast enhancement, complicating a differentiation based on solely morphological criteria.

GBMs constitute between 60% and 70% of all malignant gliomas [7]. Median survival for this highly malignant, infiltratively growing tumor is between 12 and 15 months with optimal treatment [7–9]. A higher median survival of 22 months has been reported for patients with recurrent glioblastoma undergoing a second surgical therapy [10]. The metabolism of GBM cells is adapted to the increased uptake of nutrients by utilizing aerobic glycolysis [11], thereby linking glucose metabolism to oxygen metabolism [12]. Although benefiting from a greater supply of components for cell upkeep and increased perfusion, GBMs frequently grow too fast for their vasculature, eventually resulting in the typical central necrosis with peripheral ring-enhancement [7,8].

Similarly, cMETs can also become ring-enhancing after developing a necrotic center due to them outgrowing their blood and nutrient supply [13,14]. Over 20% of cancer patients develop disseminations to the central nervous system [15]. One study identified that 55% of cMET cases had no known primary at diagnosis [16], while between 30% to 50% of cMETs have been found to first appear as solitary lesions, further complicating their correct identification [2,17,18]. Accurately discriminating between GBM and cMET is of great clinical importance because therapy approach and surgical decisions are quite different and directly affect patient outcomes [9,17,19]. The current diagnostic standard is an invasive tissue biopsy with subsequent histopathological examination [4], a procedure that is not without inherent risks with a complication rate of about 6% [20].

Hence, the discrimination of GBM and cMET has been attempted many times using a variety of different radiological approaches [4,19,21–24]. Perfusion-based studies showed relative cerebral blood volume (rCBV) in the proximal peritumoral non-enhancing T2 FLAIR hyperintensity (NET2) to be significantly higher in GBM than in cMET [21–24]. In the distal parts of NET2, GBM and cMET showed similar rCBV values which may reflect a lack of GBM cell infiltration and angiogenesis, supporting the notion that angiogenesis follows a gradient around the tumor and is highest on the surface of the contrast-enhancing tumor region (CET) [24,25]. With regards to oxygenation, tissue hypoxia is widely accepted as a predictor of therapy resistance to radiation and chemotherapy in gliomas [5]. Hypoxia has been found to stimulate the growth of new blood vessels via the induction and release of vascular endothelial growth factor [10]. This neovascularization leads to a dilated and tortuous vessel configuration, abnormal branching and arteriovenous shunts [26]. The resulting inefficiencies in the tumor vasculature were found to be correlated with greatly increased CBF, lower oxygen extraction fraction (OEF) and, in sum, a higher $CMRO_2$ [6].

In this work, tissue oxygenation was estimated using a combined QSM + qBOLD model that was introduced in 2018 and which utilizes both signal magnitude and phase of a 3D multi-gradient echo (mGRE) sequence [27–29]. We built upon an existing artificial neural network approach to perform the QSM + qBOLD analysis [29]. The artificial neural network was previously used only for mapping OEF in healthy individuals, making this study its first application in a clinical setting by employing it for the differential diagnosis of two brain tumor entities. Perfusion was measured with pseudocontinuous arterial spin labeling.

The purpose of this study was, therefore, to apply the QSM + qBOLD method for the first time to a prospectively recruited collective of GBM and cMET patients and to compare their cerebral oxygenation and perfusion. Based on the hypothesis that the infiltrative growth of GBM and the lack thereof in cMET would create differences in CET and NET2, a machine learning classifier was trained to differentiate the two entities.

2. Materials and Methods

2.1. Patients

Between December 2019 and October 2020, 15 patients with primary GBM ($n = 7$; median age: 68 years, range: 54–84 years) or cMET ($n = 8$; median age: 66 years, range: 50–78 years) before resection, radiation or chemotherapy were prospectively included in this study as a convenience sample. Ten patients were male, five were female. The cMETs, as determined by histopathology, originated from four lung carcinomas, one esophageal carcinoma and three cancers of unknown primary. Ethics committee approval was granted before recruitment of patients (reference: 2017-666N-MA). Written informed consent was obtained from every participant prior to MRI measurements and the acquired image data was anonymized before further processing. The study was conducted in compliance with the ethical standards of the Declaration of Helsinki of 2013.

2.2. Image Acquisition

All MRI scans were performed on a 3T MAGNETOM Trio system (Siemens Healthcare GmbH, Erlangen, Germany) at the Department of Neuroradiology of the Medical Faculty Mannheim. The perfusion data of the first two patients (one cMET, one GBM) were acquired with a 32-channel head coil. Due to practical reasons in clinical day-to-day MRI scans, a 12-channel head coil was used for all subsequent study participants. The MRI protocol included a 3D multi-gradient echo, an unbalanced axial two-dimensional pseudocontinuous arterial spin labeling (three-dimensional pseudocontinuous arterial spin labeling was employed for the first two patients that were scanned with a 32-channel head coil) and a magnetization prepared rapid gradient echo (MP-RAGE) sequence. The gadolinium-based contrast agent Dotarem® (Guerbet, Villepinte, France) was administered as bolus injection for the T1-weighted sequence at a dosage of 0.1 mL/kg body weight. The labeling plane for the arterial spin labeling sequence was placed circa 85 mm inferior to the anterior commissure-posterior commissure line, approximately perpendicular to the feeding arteries, in line with the consensus recommendation for arterial spin labeling imaging [30]. The sequences and specific parameters used in this study have all been described in detail in a previously published study [29].

2.3. Image Processing

The MRI images were registered using the statistical parametric mapping software SPM12 (Wellcome Centre for Human Neuroimaging, UCL, London, UK) using default values. Correct registration was verified with the open-source medical image viewer ITK-SNAP (http://www.itksnap.org/pmwiki/pmwiki.php, accessed on 13 October 2021). Post-processing was performed with MATLAB (Mathworks, Natick, MA, USA). Three regions of interest (ROI) were outlined manually (H.B.) for calculation of oxygenation and perfusion parameters: CET and central necrosis on all slices of the acquired images where the respective region was present, as well as NET2 (Figure 2). The propensity of the central necrotic region to hemorrhage and its very low to non-existent perfusion can lead to an unphysiological spike in OEF (cf. Figure 3) [5]. Therefore, the necrotic region was subtracted from CET to exclude non-vital parts of the tumor in metabolic assessment. For the peritumoral edematous area, a 15–20 mm wide ROI was defined in NET2 on three consecutive FLAIR images with the largest peritumoral hyperintensity, leaving an approximately 3 mm wide margin to avoid partial volume effects [31]. All ROIs were adapted to the tumor border, subtracting a cerebrospinal fluid mask of ventricles and sulci generated automatically in SPM12. The ROIs were audited by an experienced neuroradiologist (H.W.).

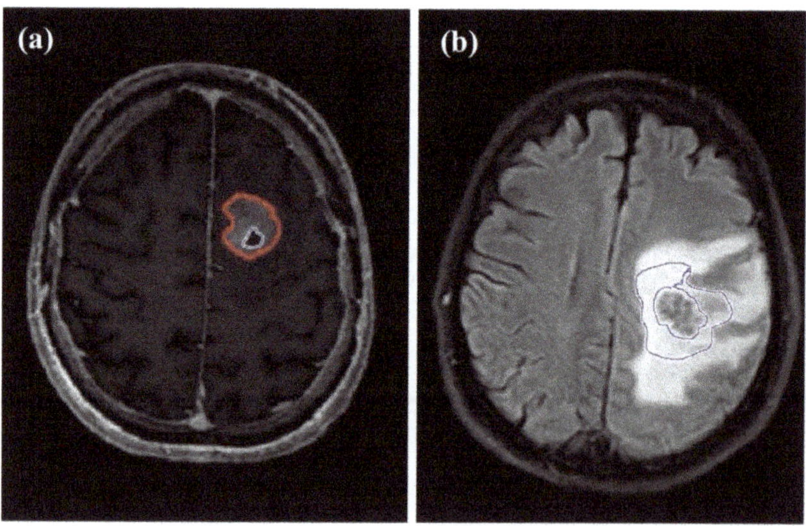

Figure 2. Examples of the different regions of interest (ROIs) assessed in this study. (**a**) Axial T1-weighted contrast-enhanced image of a 71-year-old male patient with glioblastoma in the frontal lobe of the left hemisphere. The outer ROI marks contrast-enhancing tumor while the ROI inside the tumor indicates the central necrosis. (**b**) Axial FLAIR image of a 66-year-old male patient with a left hemispheric metastasis in the perirolandic region from unknown primary. The ROI marks the proximal edema bordering the solid-appearing tumor.

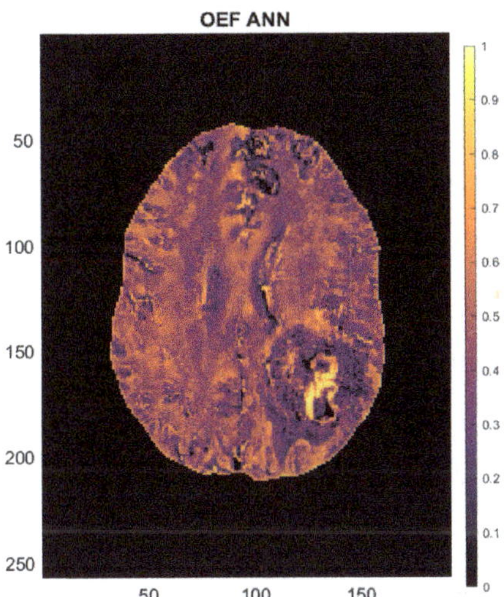

Figure 3. Oxygen extraction fraction (OEF) map of a 54-year-old female patient with left occipito-temporal glioblastoma. Axes are in mm. The scale to the right of the image indicates OEF (e.g., 0.5 = 50%).

2.4. Calculation of Perfusion and Oxygenation Parameters

In order to estimate cerebral perfusion, a quantitative perfusion map was created from the arterial spin labeling data using SPM12. Both the control and tag images from the pCASL sequence were averaged and used together with the proton density weighted image to calculate CBF in mL/100 g/min, using the consensus recommendation for implementation of arterial spin labeling perfusion MRI in clinical applications from Alsop et al. [30]. Time correction for multi-slice imaging was implemented to account for the different transit time of the labeled bolus depending on the time of slice acquisition [30].

$CMRO_2$ in µmol/100 g/min was calculated as follows:

$$CMRO_2 = CBF \cdot Y_a \cdot OEF \cdot [H], \quad (1)$$

where Υ_a is the arterial oxygen saturation, assumed to be 98%, and [H] = 7.53 µmol/mL is the heme molar concentration in tissue blood assuming a hematocrit of Hct = 0.357 in arterioles [27,32].

An artificial neural network was employed for combined QSM + qBOLD analysis of the mGRE data to estimate Υ and calculate OEF [33]:

$$OEF = 1 - \frac{Y}{Y_a}, \quad (2)$$

where Υ and Υ_a are venous and arterial blood oxygenation, the latter again assumed to be 98% [27].

2.5. Artificial Neural Network

The feed-forward artificial neural network used for estimation of Υ was designed in the Neural Network Toolbox in MATLAB (Mathworks, Natick, MA, USA), consisting of one input layer, one hidden layer with 10 nodes and one output layer. The normalized mGRE magnitude signal and the magnetic susceptibility from QSM were used as inputs. The artificial neural network has been described in detail in a previous study [26], where it was trained and used to emulate the solution of the qBOLD model for free induction decay [27], yielding an estimate of venous oxygen saturation Υ, transverse relaxation rate

R2, deoxygenated blood volume ν and non-blood susceptibility χ_{nb} [29]. After doing this for every voxel, OEF maps were created with Equation (2). In order to make the qBOLD model more robust, QSM was added by calculating magnetic susceptibility [27], using the MEDI toolbox (Cornell MRI Research Lab, Cornell University, New York, NY, USA). The resulting parameters were used for a last fitting step by giving starting values for a quasi-Newton optimization that was stopped once the relative change was smaller than 0.001 or a maximum of 50 iterations was reached (this was mostly the case after approx. 10 iterations) [34].

2.6. Statistical Analysis

The statistical analyses were descriptive and performed in MATLAB R2020b (Mathworks, Natick, MA, USA). Variables were summarized using their median, minimum–maximum and interquartile range. Outliers were included in the statistical tests. For all patients, means and standard deviations of OEF, CBF and $CMRO_2$ were calculated in the ROIs set out in Figure 2. In patients with multiple cMETs, perfusion and oxygenation parameters were not assessed on a metastasis-by-metastasis basis but averaged across all metastases present in the respective patient's brain. GBM patients were not stratified according to IDH or MGMT promoter methylation status.

Non-parametric Mann–Whitney–Wilcoxon tests for two populations were applied to properly compare the distributions of explanatory variables between the two groups. Two-tailed Wilcoxon signed-rank tests were used for intra-individual comparisons between CET and NET2. p-values < 0.05 were considered significant. Due to the explorative nature of this study, p-values were not adjusted for multiple comparisons and power analysis for determination of required sample size was not performed [35].

The primary learning objective was the binary classification of GBM vs. cMET. For this, a well-established maximum margin classifier, a linear kernel support-vector machine [36], was fitted to the following features: OEF, CBF and $CMRO_2$ in CET and NET2. Additional secondary measures, i.e., the ratios of said features in CET divided by NET2, were also calculated and used for fitting [17]. We performed five-fold cross-validation with 10 repeats to assess the robustness of these explanatory variables [37]. Receiver operating characteristic analysis was performed to calculate the area under the receiver operating characteristic curve (AUC) accuracy metrics and derivatives of the confusion matrix, averaged over the five folds [37]. In an effort to compare the classification performance of the linear kernel support-vector machine with other common binary classifiers, ROC analysis was performed for naïve Bayes, weighted k-nearest neighbor, decision trees and for quadratic as well as Gaussian kernel support-vector machines.

3. Results

Analysis of oxygenation and perfusion maps in both groups revealed OEF to be lower and CBF to be generally higher than in normal-appearing brain, leading to a net positive effect on $CMRO_2$, meaning a higher oxygen metabolism in tumor tissue than in an unaffected contralateral brain, both in GBM and in cMET patients. The OEF map of a 54-year-old female patient with left occipito-temporal GBM supplied in Figure 3 gives an illustrative example of the oxygenation values emulated by the artificial neural network. The patient's brain shows largely uniform OEF values across "healthy" brain matter. The solid-appearing region of the tumor displays a strong OEF signal compared to the surrounding tissue and the rest of the brain while OEF in the peritumoral edematous area appears to be lower than in contralateral white matter.

When comparing oxygenation and perfusion parameters between GBM and cMET, OEF in CET was found to be significantly ($p = 0.03$) lower in GBM than in cMET. No significant differences were found between CET of GBM and cMET in terms of CBF ($p = 0.33$) and $CMRO_2$ ($p = 0.15$). For cMET patients, all parameters, i.e., OEF, CBF and $CMRO_2$, were significantly ($p = 0.01$) higher in the CET region than in the NET2 region. Meanwhile, for the GBM group, neither the difference between CET and NET2 in OEF ($p = 0.11$), nor CBF

($p = 0.15$), nor $CMRO_2$ ($p = 0.08$) was significant. A visual representation of this can be found in the boxplots supplied in Figure 4a. For an overview of oxygenation and perfusion parameters on a patient-by-patient basis, please consult Table S1 (supplement).

The ratio of CET divided by NET2 was demonstrated to be another useful metric for differentiation of GBM and cMET. While OEF for CET/NET2 was not significantly different between GBM and cMET patients ($p = 0.12$), the ratio for CBF was significantly higher for cMET ($p = 0.04$), as was the ratio for $CMRO_2$ ($p = 0.01$). Boxplots depicting these quotients are displayed in Figure 4b.

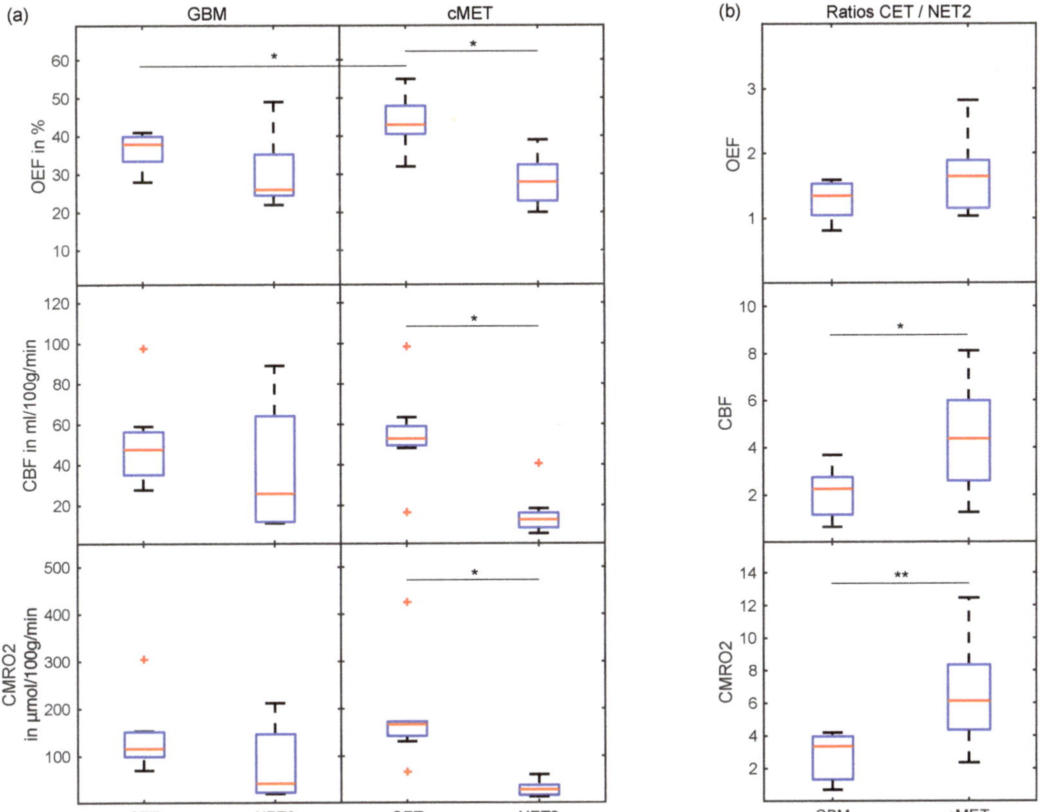

Figure 4. Boxplots (**a**) comparing OEF, CBF and $CMRO_2$ between CET and NET2 for GBM and cMET and (**b**) displaying the ratios of OEF, CBF and $CMRO_2$ in CET divided by NET2 for GBM and cMET patients. Box: first to third quartile; whiskers: 1.5 times the interquartile distance or the maximum/minimum value, if contained therein; red line: median. Outliers are displayed as red crosses. Significant differences ($p < 0.05$) are marked with an asterisk, the highly significant difference ($p = 0.01$) is marked with two asterisks. OEF: oxygen extraction fraction; CBF: cerebral blood flow; $CMRO_2$: cerebral metabolic rate of oxygen; CET: contrast-enhancing tumor; NET2: peritumoral non-enhancing T2 FLAIR hyperintensity.

After fitting a classifier support-vector machine to the oxygenation and perfusion features, different metrics for binary classification of GBM and cMET were assessed. The receiver operating characteristic curves of the ratios of OEF, CBF and $CMRO_2$ in CET divided by NET2 as well as the best overall feature combination in terms of AUC (OEF in CET and $CMRO_2$ in CET/NET2) are exhibited in Figure 5.

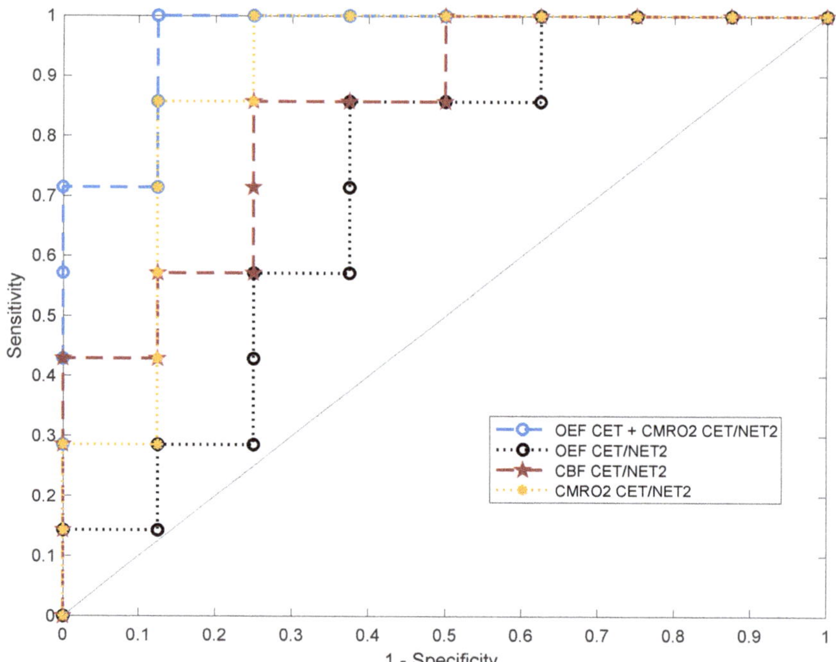

Figure 5. Five-fold cross-validated ROC curves for the support-vector machine classifier predicting binary outcome (GBM or cMET). Curves are shown for the ratios of oxygen extraction fraction ($OEF_{CET/NET2}$), cerebral blood flow ($CBF_{CET/NET2}$) and cerebral metabolic rate of oxygen ($CMRO_{2\,CET/NET2}$) in contrast-enhancing tumor (CET) divided by peritumoral non-enhancing T2 FLAIR hyperintensity (NET2) as well as for the multivariable fit to OEF_{CET} and $CMRO_{2\,CET/NET2}$. OEF_{CET}: OEF in contrast-enhancing tumor.

For each parameter, values for accuracy, optimal sensitivity and specificity and AUC are listed in Table 1. $CMRO_{2\,CET/NET2}$ emerged as the best single feature for differentiation of GBM from cMET. The resulting model had an AUC of 0.85 with an accuracy of 83% at an optimal sensitivity and specificity of 85% and 82%, respectively. The next best single classification features in terms of AUC were the ratio $CBF_{CET/NET2}$ (0.80) and OEF in CET (0.79). The highest discriminative power with the best diagnostic accuracy was achieved by combining OEF_{CET} and $CMRO_{2\,CET/NET2}$ for fitting a support-vector machine (AUC = 0.94). This allowed for an accurate classification of the tumors in 93% of cases, with a sensitivity of 99% and a specificity of 88%. Averaged over all iterations, out of 15 tumor patients included in this study, 14 were correctly diagnosed and only one case was misclassified.

We also assessed other machine learning classifiers that showed a lower discrimination performance including weighted k-nearest neighbor (AUC: 0.93, accuracy: 87%), naïve Bayes (AUC: 0.88, accuracy: 93%), and decision trees (AUC: 0.66, accuracy: 73%). Additionally, we investigated different variants of support-vector machines with quadratic (AUC: 0.89, accuracy: 87%) and Gaussian kernels (AUC: 0.86, accuracy: 87%). All of these achieved a smaller AUC with k-nearest neighbor coming closest to the linear kernel support-vector machine.

Table 1. Receiver operating characteristic analysis results of a linear kernel support-vector machine trained on oxygenation and perfusion parameters.

Region	Feature	Accuracy	Sensitivity	Specificity	AUC (Range)
CET	OEF	81%	87%	75%	0.79 (0.76–0.84)
	CBF	71%	70%	71%	0.67 (0.55–0.73)
	$CMRO_2$	63%	27%	95%	0.52 (0.41–0.68)
NET2	OEF	68%	70%	66%	0.65 (0.46–0.82)
	CBF	73%	54%	89%	0.69 (0.64–0.75)
	$CMRO_2$	73%	44%	99%	0.66 (0.61–0.71)
CET/NET2	OEF	69%	60%	78%	0.66 (0.55–0.77)
	CBF	75%	63%	86%	0.80 (0.77–0.82)
	$CMRO_2$	83%	85%	82%	0.85 (0.73–0.93)
Best combined	OEF_{CET} + $CMRO_{2\ CET/NET2}$	93%	99%	88%	0.94 (0.88–0.96)

Values are shown for contrast-enhancing tumor (CET), peritumoral non-enhancing T2 FLAIR hyperintensity (NET2) and their ratio (CET/NET2). AUC range indicates the lowest and highest values over 10 iterations. OEF: oxygen extraction fraction; CBF: cerebral blood flow; $CMRO_2$: cerebral metabolic rate of oxygen; AUC: area under the receiver operating characteristic curve.

4. Discussion

In this feasibility study, we applied for the first time the MRI-based QSM + qBOLD approach for OEF and $CMRO_2$ estimation to a prospectively recruited collective of GBM and cMET patients in order to distinguish the two entities based on their respective cerebral oxygenation and perfusion. Since the MRI protocol employed in this study for estimating cerebral tissue oxygenation and perfusion does not require the administration of an intravenous contrast agent per se, it offers the potential to facilitate metabolic imaging of cerebral tumors as well as non-invasive differential diagnosis of GBM and cMET. We demonstrated for the two patient groups that (i) OEF in the enhancing tumor was significantly lower in GBM than in cMET, (ii) the differences in perfusion and oxygenation between CET and NET2 were only significant for the cMET group, and (iii) the ratios of CBF and $CMRO_2$ in CET divided by NET2 were significantly higher for cMET patients than for GBM patients.

In order to differentiate the two entities, a support-vector machine classifier was trained on oxygenation and perfusion parameters in CET and NET2. A support-vector machine was chosen since it emerged as the best classifier with the highest accuracy, in line with previous investigations that compared it to different approaches such as naïve Bayes, weighted k-nearest neighbor and decision trees for binary classification of glioblastoma and cerebral metastasis [4,23,38]. The described procedure could identify OEF in CET and the ratio of $CMRO_2$ in CET versus NET2 as the most promising features for distinguishing GBM from cMET, achieving the highest discriminative power. With an accuracy of 93% and an AUC of 0.94, the two entities could be successfully differentiated.

Our results suggest that differentiation of GBM and cMET based solely on OEF in CET is good but not sufficient for reliably distinguishing the two groups. AUC and accuracy were higher than for any other individual parameter in CET or NET2 but lower than those obtained from using the ratios of CBF or $CMRO_2$ in CET versus NET2. This may be explained by the tumor microenvironment, in particular angiogenesis, that is similar in the contrast-enhancing part of GBM and many hematogenous cMETs [17]. In both entities, a disruption of the blood-brain barrier is caused by tumor growth and angiogenesis: the hypoxic state of the tumor and its surroundings lead to hypoxia-inducible factors being activated and their gene product VEGF being expressed [39,40]. The improved classification performance with the ratios of CET versus NET2 is likely attributable to the dissimilarities in NET2 tissue structure between GBM and cMET. Higher vascularization and neoplastic cell growth are hallmarks of the region surrounding contrast-enhancing GBM [5,6,41], while the peritumoral edema around cMET is caused purely by vasogenic edema and does not show extensive tumor cell infiltration [42]. This may explain why OEF, CBF and $CMRO_2$ were significantly higher in CET than in NET2 for cMET patients only.

We observed OEF in CET of GBM patients to be in line with the results of previous studies [26,27,43], reproducing the findings of low OEF, which is possibly a marker of less efficient oxygen extraction due to a physiologically inferior capillary configuration [27], coupled with high CBF. This gives plausibility to the parameter maps calculated with the QSM + qBOLD model that have in the past shown higher, more uniform OEF across the whole brain and more robust $CMRO_2$ estimates than QSM or qBOLD alone [27]. It was surprising, however, to find discrepancies between the OEF values computed in NET2 of GBM patients: While we found lower OEF in NET2 than in CET, reproducing the results of a previous study [43], two other studies found higher OEF in the edema surrounding the tumor [5,6]. This might either be a cause of incongruent definitions of the ROIs "edema" and "NET2" or another underlying phenomenon. A PET/MR study with a sufficiently large number of patients might be useful to shed light on true OEF in the peritumoral region of GBM.

In addition to the good classification accuracy, the $CMRO_2$ yielded by QSM + qBOLD serves as an important physiological parameter that has been shown to predict tumor response to antiangiogenic therapy as well as progression-free survival and overall survival in GBM patients [44] through association with intratumoral angiogenesis and oxygenation status [45]. Other possible applications of $CMRO_2$ that warrant further investigation consist of monitoring tumor response after initial radiation or chemotherapy and early detection of recurrence. With regard to preoperative imaging, $CMRO_2$ may also serve as a parameter to improve assessment of the surgical margins necessary for a more complete resection of glioblastomas, possibly adding complementary information to conventional MRI sequences for a "supratotal resection" [46].

This work is subject to a number of limitations. The QSM + qBOLD model requires a set of assumptions about physiologic parameters that were not measured for each patient individually. Among these assumptions are a constant tissue hematocrit and arterial oxygen saturation. Furthermore, the OEF values from QSM + qBOLD are prone to susceptibility artifacts from disturbances close to air-tissue bounds, e.g., the sinus frontalis, iron accumulations in deep grey matter or from blood degradation metabolites from hemorrhage, a phenomenon common in the necrotic centers of GBMs and cMETs, hence their exclusion from the assessment. Additional variance was introduced by changing the head coil after the first two patients. However, the perfusion values of these patients remained within a plausible range. Segmentation of ROIs was performed manually, adding a degree of intra- and inter-observer variability. Moreover, cMETs originating from various primary tumors may cause different oxygen metabolism characteristics. Subgroup analysis was not performed because of the relatively small sample size due to the explorative nature of this feasibility study. Thus, further research with larger patient populations is recommended before implementation of QSM + qBOLD into the clinical routine.

5. Conclusions

This study demonstrated that QSM + qBOLD allows for non-invasive differential diagnosis of GBM and cMET. In future studies, this differentiation could also be made without the use of an intravenous contrast agent. Going forward, our MRI approach for assessment of tissue oxygen metabolism might be helpful as a diagnostic tool that complements or replaces invasive stereotactic biopsies while yielding metabolic information about the tumor microenvironment and its surroundings, e.g., for predicting tumor response to therapy.

Supplementary Materials: The following is available online at https://www.mdpi.com/article/10.3390/app11219928/s1, Table S1: Oxygenation and perfusion parameters in CET and NET2 by individual patient.

Author Contributions: Conceptualization, H.B., A.F., L.R.S. and H.W.; methodology, H.B., S.H. and H.W.; software, H.B. and S.H.; validation, M.E.M. and S.A.M.; formal analysis, H.B., M.E.M. and S.A.M.; investigation, H.W.; resources, L.R.S. and H.W.; data curation, H.B. and S.A.M.; writing—Original draft preparation, H.B.; writing—Review and editing, S.H., S.A.M., M.E.M. and H.W.; visualization, H.B.; supervision, L.R.S. and H.W.; project administration, H.B. All authors have read and agreed to the published version of the manuscript.

Funding: This research received no external funding.

Institutional Review Board Statement: The study was conducted according to the guidelines of the Declaration of Helsinki, and approved by the Ethics Committee of the University Hospital Mannheim (reference number: 2017-666N-MA, date of approval: 6 August 2019).

Informed Consent Statement: Informed consent was obtained from all subjects involved in the study.

Data Availability Statement: The data presented in this study are available on request from the corresponding author. For a detailed breakdown of perfusion and oxygenation values by patient, please consult the supplementary material.

Acknowledgments: The authors are thankful for the technical help provided by the staff of the Institute for Computer Assisted Clinical Medicine at the Medical Faculty Mannheim.

Conflicts of Interest: The authors declare no conflict of interest.

References

1. Lee, E.J.; TerBrugge, K.; Mikulis, D.; Choi, D.S.; Bae, J.M.; Lee, S.K.; Moon, S.Y. Diagnostic value of peritumoral minimum apparent diffusion coefficient for differentiation of glioblastoma multiforme from solitary metastatic lesions. *AJR Am. J. Roentgenol.* **2011**, *196*, 71–76. [CrossRef]
2. Bauer, A.H.; Erly, W.; Moser, F.G.; Maya, M.; Nael, K. Differentiation of solitary brain metastasis from glioblastoma multiforme: A predictive multiparametric approach using combined MR diffusion and perfusion. *Neuroradiology* **2015**, *57*, 697–703. [CrossRef] [PubMed]
3. Li, X.; Wang, D.; Liao, S.; Guo, L.; Xiao, X.; Liu, X.; Xu, Y.; Hua, J.; Pillai, J.J.; Wu, Y. Discrimination between glioblastoma and solitary brain metastasis: Comparison of inflow-based vascular-space-occupancy and dynamic susceptibility contrast MR imaging. *AJNR Am. J. Neuroradiol.* **2020**, *41*, 583–590. [CrossRef]
4. Artzi, M.; Bressler, I.; Ben Bashat, D. Differentiation between glioblastoma, brain metastasis and subtypes using radiomics analysis. *J. Magn. Reason. Imaging* **2019**, *50*, 519–528. [CrossRef] [PubMed]
5. Preibisch, C.; Shi, K.; Kluge, A.; Lukas, M.; Wiestler, B.; Gottler, J.; Gempt, J.; Ringel, F.; Al Jaberi, M.; Schlegel, J.; et al. Characterizing hypoxia in human glioma: A simultaneous multimodal MRI and PET study. *NMR Biomed.* **2017**, *30*, e3775. [CrossRef] [PubMed]
6. Stadlbauer, A.; Zimmermann, M.; Kitzwogerer, M.; Oberndorfer, S.; Rossler, K.; Dorfler, A.; Buchfelder, M.; Heinz, G. MR imaging-derived oxygen metabolism and neovascularization characterization for grading and IDH gene mutation detection of gliomas. *Radiology* **2017**, *283*, 799–809. [CrossRef]
7. Wen, P.Y.; Kesari, S. Malignant gliomas in adults. *N. Engl. J. Med.* **2008**, *359*, 492–507. [CrossRef] [PubMed]
8. Noroxe, D.S.; Poulsen, H.S.; Lassen, U. Hallmarks of glioblastoma: A systematic review. *ESMO Open* **2016**, *1*, e000144. [CrossRef] [PubMed]
9. Stupp, R.; Mason, W.P.; van den Bent, M.J.; Weller, M.; Fisher, B.; Taphoorn, M.J.; Belanger, K.; Brandes, A.A.; Marosi, C.; Bogdahn, U.; et al. Radiotherapy plus concomitant and adjuvant temozolomide for glioblastoma. *N. Engl. J. Med.* **2005**, *352*, 987–996. [CrossRef]
10. Montemurro, N.; Fanelli, G.N.; Scatena, C.; Ortenzi, V.; Pasqualetti, F.; Mazzanti, C.M.; Morganti, R.; Paiar, F.; Naccarato, A.G.; Perrini, P. Surgical outcome and molecular pattern characterization of recurrent glioblastoma multiforme: A single-center retrospective series. *Clin. Neurol. Neurosurg.* **2021**, *207*, 106735. [CrossRef]
11. Stadlbauer, A.; Oberndorfer, S.; Zimmermann, M.; Renner, B.; Buchfelder, M.; Heinz, G.; Doerfler, A.; Kleindienst, A.; Roessler, K. Physiologic MR imaging of the tumor microenvironment revealed switching of metabolic phenotype upon recurrence of glioblastoma in humans. *J Cereb. Blood Flow Metab.* **2020**, *40*, 528–538. [CrossRef]
12. Montemurro, N.; Perrini, P.; Rapone, B. Clinical risk and overall survival in patients with diabetes mellitus, hyperglycemia and glioblastoma multiforme. A review of the current literature. *Int. J. Environ. Res. Public Health* **2020**, *17*, 8501. [CrossRef]
13. Pope, W.B. Brain metastases: Neuroimaging. *Handb. Clin. Neurol.* **2018**, *149*, 89–112. [PubMed]
14. Smirniotopoulos, J.G.; Murphy, F.M.; Rushing, E.J.; Rees, J.H.; Schroeder, J.W. Patterns of contrast enhancement in the brain and meninges. *Radiographics* **2007**, *27*, 525–551. [CrossRef] [PubMed]
15. Ostrom, Q.T.; Wright, C.H.; Barnholtz-Sloan, J.S. Brain metastases: Epidemiology. *Handb. Clin. Neurol.* **2018**, *149*, 27–42. [PubMed]
16. Giordana, M.T.; Cordera, S.; Boghi, A. Cerebral metastases as first symptom of cancer: A clinico-pathologic study. *J. Neurooncol.* **2000**, *50*, 265–273. [CrossRef]

17. Server, A.; Orheim, T.E.; Graff, B.A.; Josefsen, R.; Kumar, T.; Nakstad, P.H. Diagnostic examination performance by using microvascular leakage, cerebral blood volume, and blood flow derived from 3-T dynamic susceptibility-weighted contrast-enhanced perfusion MR imaging in the differentiation of glioblastoma multiforme and brain metastasis. *Neuroradiology* **2011**, *53*, 319–330.
18. Blasel, S.; Jurcoane, A.; Franz, K.; Morawe, G.; Pellikan, S.; Hattingen, E. Elevated peritumoural rCBV values as a mean to differentiate metastases from high-grade gliomas. *Acta Neurochir.* **2010**, *152*, 1893–1899. [CrossRef]
19. Lee, E.J.; Ahn, K.J.; Lee, E.K.; Lee, Y.S.; Kim, D.B. Potential role of advanced MRI techniques for the peritumoural region in differentiating glioblastoma multiforme and solitary metastatic lesions. *Clin. Radiol.* **2013**, *68*, e689–e697. [CrossRef]
20. Malone, H.; Yang, J.; Hershman, D.L.; Wright, J.D.; Bruce, J.N.; Neugut, A.I. Complications following stereotactic needle biopsy of intracranial tumors. *World Neurosurg.* **2015**, *84*, 1084–1089. [CrossRef]
21. Swinburne, N.C.; Schefflein, J.; Sakai, Y.; Oermann, E.K.; Titano, J.J.; Chen, I.; Tadayon, S.; Aggarwal, A.; Doshi, A.; Nael, K. Machine learning for semi-automated classification of glioblastoma, brain metastasis and central nervous system lymphoma using magnetic resonance advanced imaging. *Ann. Transl. Med.* **2019**, *7*, 232. [CrossRef]
22. Askaner, K.; Rydelius, A.; Engelholm, S.; Knutsson, L.; Latt, J.; Abul-Kasim, K.; Sundgren, P.C. Differentiation between glioblastomas and brain metastases and regarding their primary site of malignancy using dynamic susceptibility contrast MRI at 3T. *J. Neuroradiol.* **2019**, *46*, 367–372. [CrossRef] [PubMed]
23. Tsolaki, E.; Svolos, P.; Kousi, E.; Kapsalaki, E.; Fountas, K.; Theodorou, K.; Tsougos, I. Automated differentiation of glioblastomas from intracranial metastases using 3T MR spectroscopic and perfusion data. *Int. J. Comput. Assist. Radiol. Surg.* **2013**, *8*, 751–761. [CrossRef] [PubMed]
24. Lehmann, P.; Saliou, G.; de Marco, G.; Monet, P.; Souraya, S.E.; Bruniau, A.; Vallee, J.N.; Ducreux, D. Cerebral peritumoral oedema study: Does a single dynamic MR sequence assessing perfusion and permeability can help to differentiate glioblastoma from metastasis? *Eur. J. Radiol.* **2012**, *81*, 522–527. [CrossRef] [PubMed]
25. Asgari, S.; Rohrborn, H.J.; Engelhorn, T.; Stolke, D. Intra-operative characterization of gliomas by near-infrared spectroscopy: Possible association with prognosis. *Acta Neurochir.* **2003**, *145*, 453–459. [CrossRef] [PubMed]
26. Hardee, M.E.; Zagzag, D. Mechanisms of glioma-associated neovascularization. *Am. J. Pathol.* **2012**, *181*, 1126–1141. [CrossRef] [PubMed]
27. Cho, J.; Kee, Y.; Spincemaille, P.; Nguyen, T.D.; Zhang, J.; Gupta, A.; Zhang, S.; Wang, Y. Cerebral metabolic rate of oxygen (CMRO2) mapping by combining quantitative susceptibility mapping (QSM) and quantitative blood oxygenation level-dependent imaging (qBOLD). *Magn. Reason. Med.* **2018**, *80*, 1595–1604. [CrossRef] [PubMed]
28. Kurz, F.T.; Buschle, L.R.; Rotkopf, L.T.; Herzog, F.S.; Sterzik, A.; Schlemmer, H.P.; Kampf, T.; Bendszus, M.; Heiland, S.; Ziener, C.H. Dependence of the frequency distribution around a sphere on the voxel orientation. *Z. Med. Phys.* **2021**. [CrossRef]
29. Hubertus, S.; Thomas, S.; Cho, J.; Zhang, S.; Wang, Y.; Schad, L.R. Using an artificial neural network for fast mapping of the oxygen extraction fraction with combined QSM and quantitative BOLD. *Magn. Reson. Med.* **2019**, *82*, 2199–2211. [CrossRef]
30. Alsop, D.C.; Detre, J.A.; Golay, X.; Gunther, M.; Hendrikse, J.; Hernandez-Garcia, L.; Lu, H.; MacIntosh, B.J.; Parkes, L.M.; Smits, M.; et al. Recommended implementation of arterial spin-labeled perfusion MRI for clinical applications: A consensus of the ISMRM perfusion study group and the European consortium for ASL in dementia. *Magn. Reson. Med.* **2015**, *73*, 102–116. [CrossRef]
31. Dong, F.; Li, Q.; Jiang, B.; Zhu, X.; Zeng, Q.; Huang, P.; Chen, S.; Zhang, M. Differentiation of supratentorial single brain metastasis and glioblastoma by using peri-enhancing oedema region-derived radiomic features and multiple classifiers. *Eur. Radiol.* **2020**, *30*, 3015–3022. [CrossRef]
32. Ma, Y.; Mazerolle, E.L.; Cho, J.; Sun, H.; Wang, Y.; Pike, G.B. Quantification of brain oxygen extraction fraction using QSM and a hyperoxic challenge. *Magn. Reason. Med.* **2020**, *84*, 3271–3285. [CrossRef]
33. Hubertus, S.; Thomas, S.; Cho, J.; Zhang, S.; Wang, Y.; Schad, L.R. Comparison of gradient echo and gradient echo sampling of spin echo sequence for the quantification of the oxygen extraction fraction from a combined quantitative susceptibility mapping and quantitative BOLD (QSM+qBOLD) approach. *Magn. Reason. Med.* **2019**, *82*, 1491–1503. [CrossRef]
34. Cho, J.; Zhang, S.; Kee, Y.; Spincemaille, P.; Nguyen, T.D.; Hubertus, S.; Gupta, A.; Wang, Y. Cluster analysis of time evolution (CAT) for quantitative susceptibility mapping (QSM) and quantitative blood oxygen level-dependent magnitude (qBOLD)-based oxygen extraction fraction (OEF) and cerebral metabolic rate of oxygen (CMRO2) mapping. *Magn. Reason. Med.* **2020**, *83*, 844–857. [CrossRef] [PubMed]
35. Wenz, H.; Maros, M.E.; Meyer, M.; Forster, A.; Haubenreisser, H.; Kurth, S.; Schoenberg, S.O.; Flohr, T.; Leidecker, C.; Groden, C.; et al. Image quality of 3rd generation spiral cranial dual-source CT in combination with an advanced model iterative reconstruction technique: A prospective intra-individual comparison study to standard sequential cranial CT using identical radiation dose. *PLoS ONE* **2015**, *10*, e0136054. [CrossRef] [PubMed]
36. Cortes, C.; Vapnik, V. Support-vector networks. *Mach. Learn.* **1995**, *20*, 273–297. [CrossRef]
37. Maros, M.E.; Capper, D.; Jones, D.T.W.; Hovestadt, V.; von Deimling, A.; Pfister, S.M.; Benner, A.; Zucknick, M.; Sill, M. Machine learning workflows to estimate class probabilities for precision cancer diagnostics on DNA methylation microarray data. *Nat. Protoc.* **2020**, *15*, 479–512. [CrossRef]

38. Shrot, S.; Salhov, M.; Dvorski, N.; Konen, E.; Averbuch, A.; Hoffmann, C. Application of MR morphologic, diffusion tensor, and perfusion imaging in the classification of brain tumors using machine learning scheme. *Neuroradiology* **2019**, *61*, 757–765. [CrossRef]
39. Fanelli, G.N.; Grassini, D.; Ortenzi, V.; Pasqualetti, F.; Montemurro, N.; Perrini, P.; Naccarato, A.G.; Scatena, C. Decipher the glioblastoma microenvironment: The first milestone for new groundbreaking therapeutic strategies. *Genes* **2021**, *12*, 445. [CrossRef]
40. Guan, Z.; Lan, H.; Cai, X.; Zhang, Y.; Liang, A.; Li, J. Blood-brain barrier, cell junctions, and tumor microenvironment in brain metastases, the biological prospects and dilemma in therapies. *Front. Cell Dev. Biol.* **2021**, *9*, 722917. [CrossRef] [PubMed]
41. Sunwoo, L.; Yun, T.J.; You, S.H.; Yoo, R.E.; Kang, K.M.; Choi, S.H.; Kim, J.H.; Sohn, C.H.; Park, S.W.; Jung, C.; et al. Differentiation of glioblastoma from brain metastasis: Qualitative and quantitative analysis using arterial spin labeling MR imaging. *PLoS ONE* **2016**, *11*, e0166662. [CrossRef] [PubMed]
42. Fink, K.R.; Fink, J.R. Imaging of brain metastases. *Surg. Neurol. Int.* **2013**, *4*, S209–S219. [CrossRef] [PubMed]
43. Hubertus, S.; Thomas, S.; Cho, J.; Zhang, S.; Kovanlikaya, I.; Wang, Y.; Schad, L.R. In MRI-based oxygen extraction fraction and cerebral metabolic rate of oxygen mapping in high-grade glioma using a combined quantitative susceptibility mapping and quantitative blood oxygenation level-dependent approach. In Proceedings of the International Society for Magnetic Resonance in Medicine, Montréal, QC, Canada, 11–16 May 2019; p. 0391.
44. Bonekamp, D.; Mouridsen, K.; Radbruch, A.; Kurz, F.T.; Eidel, O.; Wick, A.; Schlemmer, H.P.; Wick, W.; Bendszus, M.; Ostergaard, L.; et al. Assessment of tumor oxygenation and its impact on treatment response in bevacizumab-treated recurrent glioblastoma. *J. Cereb. Blood Flow Metab.* **2017**, *37*, 485–494. [CrossRef] [PubMed]
45. Kickingereder, P.; Brugnara, G.; Hansen, M.B.; Nowosielski, M.; Pfluger, I.; Schell, M.; Isensee, F.; Foltyn, M.; Neuberger, U.; Kessler, T.; et al. Noninvasive characterization of tumor angiogenesis and oxygenation in bevacizumab-treated recurrent glioblastoma by using dynamic susceptibility MRI: Secondary analysis of the European Organization for Research and Treatment of Cancer 26101 trial. *Radiology* **2020**, *297*, 164–175. [CrossRef] [PubMed]
46. Yordanova, Y.N.; Duffau, H. Supratotal resection of diffuse gliomas—An overview of its multifaceted implications. *Neurochirurgie* **2017**, *63*, 243–249. [CrossRef] [PubMed]

Article

Choroidal Neovascularization Screening on OCT-Angiography Choriocapillaris Images by Convolutional Neural Networks

Kawther Taibouni [1], Alexandra Miere [2], Abdourahmane Samake [1], Eric Souied [2], Eric Petit [1] and Yasmina Chenoune [1,3,*]

1 Laboratory of Images, Signals and Intelligent Systems (LISSI, EA N° 3956), University Paris-Est Créteil, Vitry sur Seine, 94400 Paris, France; kawther.taibouni@univ-paris-est.fr (K.T.); abdou.samake0908@gmail.com (A.S.); petit@u-pec.fr (E.P.)
2 Centre Hospitalier Intercommunal de Créteil, Department of Ophthalmology, 40, Avenue de Verdun, Créteil, 94010 Paris, France; alexandra.miere@chicreteil.fr (A.M.); Eric.Souied@chicreteil.fr (E.S.)
3 ESME Sudria Research Lab., 34 Rue de Fleurus, 75006 Paris, France
* Correspondence: yasmina.chenoune@esme.fr

Abstract: Choroidal Neovascularization (CNV) is the advanced stage of Age-related Macular Degeneration (AMD), which is the leading cause of irreversible visual loss for elder people in developed countries. Optical Coherence Tomography Angiography (OCTA) is a recent non-invasive imaging technique widely used nowadays in diagnosis and follow-up of CNV. In this study, an automatic screening of CNV based on deep learning is performed using OCTA choriocapillaris images. CNV eyes (advanced wet AMD) are diagnosed among healthy eyes (no AMD) and eyes with drusen (intermediate AMD). An OCTA dataset of 1396 images is used to train and evaluate the model. A pre-trained convolutional neural network (CNN) is fine-tuned and validated on 80% of the dataset while the remaining 20% is used independently for predictions. The model can accurately detect CNV on the test set with an accuracy of 89.74%, precision of 0.96 and 0.99 area under the curve of the receiver operating characteristic. A good overall classification accuracy of 88.46% is obtained on a balanced test set. Detailed analysis of misclassified images shows that they are also considered ambiguous images for expert clinicians. This novel CNN-based application is truly a breakthrough to assist clinicians in the challenging task of screening for neovascular complications.

Keywords: age-related macular degeneration; choroidal neovascularization; convolutional neural networks; image classification; optical coherence tomography angiography

Citation: Taibouni, K.; Miere, A.; Samake, A.; Souied, E.; Petit, E.; Chenoune, Y. Choroidal Neovascularization Screening on OCT-Angiography Choriocapillaris Images by Convolutional Neural Networks. *Appl. Sci.* **2021**, *11*, 9313. https://doi.org/10.3390/app11199313

Academic Editors: Leonardo Rundo, Carmelo Militello and Andrea Tangherloni

Received: 15 September 2021
Accepted: 2 October 2021
Published: 8 October 2021

1. Introduction

Age-related Macular Degeneration (AMD) is the leading cause of irreversible blindness in the elderly population of developed countries. AMD is characterized by changes in the Retinal Pigment Epithelium (RPE), Bruch's Membrane (BM), or Choriocapillaris (CC) complex [1]. There are several staging systems for AMD, but the most widely used is the AREDS (Age Related Eye Disease Study) classification, distinguishing between early, intermediate and late AMD (see Figure 1) [2]. Early and intermediate AMD are characterized by the presence of drusen and pigmentary changes. Late AMD consists of wet AMD, characterized by choroidal neovascularization (CNV), and dry AMD, characterized by geographic atrophy (GA) in the macular area [3,4]. While both dry and wet AMD are visually threatening, in the particular case of wet AMD, CNV progression can result in rapidly deteriorating visual acuity, leading to scarring and irreversible visual loss [4,5]. Moreover, as hallmarks of early and intermediate AMD, drusen precede the progression to late AMD [6,7]. Hence, distinguishing between the early, intermediate, and late AMD plays a key role in both follow-up and treatment decisions, in order to preserve the visual prognosis.

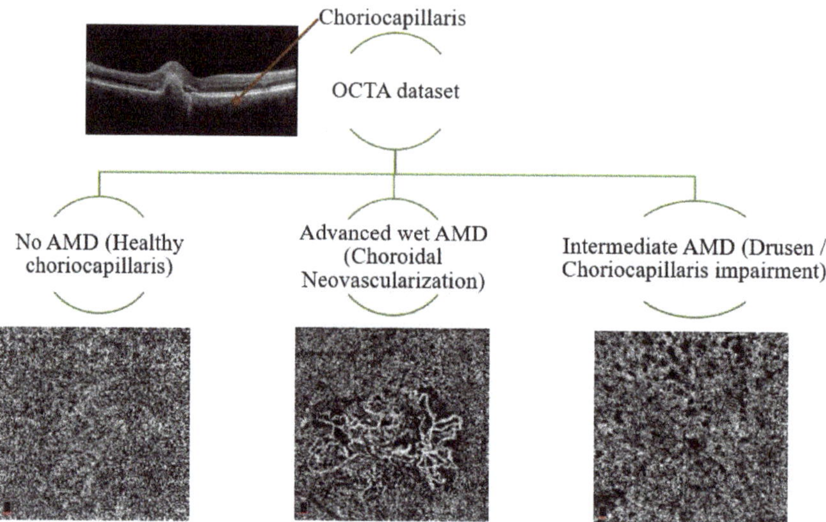

Figure 1. Classification of three types of OCTA choriocapillaris images (no AMD, advanced wet AMD, intermediate AMD). On the top-left: the OCTA cross-sectional B-scan showing retinal layers.

Novel imaging techniques, such as optical coherence tomography angiography (OCTA), contribute to the accurate diagnosis of early, intermediate and late AMD in a depth-resolved and non-invasive manner [8–10]. Besides the accurate detection of CNV or GA in the CC segmentation [11], OCTA has demonstrated CC alterations, i.e., flow deficits (FD) and signal attenuation associated with drusen, in early and intermediate AMD [12]. Moreover, recent literature has shown that choriocapillaris FD predict drusen enlargement, therefore being a significant risk factor for late forms of AMD [7]. Therefore, earlier diagnosis could ensure better follow-up of patients at high risk of conversion to advanced wet AMD. A computer-aided diagnosis (CAD) tool in ophthalmic clinical routine could be of significant assistance for clinicians at daily diagnosis and during follow-up.

The detection of CNV in the context of wet AMD on OCTA images is very challenging due to OCTA various artifacts [8,9]. Many recent papers dealt with the contributions of artificial intelligence (AI) and deep learning (DL) approaches in ophthalmology [13–15]. However, only few works have addressed CNV screening on OCTA images using DL methods [16]. Most of the existing works focused on the diagnosis of AMD using OCT or Color Fundus Photography. Moreover, published works on OCTA images involved other retinal diseases such as Diabetic Retinopathy (DR) [17].

In 2017, Rasti et al. [18] proposed a CAD system based on a multi-scale convolutional mixture of expert model to identify accurately dry AMD and diabetic macular edema using OCT images. Two different macular OCT datasets of 4142 and 3247 B-scans were used for the training step. A very good classification score of 99.85% was derived from the receiver operating characteristic curve (ROC-AUC).

In 2018, Burlina et al. [19] used a set of 67,401 color fundus images of AMD patients to estimate 5-year risk of progression to advanced stages of AMD by DL techniques. Two AMD severity scales (4-step and 9-step) were considered, and a human versus machine comparison was carried out. This study achieved weighted k scores of 0.77 for the 4-step and 0.74 for the 9-step AMD severity scales. The same year, Grassmann et al. [20] exploited a database of 120,656 color fundus images, manually graded in 13 AMD severity levels, to train several CNN architectures (AlexNet, GoogLeNet, VGG, Inception-V3, ResNet and I-ResNet-V2). A very good-weighted k of 92% was obtained. Govindaiah et al. [19] have also shown that deep CNN could be efficient to grade color fundus images in four classes: no AMD, early AMD, intermediate AMD, and advanced AMD. The study included a comparison between training using the transfer learning approach and training from scratch. The obtained accuracies were 78.1% for transfer learning and 83% without transfer learning.

More recently, Russakoff et al. [21] developed a DL architecture so called "AMDnet" on OCT images to predict the conversion from early/intermediate AMD to advanced wet AMD. The study included 71 patients with confirmed early/intermediate AMD that were imaged with OCT three times over 2 years. Results showed a ROC-AUC of 0.89 at the B-scan levels and 0.91 for volumes. Hwang et al. [22] used 35,900 labeled OCT images from AMD patients to train three types of Convolutional Neural Networks (CNNs), VGG19, InceptionV3 and ResNet50, to perform AMD diagnosis. The authors developed an AI and cloud-based telemedicine interaction tool dedicated to diagnosis and therapeutic of AMD. The image discrimination rates obtained by expert clinicians (92.73% and 91.90%) and provided by the AI-based platform (above 90%) were almost the same.

Further works on OCT imaging, such as the recent study of Romo-Bucheli et al. [23], proposed a treatment predictive model using a densely connected neural network (DenseNet) and a recurrent neural network (RNN) on longitudinal OCT scans for neovascular AMD patients (281 patients for training and 69 for tests). The CNN model achieved 0.85 AUC in detecting patients with low treatment requirements and 0.81 AUC for patients with high treatment requirements.

In what concerns the use of OCTA in DL, Le et al. [17] tested the feasibility of using DL for DR detection from OCTA including 77 patients and 20 control subjects. The authors applied transfer learning on a VGG16 network for robust OCTA classification. The obtained results showed an accuracy of 87.27% in differentiating healthy, no DR and DR eyes. In the same period, Wang et al. [16] developed an algorithm based on two CNNs to classify input OCTA images (using structural volumes and enface retinal angiograms) as CNV or Non-CNV and then segment the CNV membrane when present. The proposed neural network included a cutoff threshold for CNV area to overcome the residual artifacts limitation that could be confounded with CNV. CNV binary classification ROC-AUC was 0.997.

In this work, we aim to fill the gap of CNV screening on OCTA images using DL by promoting a novel application of CNNs on OCTA images using the choriocapillaris slab. The main contribution of this paper is the deep learning-based solution to classify AMD on two major forms: advanced wet AMD (CNV) and intermediate AMD (drusen/pigmentary changes) including a healthy control group (no AMD) using choriocapillaris OCTA images. A second contribution is the adaptation of a pre-trained VGG19 model on non-medical ImageNet dataset to medical domain using an adapted densely connected classifier on our limited OCTA data. Additionally, class activation mapping is used to interpret the CNN prediction on choriocapillaris OCTA images, which is a promising DL application for CAD systems in retinal clinical routine.

2. Materials and Methods

2.1. Dataset and Study Population

Data from patients with AMD is collected from the Ophthalmology Department of Intercommunal Hospital Center of Créteil, France, between September 2014 and July 2019. A database of 1396 choriocapillaris OCTA images of size 304 × 304 with a pixel size of 9.87 × 9.87 µm is built from 787 eyes related to 508 patients (mean age 70.67 ± 17.74 years). All patients underwent a 3 × 3 mm OCTA examination (AngioVue, Optovue, Freemont, CA, USA). The choriocapillaris slab is extracted and there are no excluded images due to motion or projection artifacts.

A retina specialist (A.M.) classified the OCTA images into three classes (391 with no AMD images from healthy eyes of 156 subjects, 457 images with CNV from 187 AMD patients, and 548 images with intermediate AMD from 274 patients). Multiple images per patient are included in this database. On one hand, follow-up images acquired at different dates are considered in this study as they show notable and significant changes in the CNV progression or in the number and size of drusen. On the other hand, both eyes are considered for some of the patients when both eyes' examinations are available. This study is performed in accordance with the Declaration of Helsinki and current French legislation and with approval of our local ethics committee.

2.2. CNV Screening on OCTA Images

Our goal is to discriminate from OCTA images the three predefined classes: No AMD (healthy CC), advanced wet AMD (CNV) and intermediate AMD (drusen or CC impairment) (see Figure 1). OCTA imaging allows physicians to visualize blood vessels in the individual layers of the retina and choroid without dye injection. Thus, CNV, drusen and impairment within the CC (pigmentary changes) can be clearly identified on OCTA images. The healthy choriocapillaris appears on OCTA images as a grainy texture with bright and dark spots corresponding to blood flow and flow deficits, respectively [7]. Drusen and CC impairment are characterized by black nonflow areas of different sizes related to flow deficits that can appear anywhere on the OCTA image surrounded by the grainy texture of the choriocapillaris. Regarding CNV, neovascular membranes harbor the aspect of a vascular branching, surrounded by the grainy texture of the choriocapillaris.

Nevertheless, these images are corrupted by speckle noise due to the physical principles of OCT, in addition to the image acquisition process and artifacts [24]. Moreover, included CNV lesions could have different sizes and locations with irregular shapes of neovascular membranes, thus small ones may be confused with the grainy texture of the choriocapillaris. This makes CNV detection on OCTA images a very challenging task. Figure 1 illustrates the OCTA choriocapillaris images of the three classes used for the classification in this work.

2.3. CNN Architecture and Transfer Learning

As depicted on Figure 2, our methodological approach consists of two parts: the VGG19 deep network [25] that provides the features extraction process on the OCTA image and a personalized densely connected network that represents the classification part.

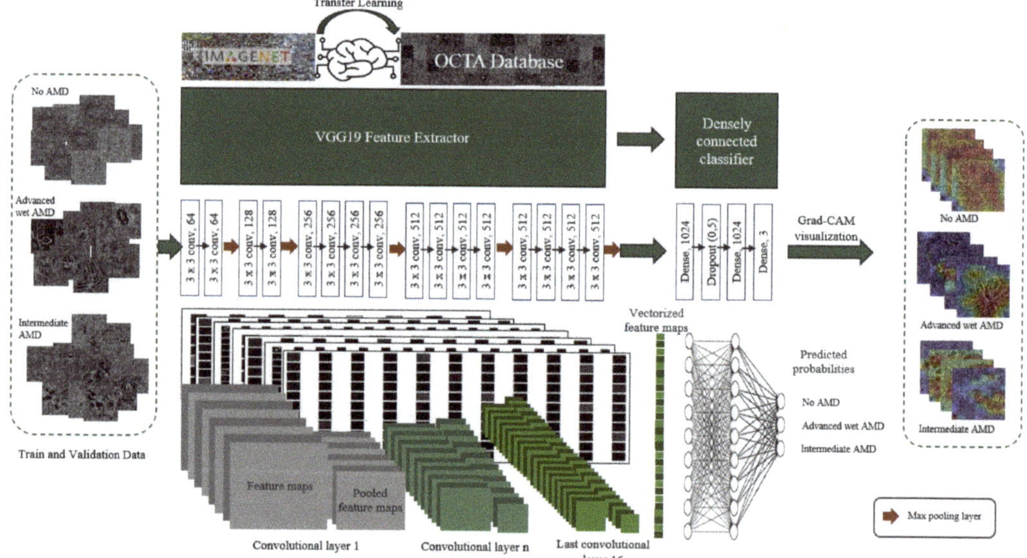

Figure 2. Our proposed modified VGG19 architecture. OCTA images are mapped through the VGG19 feature extractor to build discriminant 9 × 9 feature maps that are used by the densely connected classifier to generate predicted probabilities and Grad-CAM heatmaps to visualize feature attribution for each class: No AMD (healthy choriocapillaris), advanced wet AMD (CNV), intermediate AMD (drusen/choriocapillaris impairment).

VGG19 is a competition-winning model of the *ImageNet Large-Scale Visual Recognition Challenge* (ILSVRC) [26], that has a sequential pipeline architecture consisting of 16 convolutional and 3 fully connected trainable layers including five max-pooling layers. In this work, only the VGG19 convolutional part is used as a feature extractor where convolution layers include 3 × 3 convolution filters. A rectified linear unit (ReLU) activation function and a downsampling 2 × 2 max-pooling operation follow each convolutional stack. This convolutional part provides vectorized feature maps used as input to the densely connected classifier.

The densely connected classifier contains three fully connected layers, layers 1 and 2 are composed of 1024 nodes each. The last one consists of three nodes that provide the classification result into three types of OCTA images. A regularization dropout layer is included after the first dense layer to overcome overfitting the model by randomly dropping out 50% of the activations at that layer. ReLU activation function is used on the two first dense layers, whereas a softmax activation function is used on the last one.

According to the huge number of learnable VGG19 parameters (144 million) and the limited amount of OCTA data in our training dataset, transfer learning from non-medical data is applied in our approach [27]. Therefore, learned knowledge from the ImageNet dataset [26,28] is transferred to the model and adapted to our application by fine-tuning the convolutional part using OCTA images. The densely connected layers are trained from scratch on our OCTA data to classify OCTA images. Finally, feature maps from the last convolutional layer are mapped through the densely connected classifier to generate Gradient-weighted Class Activation Mapping (Grad-CAM) visualization [29]. The Grad-CAM produces a localization map that highlights the image's important features to the CNN for class predictions.

Additionally, to assess the impact of our approach (transfer learning on modified VGG19 model) on CNV detection accuracy, we trained the original VGG19 model independently from scratch with random initialization on our OCTA data.

2.4. Implementation Details

The dataset is divided into two independent subsets for training and testing. From the whole dataset, 80% (1115 images) is dedicated to fine-tuning, training, and validation. Subsequently, this first partition is further separated into 80% (892 images) for train and 20% (223 images) for validation. The remaining 20% (281 images) of the whole dataset is used for the performance evaluation and tests.

The whole network is trained end-to-end on 100 epochs for transfer learning and 200 epochs for training original VGG19. Stochastic Gradient Descent (SGD) optimization algorithm [30] and categorical cross entropy loss function are used. The learning-rate is set to 10^{-5} for transfer learning and to 10^{-4} for training original VGG19. Data augmentation is applied during training to reduce overfitting. Only random zoom is used in transfer learning to generate 16 OCTA images at each batch while rotation, horizontal and vertical flip are used in addition in training from scratch the original VGG19 model to generate 8 OCTA images at each batch.

The pipeline is implemented in Python with the Keras-TensorFlow library [31,32]. Training and testing are performed on a NVIDIA Corporation GP104GL [Quadro P4000] Graphics Processing Unit.

2.5. Performance Evaluation

Performance is evaluated using an independent balanced test set (78 images for each class) and there are no excluded images due to motion and projection artifacts or to image quality. The CNN prediction output is compared to the ground truth set by the expert reader (A.M.). Four statistical metrics are used to report classification performance [33]:

First, the accuracy that generally describes how the model performs across all classes. It is obtained as the ratio between the number of correct predictions to the total number of predictions:

$$\text{Accuracy} = \frac{\text{True}_{\text{positive}} + \text{True}_{\text{negative}}}{\text{True}_{\text{positive}} + \text{True}_{\text{negative}} + \text{False}_{\text{positive}} + \text{False}_{\text{negative}}}$$

Then the precision, that is calculated as the ratio between the number of positive samples correctly classified to the total number of samples classified as positive. The precision measures the model's accuracy in classifying a sample as positive:

$$\text{Precision} = \frac{\text{True}_{\text{positive}}}{\text{True}_{\text{positive}} + \text{False}_{\text{positive}}}$$

The recall is calculated as the ratio between the number of positive samples correctly classified as Positive to the total number of positive samples. The recall measures the model's ability to detect positive samples. The higher the recall, the more positive samples detected:

$$\text{Recall} = \frac{\text{True}_{\text{positive}}}{\text{True}_{\text{positive}} + \text{False}_{\text{negative}}}$$

Finally, the F1-score is a way of combining the precision and recall:

$$\text{F1-score} = \frac{2 \times \text{Precision} \times \text{Recall}}{\text{Precision} + \text{Recall}}$$

Confusion matrix and area under the curve (AUC) of the receiver operating characteristic (ROC) and precision-recall (PRC) curves (ROC-AUC and PRC-AUC) are supplied. Additionally, multiple class activation maps are generated to analyze the feature attribution and understand the CNN predictions.

3. Results

CNV screening evaluation, reported in Table 1 and Figure 3, shows that CNV detection on OCTA images achieves the best performance with a precision of 96%, recall of 90%, F1-score of 0.93 and an accuracy of 89.74%. In addition, ROC-AUC and PRC-AUC are 0.99 each. No AMD (healthy CC) OCTA images are also well classified by the proposed VGG19 modified model with a very good accuracy of 94.87% and F1-score of 0.90 (precision 0.85, recall 0.95). Regarding intermediate AMD (drusen/CC impairment) class, the images are, in some cases, confused with no AMD class. Sixty-three images from the intermediate AMD test dataset are correctly classified while 12 images are predicted as no AMD images and 3 as CNV images. This is summarized by the confusion matrix in Table 2.

Table 1. CNV screening performance of the modified VGG19. Accuracy, precision, recall, F1-score, area under the curve (AUC) for precision-recall (PRC) and receiver operation characteristic (ROC) curves for the three classes of OCTA images (HCC: healthy choriocapillaris, DCCI: drusen/choriocapillaris impairment, CNV: choroidal neovascularization).

	Accuracy (%)	Precision	Recall	F1-Score	PRC-AUC	ROC-AUC
HCC	94.87	0.85	0.95	0.90	0.97	0.99
DCCI	80.77	0.85	0.81	0.83	0.94	0.97
CNV	**89.74**	**0.96**	**0.90**	**0.93**	**0.99**	**0.99**

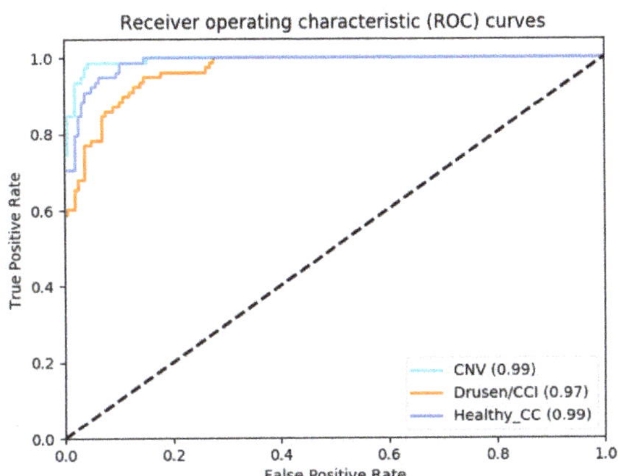

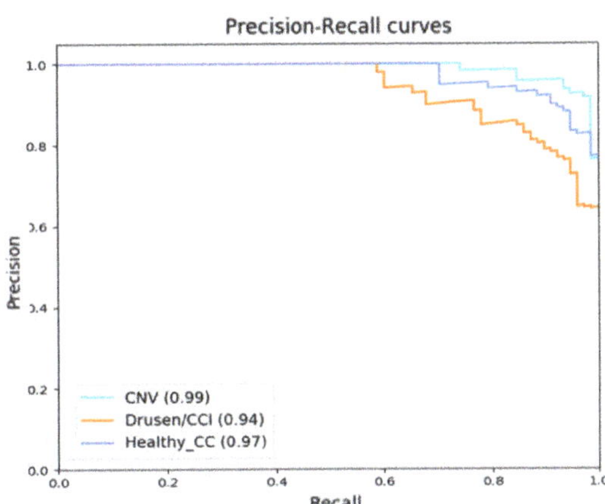

Figure 3. Receiver operating characteristic (ROC) and precision-recall (PRC) curves for the three classes: advanced wet AMD (CNV), intermediate AMD (drusen/CCI: choriocapillaris impairment), no AMD (healthy choriocapillaris). Area under the curve (AUC) values for each class.

Table 2. Confusion matrix of the modified VGG19 prediction on a balanced test dataset of 234 OCTA images (78 images for each class). CNV: choroidal neovascularization, DCCI: drusen/choriocapillaris impairment, HCC: healthy choriocapillaris.

Ground Truth Class	Predicted Class		
	CNV	**DCCI**	**HCC**
CNV	70	7	1
DCCI	3	63	12
HCC	0	4	74

The overall classification accuracy using transfer learning on our modified VGG19 is 88.46% with a loss of 0.089, while the overall classification accuracy using original VGG19 is 83.76% with a loss of 0.37. The proposed approach on the modified VGG19 achieved better performance than original VGG19 for CNV detection. This is reported in classification reports (Tables 1 and 3) and confusion matrices (Tables 2 and 4) where CNV screening accuracy is 74.36% for original VGG19 against 89.74% for our proposed approach. In addition, PRC-AUC and ROC-AUC are 0.99 each using our modified VGG19, while PRC-AUC is 0.95 and ROC-AUC is 0.97 using original VGG19.

Table 3. CNV screening performance of the original VGG19. Accuracy, precision, recall, F1-score, area under the curve (AUC) for precision-recall (PRC) and receiver operation characteristic (ROC) curves for the three classes of OCTA images (HCC: healthy choriocapillaris, DCCI: drusen/choriocapillaris impairment, CNV: choroidal neovascularization).

	Accuracy (%)	Precision	Recall	F1-Score	PRC-AUC	ROC-AUC
HCC	97.44	0.84	0.97	0.90	0.96	0.99
DCCI	79.49	0.75	0.79	0.77	0.83	0.93
CNV	**74.36**	**0.97**	**0.74**	**0.84**	**0.95**	**0.97**

Table 4. Confusion matrix of the original VGG19 prediction on a balanced test dataset of 234 OCTA images (78 images for each class). CNV: choroidal neovascularization, DCCI: drusen/choriocapillaris impairment, HCC: healthy choriocapillaris.

Ground Truth Class	Predicted Class		
	CNV	DCCI	HCC
CNV	58	19	1
DCCI	2	62	14
HCC	0	2	76

Figures 4 and 5 display Grad-CAM visualizations for correct predictions and incorrect predictions of the three classes, respectively, (no AMD, intermediate AMD, and advanced wet AMD). Grad-CAM heatmaps are superimposed on original OCTA choriocapillaris images with warm colors (red, orange, and yellow) for discriminant features and cold colors (blue, cyan, and green) for non-discriminant features.

In Figure 4, expected discriminant features are correctly highlighted by Grad-CAM heatmaps for each class: grainy texture throughout the whole OCTA image for no AMD images (Figure 4A,B), flow deficits/nonflow areas for intermediate AMD images (Figure 4C,D), and high flow vascular networks (CNV) for advanced wet AMD images (Figure 4E,F,G,H). This is further supported by the CNN predicted probabilities for each image. Regarding no AMD images (Figure 4A,B), the CNN predicted high probabilities were 0.94 and 0.97, respectively, and were attributed to no AMD class. Predicted probabilities for intermediate AMD images (Figure 4C,D) were 0.70 and 0.99, respectively, and were attributed to the correct class. Finally, regarding advanced wet AMD images (Figure 4E,F,G,H) predicted probabilities were 0.99 for images Figure 4E,G,H and 0.63 for image Figure 4F, correctly attributed to CNV.

On the other hand, non-discriminant CNV features are highlighted by Grad-CAM heatmaps in Figure 5I,J, including the flow deficits/nonflow areas or regions in Figure 5I and grainy texture in Figure 5J.

The CNN predicted probabilities reinforce this observation where image Figure 5I is predicted as intermediate AMD with 0.84 probability and image Figure 5J as no AMD image with 0.90 probability. Regarding the Figure 5K, non-discriminant drusen features are highlighted by Grad-CAM heatmap, grainy texture is highlighted as discriminant features showing the CNN prediction as no AMD with 0.59 of probability against 0.41 for intermediate AMD. Conversely, Figure 5L represents a healthy CC (no AMD) image predicted as intermediate AMD with 0.58 of probability against 0.42 for no AMD. The Grad-CAM heatmap supports these probabilities by highlighting only drusen features as discriminant rather than those from grainy texture of no AMD images.

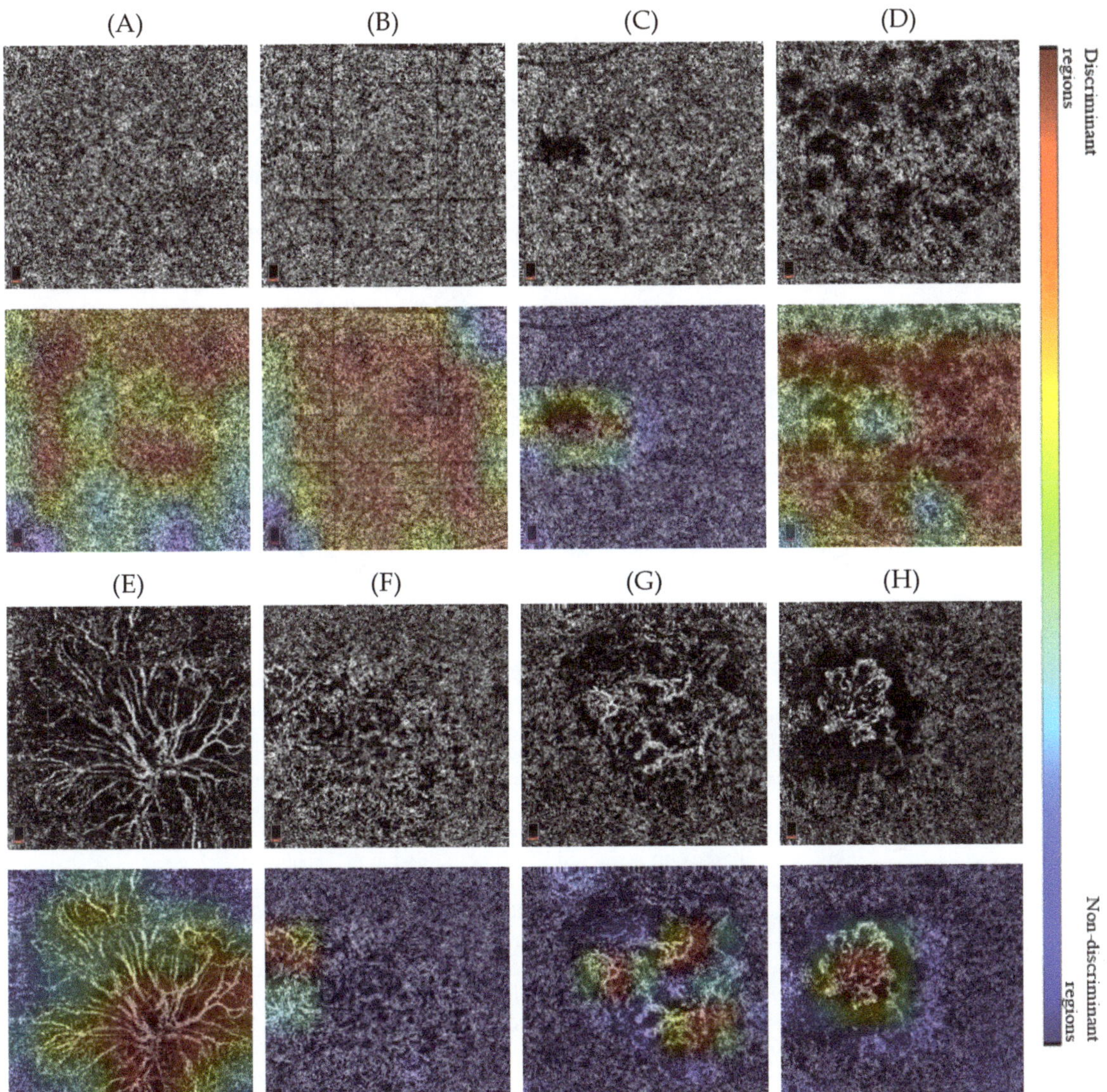

Figure 4. Grad-CAM visualizations for correct predictions by the proposed VGG19 modified model. Below each OCTA image, the corresponding Grad-CAM visualization of the CNN prediction. (**A**,**B**) Healthy choriocapillaris—no AMD. (**C**,**D**) Drusen/Choriocapillaris impairment-intermediate AMD. (**E**–**H**) Choroidal Neovascularization—advanced wet AMD.

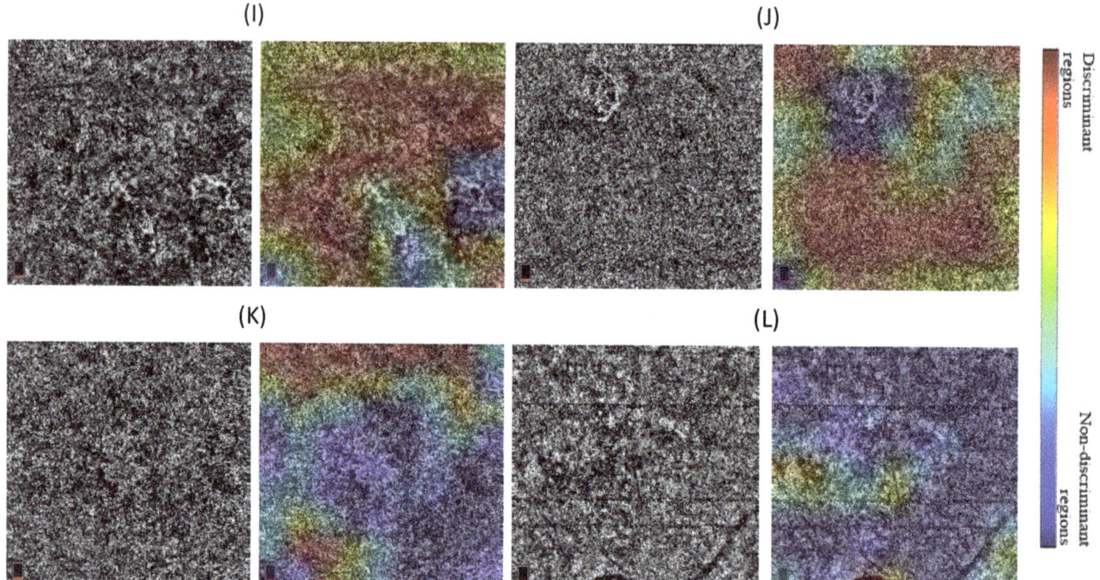

Figure 5. Grad-CAM visualizations for incorrect predictions by the proposed VGG19 modified model. To the right of each OCTA image, the corresponding Grad-CAM visualization of the CNN prediction. (**I**,**J**) advanced wet AMD (CNV) images, (**K**) intermediate AMD image and (**L**) No AMD (healthy CC) image.

4. Discussion and Conclusions

CNV screening on OCT-angiography choriocapillaris images is challenging due to the high variance in neovascular membranes sizes, shapes, and locations. Additionally, the speckle noise and the image acquisition process makes the detection task even more difficult, particularly to differentiate CNV or early drusen from the grainy texture of the choriocapillaris. Our proposed VGG19 modified model achieved very good CNV detection performance with an accuracy of 89.74% and F1-score of 0.93 (as reported in Table 1 [0.96 of precision, 0.90 of recall]). The ROC-AUC and PRC-AUC were 0.99 each (Table 1 and Figure 3). The confusion matrix in Table 2 shows that the CNN could accurately detect the CNV lesion on 70 images out of 78 independent test images. Seven CNV images are confused with intermediate AMD images, only one CNV image is misclassified as healthy CC, and 3 images of intermediate AMD are misclassified as CNV.

Concerning the no AMD (healthy CC) classification performance by our model, the statistical analysis reported in Table 1 and Figure 3 shows that it is the second-best classified class after CNV. Precision and recall are 0.85 and 0.95, respectively, with a ROC-AUC of 0.99, a PRC-AUC of 0.97 and an accuracy of 94.87%. The confusion matrix in Table 2 indicates that 74 no AMD images are correctly classified, only 4 images are confused with intermediate AMD images and no healthy CC image is misclassified as CNV.

Finally, for intermediate AMD detection, statistics show a F1-score of 0.83 (precision of 0.85 and recall of 0.81), an accuracy of 80.77%, a ROC-AUC of 0.97 and a PRC-AUC of 0.94 (reported in Table 1 and Figure 3). These results are better explained by the confusion matrix attributions in Table 2, where 63 of the 78 intermediate AMD images are correctly classified, while 12 images are predicted as no AMD and 3 images confused with CNV.

Furthermore, the CNN predicted probabilities demonstrate, on one hand, the CNN certainty when predicting correct classes (probability higher than 0.90 for the images in Figure 4A,B,D,E,G,H), and on the other hand, the CNN uncertainty for more difficult cases, such as images C and F of Figure 4. Image C illustrates a flow impairment clustered only on one region of the image and surrounded by the grainy texture of the CC on the whole image. Therefore, the CNN predicted probabilities for image C are 0.70 as intermediate AMD and 0.29 as no AMD. Regarding image F, the CNV membrane is covered by different drusen surrounded by the grainy texture of the CC. Consequently, the CNN predicted probabilities are 0.63 for CNV and 0.36 for intermediate AMD. This shows the CNN's high ability to discriminate wet AMD eyes from healthy and intermediate AMD eyes and to show uncertainty for ambiguous and confusing cases.

The Grad-CAM visualization (see Figure 4) provides a better understanding of these results. No matter the shape, size, and location of the CNV on images of Figure 4E–H; the CNN's predicted high probability is attributed to the correct region with high discriminative CNV features. Conversely, the CNN's low probability is attributed to the non-discriminative regions. Similarly, regardless of the presence of artifacts on images of Figure 4B,C); the CNN feature attribution is correctly highlighted by Grad-CAM heatmaps. This proves that the CNN prediction is based on relevant regions of the OCTA choriocapillaris images related to the three classes: no AMD, intermediate AMD, and advanced wet AMD, that are also considered by the expert reader to detect CNV on OCTA images.

The most difficult step of this classification problem is, on one hand, the detection of some confusing CNV where a small, indefinite vascular shape is visible and for which very small amount of OCTA images are available in our training dataset. On the other hand, the discrimination of early small drusen from healthy choriocapillaris remains problematic. These difficulties are addressed in Figure 5 that illustrates two misclassified CNV images (Figure 5I,J), a misclassified intermediate AMD image (Figure 5K) and a misclassified no AMD image (Figure 5L).

Image I is misclassified as intermediate AMD with 0.84 of probability and image J is the only CNV image of our test dataset misclassified as no AMD with 0.90 of probability. Feature attribution visualization for these two misclassified images illustrated by Grad-CAM heatmaps in Figure 5 helps understanding the CNN prediction for these cases. The Grad-CAM heatmap of image Figure 5I shows that discriminant features are those of drusen by only black nonflow regions which explains the CNN predicted high probability attributed to intermediate AMD class rather than CNV class. Only one small CNV is visible in Figure 5I and is considered as non-discriminant feature. The CNN prediction is thus based on the most present features on the image. Such images are also ambiguous for clinicians and additional imaging modalities are needed to establish a clear diagnosis.

The Grad-CAM heatmap of image Figure 5J explains again the CNN misclassification. Only grainy regions present throughout the whole image are considered as discriminant features by the CNN. The tiny CNV membrane visible on this image is considered as non-discriminant as it is hidden in the grainy texture. This explains the CNN's prediction with high probability (0.90) for image Figure 5J, attributed to no AMD class.

To overcome these classification errors, we should supply our training dataset with more ambiguous OCTA choriocapillaris images such as images I and J of Figure 5 to train the model to detect tiny CNV membranes when they are hidden and confused with drusen or CC grainy texture.

Drusen or significant flow impairment on OCTA choriocapillaris images appear as nonflow areas/flow deficits that are generally surrounded by the grainy texture of CC. When these areas are small and less important than the CC texture, the OCTA image is considered as ambiguous and confusing even by clinicians. This is the case of the 12 OCTA images misclassified by the CNN and predicted as no AMD images. Early small drusen are hardly visible on OCTA images and can be considered as flow deficits related to OCTA image acquisition process. Thus, the early small drusen manually classified as intermediate AMD images are, in some cases, predicted as no AMD by the CNN. These cases are illustrated by typical images in Figure 5K,L.

The above analysis is supported by the predicted probabilities for images K and L of Figure 5. Figure 5K represents an intermediate AMD image predicted as no AMD by the CNN with close probabilities for both classes (0.59 for no AMD against 0.41 for intermediate AMD). Similarly, Figure 5L is a no AMD image predicted as intermediate

AMD by the CNN with close probabilities (0.58 for intermediate AMD against 0.42 for no AMD). This last result reveals again the CNN's ability to show uncertainty in cases of confusing features. To resolve this ambiguity, it would be relevant to classify these images as uncertain images for which the clinician should use additional information from other imaging modalities or patient history to decide and make a diagnosis.

Despite these few misclassification errors, the CNN showed a great ability to screen and detect CNV on OCTA choriocapillaris images. This main finding is achieved through the transfer learning approach that is used to train the proposed VGG19 modified model to overcome the limitation of the small amount of training data. Fine-tuning the proposed modified VGG19 improved the overall classification accuracy compared to that obtained from training the original VGG19 from random initialization. The overall accuracy increased from 83.76% using original VGG19 to 88.46% using transfer learning on the modified VGG19 although it was applied from non-medical data.

This study is one of the few works dealing with CNV screening on OCTA data using only images at the choriocapillaris slab. Obtained results revealed a promising application of CAD systems to diagnose CNV on OCTA choriocapillaris images in clinical routine using DL-based methods. In order to produce more reliable results to clinicians and to help them quantify the CNN uncertainty, we aim to measure the CNN prediction uncertainty in further works to identify how much a CNN could be trusted in diagnosis [34,35] and to avoid using images not suitable for diagnosis when high uncertainty is detected [35].

Further studies on this topic will focus on data augmentation, as well as including more CNV images to work on larger datasets and different imaging modalities to improve classification performance for ambiguous cases.

Author Contributions: Conceptualization, K.T., A.M., E.S., E.P. and Y.C.; methodology, K.T.; validation, E.S., E.P. and Y.C.; investigation, K.T., A.M. and A.S.; resources, A.M., A.S. and E.S.; writing—original draft preparation, K.T.; writing—review and editing, K.T., A.M., E.S., E.P. and Y.C.; supervision, E.S., E.P. and Y.C. All authors have read and agreed to the published version of the manuscript.

Funding: This research received no external funding.

Institutional Review Board Statement: The study was conducted according to the guidelines of the Declaration of Helsinki and current French legislation and with approval of our local ethics committee.

Informed Consent Statement: Patient consent was waived due to the retrospective nature of the study.

Data Availability Statement: The data presented in this study are available on reasonable request from the corresponding author.

Conflicts of Interest: The authors declare no conflict of interest.

References

1. Biesemeier, A.; Taubitz, T.; Julien, S.; Yoeruek, E.; Schraermeyer, U. Choriocapillaris breakdown precedes retinal degeneration in age-related macular degeneration. *Neurobiol. Aging* **2014**, *35*, 2562–2573. [CrossRef] [PubMed]
2. Age-Related Eye Disease Study Research Group A Randomized, Placebo-Controlled, Clinical Trial of High-Dose Supplementation With Vitamins C and E, Beta Carotene, and Zinc for Age-Related Macular Degeneration and Vision Loss. *Arch. Ophthalmol.* **2001**, *119*, 1417–1436. [CrossRef]
3. Colijn, J.M.; Buitendijk, G.H.S.; Prokofyeva, E.; Alves, D.; Cachulo, M.L.; Khawaja, A.P.; Cougnard-Gregoire, A.; Merle, B.M.J.; Korb, C.; Erke, M.G.; et al. Prevalence of age-related macular degeneration in Europe: The past and the future. *Ophthalmology* **2017**, *124*, 1753–1763. [CrossRef] [PubMed]
4. Abramoff, M.D.; Garvin, M.K.; Sonka, M. Retinal Imaging and Image Analysis. *IEEE Rev. Biomed. Eng.* **2010**, *3*, 169–208. [CrossRef] [PubMed]
5. De Jong, P.T. Age-related macular degeneration. *N. Engl. J. Med.* **2006**, *355*, 1474–1485. [CrossRef]
6. Lipecz, A.; Miller, L.; Kovacs, I.; Czakó, C.; Csipo, T.; Baffi, J.; Csiszar, A.; Tarantini, S.; Ungvari, Z.; Yabluchanskiy, A.; et al. Microvascular contributions to age-related macular degeneration (AMD): From mechanisms of choriocapillaris aging to novel interventions. *GeroScience* **2019**, *41*, 813–845. [CrossRef] [PubMed]

7. Nassisi, M.; Tepelus, T.; Nittala, M.G.; Sadda, S.R. Choriocapillaris flow impairment predicts the development and enlargement of drusen. *Graefe's Arch. Clin. Exp. Ophthalmol.* **2019**, *257*, 2079–2085. [CrossRef]
8. de Carlo, T.E.; Romano, A.; Waheed, N.K.; Duker, J.S. A review of optical coherence tomography angiography (OCTA). *Int. J. Retin. Vitr.* **2015**, *1*, 1–15. [CrossRef]
9. Spaide, R.F.; Fujimoto, J.G.; Waheed, N.K. Image artifacts in optical coherence angiography. *Retina* **2015**, *35*, 2163. [CrossRef]
10. Ma, J.; Desai, R.; Nesper, P.; Gill, M.; Fawzi, A.; Skondra, D. Optical Coherence Tomographic Angiography Imaging in Age-Related Macular Degeneration. *Ophthalmol. Eye Dis.* **2017**, *9*, 1179172116686075. [CrossRef]
11. Miere, A.; Butori, P.; Cohen, S.Y.; Semoun, O.; Capuano, V.; Jung, C.; Souied, E.H. Vascular Remodeling of Choroidal Neovascularization after Anti–Vascular Endothelial Growth Factor Therapy Visualized on Optical Coherence Tomography Angiography. *Retina* **2019**, *39*, 548–557. [CrossRef] [PubMed]
12. Alten, F.; Lauermann, J.L.; Clemens, C.R.; Heiduschka, P.; Eter, N. Signal reduction in choriocapillaris and segmentation errors in spectral domain OCT angiography caused by soft drusen. *Graefe's Arch. Clin. Exp. Ophthalmol.* **2017**, *255*, 2347–2355. [CrossRef]
13. Schmidt-Erfurth, U.; Sadeghipour, A.; Gerendas, B.S.; Waldstein, S.M.; Bogunović, H. Artificial intelligence in retina. *Prog. Retin. Eye Res.* **2018**, *67*, 1–29. [CrossRef] [PubMed]
14. Ting, D.S.W.; Pasquale, L.R.; Peng, L.; Campbell, J.P.; Lee, A.Y.; Raman, R.; Tan, G.S.W.; Schmetterer, L.; Keane, P.A.; Wong, T.Y. Artificial intelligence and deep learning in ophthalmology. *Br. J. Ophthalmol.* **2018**, *103*, 167–175. [CrossRef] [PubMed]
15. Sarhan, M.H.; Nasseri, M.A.; Zapp, D.; Maier, M.; Lohmann, C.P.; Navab, N.; Eslami, A. Machine Learning Techniques for Ophthalmic Data Processing: A Review. *IEEE J. Biomed. Health Inform.* **2020**, *24*, 3338–3350. [CrossRef]
16. Wang, J.; Hormel, T.T.; Gao, L.; Zang, P.; Guo, Y.; Wang, X.; Bailey, S.T.; Jia, Y. Automated diagnosis and segmentation of choroidal neovascularization in OCT angiography using deep learning. *Biomed. Opt. Express* **2020**, *11*, 927–944. [CrossRef] [PubMed]
17. Le, D.; Alam, M.; Yao, C.K.; Lim, J.I.; Hsieh, Y.-T.; Chan, R.V.P.; Toslak, D.; Yao, X. Transfer Learning for Automated OCTA Detection of Diabetic Retinopathy. *Transl. Vis. Sci. Technol.* **2020**, *9*, 35. [CrossRef]
18. Rasti, R.; Rabbani, H.; Mehridehnavi, A.; Hajizadeh, F. Macular OCT Classification Using a Multi-Scale Convolutional Neural Network Ensemble. *IEEE Trans. Med. Imaging* **2018**, *37*, 1024–1034. [CrossRef]
19. Burlina, P.M.; Joshi, N.; Pacheco, K.D.; Freund, D.E.; Kong, J.; Bressler, N.M. Use of Deep Learning for Detailed Severity Characterization and Estimation of 5-Year Risk Among Patients With Age-Related Macular Degeneration. *JAMA Ophthalmol.* **2018**, *136*, 1359–1366. [CrossRef]
20. Grassmann, F.; Mengelkamp, J.; Brandl, C.; Harsch, S.; Zimmermann, M.E.; Linkohr, B.; Peters, A.; Heid, I.M.; Palm, C.; Weber, B.H. A Deep Learning Algorithm for Prediction of Age-Related Eye Disease Study Severity Scale for Age-Related Macular Degeneration from Color Fundus Photography. *Ophthalmology* **2018**, *125*, 1410–1420. [CrossRef]
21. Russakoff, D.B.; Lamin, A.; Oakley, J.D.; Dubis, A.M.; Sivaprasad, S. Deep Learning for Prediction of AMD Progression: A Pilot Study. *Investig. Opthalmol. Vis. Sci.* **2019**, *60*, 712–722. [CrossRef]
22. Hwang, D.-K.; Hsu, C.-C.; Chang, K.-J.; Chao, D.; Sun, C.-H.; Jheng, Y.-C.; Yarmishyn, A.A.; Wu, J.-C.; Tsai, C.-Y.; Wang, M.-L.; et al. Artificial intelligence-based decision-making for age-related macular degeneration. *Theranostics* **2019**, *9*, 232–245. [CrossRef]
23. Romo-Bucheli, D.; Erfurth, U.S.; Bogunovic, H. End-to-End Deep Learning Model for Predicting Treatment Requirements in Neovascular AMD From Longitudinal Retinal OCT Imaging. *IEEE J. Biomed. Health Inform.* **2020**, *24*, 3456–3465. [CrossRef]
24. Baghaie, A.; Yu, Z.; D'Souza, R.M. State-of-the-art in retinal optical coherence tomography image analysis. *Quant. Imaging Med. Surg.* **2015**, *5*, 603–617. [CrossRef]
25. Simonyan, K.; Zisserman, A. Very deep convolutional networks for large-scale image recognition. *arXiv* **2014**, arXiv:1409.1556.
26. Russakovsky, O.; Deng, J.; Su, H.; Krause, J.; Satheesh, S.; Ma, S.; Huang, Z.; Karpathy, A.; Khosla, A.; Bernstein, M.; et al. Imagenet large scale visual recognition challenge. *Int. J. Comput. Vis.* **2015**, *115*, 211–252. [CrossRef]
27. Pan, S.J.; Yang, Q. A Survey on Transfer Learning. *IEEE Trans. Knowl. Data Eng.* **2010**, *22*, 1345–1359. [CrossRef]
28. Morid, M.A.; Borjali, A.; Del Fiol, G. A scoping review of transfer learning research on medical image analysis using ImageNet. *Comput. Biol. Med.* **2021**, *128*, 104115. [CrossRef] [PubMed]
29. Selvaraju, R.R.; Cogswell, M.; Das, A.; Vedantam, R.; Parikh, D.; Batra, D. Grad-CAM: Visual Explanations from Deep Networks via Gradient-Based Localization. *Int. J. Comput. Vis.* **2020**, *128*, 336–359. [CrossRef]
30. Ruder, S. An overview of gradient descent optimization algorithms. *arXiv* **2016**, arXiv:1609.04747.
31. Chollet, F. Keras. 2015. Available online: https://keras.io/ (accessed on 11 May 2021).
32. Abadi, M.; Agarwal, A.; Barham, P.; Brevdo, E.; Chen, Z.; Citro, C.; Corrado, C.S.; Davis, A.; Dean, J.; Devin, M.; et al. Tensorflow: Large-scale machine learning on heterogeneous distributed systems. *arXiv Prepr.* **2016**, arXiv:1603.04467.
33. Hossin, M.; Sulaiman, M.N. A review on evaluation metrics for data classification evaluations. *Int. J. Data Min. Knowl. Manag. Process.* **2015**, *5*, 1.
34. Araújo, T.; Aresta, G.; Mendonça, L.; Penas, S.; Maia, C.; Carneiro, Â.; Mendonça, A.M.; Campilho, A. DR|GRADUATE: Uncertainty-aware deep learning-based diabetic retinopathy grading in eye fundus images. *Med. Image Anal.* **2020**, *63*, 101715. [CrossRef]
35. Laves, M.-H.; Ihler, S.; Ortmaier, T. Uncertainty Quantification in Computer-Aided Diagnosis: Make Your Model say "I don't know" for Ambiguous Cases. *arXiv* **2019**, arXiv:1908.00792.

applied sciences

MDPI

Article

Deep Learning-Based Segmentation of Various Brain Lesions for Radiosurgery

Siangruei Wu [1,†], Yihong Wu [2,†], Haoyun Chang [3], Florence T. Su [4], Hengchun Liao [2], Wanju Tseng [5], Chunchih Liao [6], Feipei Lai [7], Fengming Hsu [8] and Furen Xiao [9,*]

1 Graduate Institute of Communication Engineering, National Taiwan University, Taipei 106, Taiwan; raywu0@gmail.com
2 School of Medicine, National Taiwan University, Taipei 100, Taiwan; patrickwu97@gmail.com (Y.W.); b05401014@ntu.edu.tw (H.L.)
3 Graduate Institute of Electrical Engineering, National Taiwan University, Taipei 106, Taiwan; hyzhang@eda.ee.ntu.edu.tw
4 College of Arts and Sciences, Santa Clara University, Santa Clara, CA 95053, USA; fsu1@scu.edu
5 Data Intelligence and Application Division, QNAP Systems, Inc., New Taipei 221, Taiwan; rowantseng@qnap.com
6 Department of Neurosurgery, Taipei Hospital, New Taipei 242, Taiwan; ns00360@tph.mohw.gov.tw
7 Department of Computer Science & Information Engineering, National Taiwan University, Taipei 106, Taiwan; flai@csie.ntu.edu.tw
8 Department of Oncology, National Taiwan University Hospital, Taipei 100, Taiwan; hsufengming@ntuh.gov.tw
9 Department of Neurosurgery, National Taiwan University Hospital, Taipei 100, Taiwan
* Correspondence: xiao@ntuh.gov.tw; Tel.: +886-2-23123456
† These authors contributed equally to this work.

Citation: Wu, S.; Wu, Y.; Chang, H.; Su, F.T.; Liao, H.; Tseng, W.; Liao, C.; Lai, F.; Hsu, F.; Xiao, F. Deep Learning-Based Segmentation of Various Brain Lesions for Radiosurgery. *Appl. Sci.* **2021**, *11*, 9180. https://doi.org/10.3390/app11199180

Academic Editors: Leonardo Rundo, Carmelo Militello, Andrea Tangherloni and Donato Cascio

Received: 27 July 2021
Accepted: 24 September 2021
Published: 2 October 2021

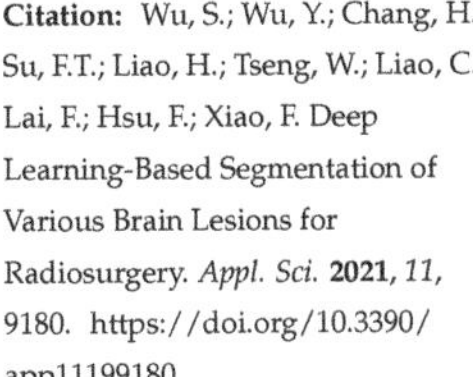

Featured Application: This study implemented deep learning methods to the task of segmentation of various brain lesions, facilitating the treatment planning process of neurosurgery and radiation oncology.

Abstract: Semantic segmentation of medical images with deep learning models is rapidly being developed. In this study, we benchmarked state-of-the-art deep learning segmentation algorithms on our clinical stereotactic radiosurgery dataset. The dataset consists of 1688 patients with various brain lesions (pituitary tumors, meningioma, schwannoma, brain metastases, arteriovenous malformation, and trigeminal neuralgia), and we divided the dataset into a training set (1557 patients) and test set (131 patients). This study demonstrates the strengths and weaknesses of deep-learning algorithms in a fairly practical scenario. We compared the model performances concerning their sampling method, model architecture, and the choice of loss functions, identifying suitable settings for their applications and shedding light on the possible improvements. Evidence from this study led us to conclude that deep learning could be promising in assisting the segmentation of brain lesions even if the training dataset was of high heterogeneity in lesion types and sizes.

Keywords: deep learning; image segmentation; brain tumors; radiosurgery; magnetic resonance imaging

1. Introduction

Stereotactic radiosurgery (SRS) is a treatment modality using ionizing radiation, focusing on precisely selected areas of tissue. It is usually delivered in a single session, but the radiation dose can also be fractionated. Targeting accuracy and anatomic precision are critical to successful SRS, but are historically secondary concerns in other types of radiation therapy [1]. Undoubtedly, as technology evolves, standards in this area will have to change. Nevertheless, when root mean square errors can be reduced to approximately 1 mm, a threshold of surgical possibilities is reached both in the brain and throughout the rest of the body. As the ACR-ASTRO guidelines suggest, a targeting accuracy is approximately 1 mm [2–4]. Although SRS can be performed in many parts of the body, it is best known to treat intracranial lesions. The common indications for intracranial SRS include many different types of brain tumors, vascular malformations (including arteriovenous malformation, AVM), and functional diseases such as trigeminal neuralgia (TN). Brain metastases, vestibular schwannomas, meningiomas, and pituitary adenomas are common tumor types treated by SRS.

Before the delivery of SRS to the target (e.g., a brain tumor), detailed treatment planning with precise contouring of the target is conducted by a neurosurgeon or a radiation oncologist. The contouring is performed on computed tomography (CT) or magnetic resonance images (MRI). Sometimes, both CT and MRI are used, depending on the devices and diseases. Normal organs or tissues sensitive to radiation are also contoured so that radiation dose and risk of injury can be estimated. These normal organs are called critical organs or organs at risk (OARs). In terms of image analysis, "precise" segmentation of targets and OARs is mandatory for SRS treatment planning. In current clinical practice, the segmentation is performed by professional personnel. The manual contouring process is time-consuming and prone to substantial inter-practitioner variability, even amongst experts, and may lead to large variation in care quality. Several pieces of research suggest computer assistance [5–10]. We expect an AI-based assistive tool could improve tumor detection, shorten mean contouring time, and increase inter-clinician agreement [11].

As convolutional neural networks (CNNs), the dominant deep learning models, are leading the breakthrough in computer vision recently, they also dominate MRI segmentation tasks. Havaei et al. (2017) proposed the idea of using a deep learning model to perform brain tumor segmentation tasks on MRI images [12]. They pointed out that both local and global representations are essential to produce better results, and this intuition was later realized in various ways. Kamnitsas et al. (2017) later perfected this idea and achieved state-of-the-art performance with a two-path model [13]. On the other hand, U-Net was first proposed for the cell tracking task [14], but then became widely used in many other segmentation tasks [15,16]. In MICCAI BraTS 2017 competition [17], most participants used U-Net variants, as the winner [18] simply ensembled three kinds of the most common deep learning models, namely FCN (fully convolutional network) [19], V-Net [20], and DeepMedic [13]. Other than deep learning, some studies on brain cancer segmentation took advantage of fuzzy c-means clustering [21–23], cellular automata [24], random walker [8], and so on [5,7,25,26]. However, they are not deep learning by not possessing over two hidden layers and will not be further discussed.

However, few studies apply deep learning methods to the actual SRS datasets. Unlike the BraTS competitions, real applicable models may need to handle much more diversity rather than a single type of disease. Liu et al. (2017) proposed a modification of DeepMedic that outperformed its parent method in segmentation during SRS treatment planning by adding a subpath, with a dice score reaching 0.67 in a cohort of 240 patients [27]. Lu et al. (2019) ensembled two neural networks, namely 3D U-Net and DeepMedic, which were trained with different hyper-parameters so that one neural network focused on small metastases with high sensitivity while the other one addressed overall tumor segmentation with high specificity, yielding a good performance on segmentation within 305 patients, with a median dice score of 0.74 [28]. Fong et al. (2019) trained the convoluted neural network with multiplanar slices, reducing false-positive predictions and yielding a dice score of 0.77 on the 248-patient dataset while maintaining competent 80% isodose coverage [29]. Lu et al. (2021) implemented deep learning in the treatment planning process, decreasing the plethora of time consumed during the planning process as well as enhancing the prediction overlap with ground truth significantly especially in the subgroup of non-experts. However, the cohort size was rather small [11]. Heterogeneity could have been considered a major problem for machine learning decades ago; however, it should be considered a real-world situation. A heterogeneous dataset could help the generalizability and transferability of trained models [11,23,25,30–32]. However, in the previously-mentioned studies, small sample sizes were important contributors to the lack of confidence to infer the generalization of deep-learning models in clinical practices with heterogeneous lesion types. For the technology to achieve satisfactory performance, we explored the behavior of deep-learning models in a realistic scenario. Therefore, we collected a relatively large dataset with 1688 patients and analyzed the performance of models with various types of settings and architectures. More specifically, we benchmarked the performance of different segmentation models previously proposed for other tasks and also compared the effectiveness of various sampling methods and the choice of loss functions. We used the BRATS dataset to evaluate whether our implementations of deep learning models were correct and comparable to their original implementations.

2. Materials and Methods

2.1. Dataset

2.1.1. NTUH (National Taiwan University Hospital) Dataset

The data were extracted from a medical center in northern Taiwan. The SRS device used was CyberKnife (Accuray, Sunnyvale, CA, USA) and commenced operation in January 2008. In the decade until December 2017, there were 2578 treatment courses completed in 2411 patients. Among these, 2036 treatment courses of 1921 patients were intracranial.

We only selected patients undergoing first SRS with contrast-enhanced T1-weighted (T1+C) MRI images available. Finally, there were 1688 patients included in our dataset. Their data were randomly divided into training and test sets (Table 1). However, because treatment targets for patients with trigeminal neuralgia are neither tumor nor vascular malformation, their data were all assigned to the training set.

Table 1. Clinical diagnoses of 1688 patients in the final dataset.

Types of Brain Lesions	Train	Test
Metastases	504	53
Meningioma	314	29
Schwannoma	305	20
Pituitary tumor	147	8
Arteriovenous malformation	80	6
Trigeminal neuralgia	38	0
Other tumors	169	15
Total	1557	131

For each patient, the target was extracted from the treatment planning system together with axial T1+C MRI. Most of the time, the targets were contoured by a neurosurgeon and then reviewed by a radiation oncologist. Occasionally only one radiation oncologist contoured the targets without review by another physician. An image volume could contain more than one target, particularly in patients with brain metastases. NTUH volumes were registered to CT volumes before contouring, with tumor contours stored in CT coordinates. We had to make an inverse transformation to put the tumor labels in MRI coordinates. In other words, volumes were resampled on a common voxel grid instead of directly cropped on grids of different voxel numbers. The images were retrieved from DICOM format and saved in NIfTI-1 data format, where names, birth dates, and geographic data were removed. After registration and de-identification, these image/label pairs were used for the training and evaluation of deep neural networks. The images were presented in native axial slices, with 1–2 mm slice thicknesses. The number of slices varied from 30 to 233 since the slices did not necessarily cover the full cranial regions, instead, they could only include the region of interest. The in-plane resolution was usually 512×512, and the smallest resolution was 197×197. The field of view in the x–y plane was usually 300 mm, ranging from 200 to 350 mm. The pixel size was mostly 0.5859×0.5859 mm^2 and was 1.1719×1.1719 mm^2 for some images with lower resolution.

There were a total of 2568 distinct targets in these 1688 image sets. The target volumes ranged from 20 to 72,646 mm^3, with a median of 1236 mm^3 and a mean of 3696 $\pm$ 6637 mm^3. In 1013 image sets, there was only one target. The number of targets may reach up to 34 in a single image set.

2.1.2. BraTS Dataset

The BraTS 2015 dataset is a standard benchmark dataset for MRI segmentation tasks. It includes 220 multi-modal scans of patients with high-grade glioma (HGG) and 54 with low-grade glioma (LGG). T1-weighted, contrast-enhanced T1+C, T2-weighted, and FLAIR images are available. The data had a common dimension of 240 $\times$ 240 $\times$ 155 with 1 mm^3 resolution. The annotation contains five classes: 0 for background, 1 for necrotic core (NC), 2 for edema (OE), 3 for non-enhancing core, and 4 for enhancing core. The evaluation follows the rules of the competition by merging the predictions into three sets: whole tumor (classes 1,2,3,4), core (classes 1,3,4), and enhancing core (class 4). The train to test ratio was 10:1 in each of the experiments we conducted.

2.2. *Preprocessing*

The raw data of the NTUH dataset contains images of different resolutions and fields of view (FOVs). We first used the skull stripping function of Brain-Suite [33] to locate the brain, then utilized the information possessed by the brain masks for centration and cropping of the MRI to make sure the images contain fewer extracranial areas. Brain-Suite was used only to locate the brain center for better cropping. Everything from the scalp to skull remained on the cropped images for reasons that some of our lesions may locate extra-axially. The final input images size was 200 $\times$ 200 $\times$ 200 mm^3. Finally, we normalized them by the z-scores.

Images in the BraTS dataset were already registered, cropped, and normalized with bias field corrections. We only normalized the data by the z-scores for every pulse sequence (T1, T2, T1+C, FLAIR).

2.3. *Data Augmentation*

To perform a fair comparison of the model architectures, we established the following standard data augmentation in the training phase. For 2D models, we performed data augmentation with translation, rotation, shear, zoom, brightness, and elastic distortion [15]. For 3D models, since data augmentation did not yield higher performances of segmentation in the preliminary experiment we performed, we did not perform any type of data augmentation.

2.4. *Deep Learning Models*

The design of the models we employed could be found online [34]. We will discuss the rationale and the architecture below.

2.4.1. DeconvNet

DeconvNet is an architecture adopted from VGG16, a 16-layered CNN by the Visual Geometry Group, and is rather simple to implement [35]. The objective of this design is to overcome the limitations of FCN, which cannot detect objects that are bigger or smaller than a specific size. In this case, the object may be fragmented or mislabeled. Furthermore, FCN only uses one convolution transpose layer to construct its output, so the output loses much detail. As a consequence, DeconvNet uses several layers of transpose-convolution and up-pooling.

The model can be divided into two parts: the encoder and the decoder, which are formed by convolution and deconvolution operations, respectively. It is worth noting that we replaced the max-pooling and up-sampling operations by setting the stride of Conv and Deconv to 2 in our implementation. This is inspired by the recent proposition of generative adversarial networks.

2.4.2. DeepMedic

DeepMedic is another kind of 3D CNN [13]. It is special for taking two inputs, high resolution, and low resolution. This design seeks to balance fine structures and high-level information. High-resolution inputs for DeepMedic are patches from our preprocessed data. Low-resolution inputs are downsampled using 3D average pooling from each corresponding high-resolution patch. Both inputs go through a series of convolution layers with skip connection, and then it constructs the output by fusing the features of both pathways. This is a state-of-the-art model, and we expect the model to perform well on segmentation of brain lesions based on previous benchmarks [27,36–38].

2.4.3. PSPNet

Pyramid scene parsing network, or PSPNet, is a state-of-the-art model in scene parsing tasks [39]. We included it because it is also suitable for our segmentation task. The PSPNet utilizes the high-level representations extracted by a pretrained network, and a novel design of the pyramid pooling module serves as a backend to predict the segmentations. The pyramid pooling modules pool the extracted feature maps to obtain features of different scales. The pretrained model is typically a ResNet trained on an ImageNet dataset [39]. However, on an MRI dataset, the features are not transferable due to the large consistency and the absence of common pretrained models to process MRI images. In our implementation, we randomly initialized the ResNet backend and also removed the deep supervision loss.

2.4.4. U-Net

U-Net tries to improve the fine structure of segmentations and increase the amount of context used [14]. Traditionally, when a certain amount of pooling is required, if one is intending to train with large patches, it unfortunately degrades the performance such as in FCN and DeconvNet. Hence, the U-Net model utilizes skip connections to forward the unpooled features, thus the model can utilize the information of various scales. In our implementation, we abandoned the max-pooling and up-sampling operations for the same reason as in DeconvNet.

2.4.5. V-Net

V-Net is the adaption of U-Net for 3-dimensional data to capture the relationships in consecutive slices, which were omitted in the 2D models, addressing contiguity problems and yielding better results in the segmentation of various 3D images [20]. It replaces the convolution and pooling operations with 3D versions.

2.5. Sampling Method

Batch samplers were defined in the source code (see Supplementary Material) [40]. We had four batch sampler designs for our models to experiment with the most efficient and accurate settings. We tried each kind of sampling method for each model in our preliminary experiment, but only those sampling methods that did not cause much overfitting, excessive memory consumption, or lower performance than other sampling methods were benchmarked.

2.5.1. Two Dimensional

For two-dimensional models, we split the MRI data slice by slice and performed predictions separately. This may result in noises along the sliced axis due to the loss of spatial contiguity information.

2.5.2. Three Dimensional

For three-dimensional models, the basic strategy is to feed the whole brain image data directly. In the preliminary experiment, three-dimensional patch resulted in high memory consumption when we employed the DeepMedic. Thus, we did not perform a three-dimensional patch for the DeepMedic. Furthermore, while we experimented with this setting on the BraTS2015 dataset, we found it caused overfitting, and we suspect that this is because many of the voxels are irrelevant and redundant for the prediction. Thus, we added two more three-dimensional sampling methods described below.

2.5.3. Uniform Patch

To reduce the redundant voxels and save memory usage, we sampled small patches within the brain regions. While inferencing, we simply reassembled the patch predictions together. The patch size used was 152 × 128 × 128. Worth noting, for the DeepMedic, uniform patch resulted in generally lower performance in segmentation, sensitivity, and precision compared with the center patch. Hence this method was not employed when training the DeepMedic.

2.5.4. Center Patch

It has been suggested that patches containing foreground regions are crucial to the training [35]. We thus deployed this sampling strategy, which guarantees at least one foreground voxel in the patch. The patch size was default to 64 × 64 × 64. However, we could set the patch size to 96 × 96 × 96 or the same as the size of the uniform patch when we applied the sampling method. When we experimented with the settings in the preliminary test, center patch sizes lower than 190 × 190 would result in tremendously low segmentation performance for V-Net in the NTUH dataset, so this sampling method was not used in the formal benchmark analysis.

2.6. Hyperparameters

We used fixed optimizer settings across all experiments. The optimizer chosen was Adam. The learning rate was initially 1×10^{-4}, with step decay of factor 0.1 at 50 and 70 epochs. Patch-wise methods were inferenced with patches cropped without overlap and excessive boundaries were filled with zero paddings. Samples were generated on the fly in a patch-wise method.

2.7. Loss Functions

Class imbalance is a major problem in most tumor segmentation problems, and it is even more severe in our task compared to the BraTS glioma dataset because of small target volumes. The imbalance would most likely lead the model to a trivial solution, which predicts all voxels as background. There are several ways to deal with this problem by modifying the loss function.

2.7.1. Weighted Cross-Entropy

Re-weighting the sparse class is the most common solution to the class imbalance problem. In this study, we set the class weights inversely proportional to the ratio of the class. In particular:

$$C = -\sum_{c=1}^{M} \frac{g_{oc} log(p_{oc})}{r_c} \quad (1)$$

where M is the number of classes and r_c is the ratio of class c in the whole volume/dataset (as an implementation choice); g_{oc} is the ground-truth label of a voxel; and p_{oc} is the predicted label probability of a voxel of class c.

2.7.2. Soft-Dice

Milletari et al. (2016) suggest using the differentiable version of the dice score, namely soft-dice, directly as the objective due to its resistibility to class imbalance [20]. It is fairly natural to use this loss function because the dice score is the most common evaluation metric in related tasks. There are two implementations of the soft-dice loss function.

Regarding the cardinality of sets, one can perform summation directly or with squaring. In particular:

$$D1 = \frac{2\sum_i^N p_i g_i}{\sum_i^N p_i^2 + \sum_i^N g_i^2} \text{ or } D2 = \frac{2\sum_i^N p_i g_i}{\sum_i^N p_i + \sum_i^N g_i} \quad (2)$$

where p_i is the predicted label probability and g_i is the ground-truth label. We found the two versions producing almost identical performances. In this study, we refer to the second version as the soft-dice loss function.

2.8. Evaluation Metrics

2.8.1. Dice Score (Hard Dice)

Dice score is the standard metric for evaluating segmentation results. It is defined as

$$\mathrm{D} = \frac{2|\mathrm{X} \cap \mathrm{Y}|}{|\mathrm{X}| + |\mathrm{Y}|} \quad (3)$$

where X and Y are the sets of predicted and labeled lesion voxels.

The previously mentioned soft-dice loss is a modified differentiable version of the dice score. We, therefore, refer to the dice score metric as hard-dice to distinguish the two.

2.8.2. Precision and Sensitivity

Precision and sensitivity (also known as recall) are standard metrics of binary classification, which is a more general scheme for segmentation. Precision quantifies the volume ratio of correctly predicted lesion voxels (TP) to all predicted lesion voxels (TP + FP). It is defined as

$$\text{Precision} = \frac{\mathrm{TP}}{\mathrm{TP} + \mathrm{FP}} \quad (4)$$

Sensitivity quantifies the volume ratio of correctly predicted lesion voxels (TP) to all labeled lesion voxels (TP + FN). It is defined as

$$\text{Sensitivity} = \frac{\mathrm{TP}}{\mathrm{TP} + \mathrm{FN}} \quad (5)$$

2.9. Experiments

In all our experiments, the training and testing were conducted under a 10:1 train-test ratio.

2.9.1. Performances of Models on Segmentation of Brain Lesions in NTUH Dataset

We first experimented with the performances of different models on the segmentation of brain tumors after training with various brain lesions. Different batch samplers and loss functions were employed based on the models used. The hard dice, precision, and sensitivity were the outcomes we were interested in. The higher dice score, precision, and sensitivity were deemed as better segmentation performance.

2.9.2. Performances of Models on Segmentation of Brain Tumors in BraTS Dataset

To compare the variables contributing to the performances of the models trained with the NTUH dataset, we experimented with the segmentation of the brain tumors in the BraTS dataset. We trained our models with either 4-channel or only T1+C inputs to compare the performances of the models trained with the same imaging modality in the NTUH dataset. During training, the labels encompassed either five classes or only tumor cores. The segmentation performances were measured using the hard dice. The evaluation was based on predictions of the tumor cores.

3. Results

Three cases from the NTUH dataset showing representative results of different models were shown in Tables 2–4. The overall dice scores of these networks on the NTUH dataset ranged from 0.33 (DeepMedic) to 0.51 (V-Net). Table 5 shows the detailed performance of each network tested with the NTUH dataset.

Table 2. Predictions with low dice scores.

Ground truth	DeconvNet	DeepMedic
	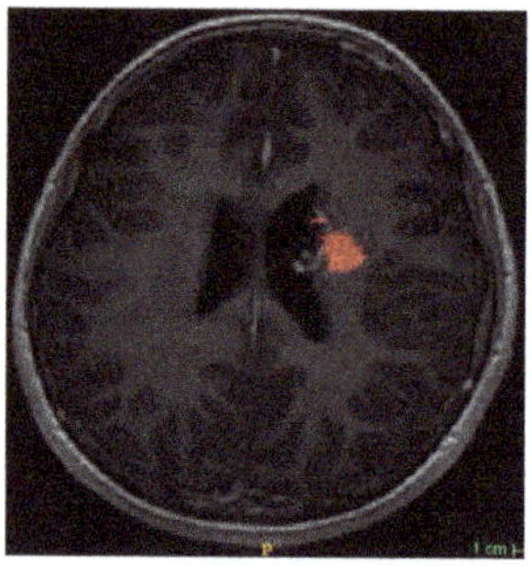	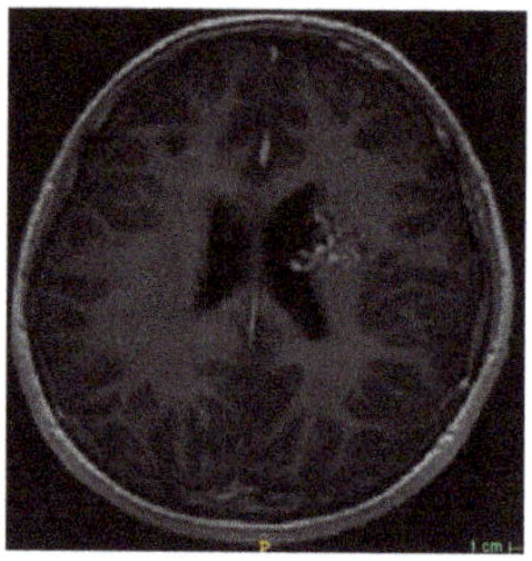
PSPNet	**U-Net**	**V-Net**
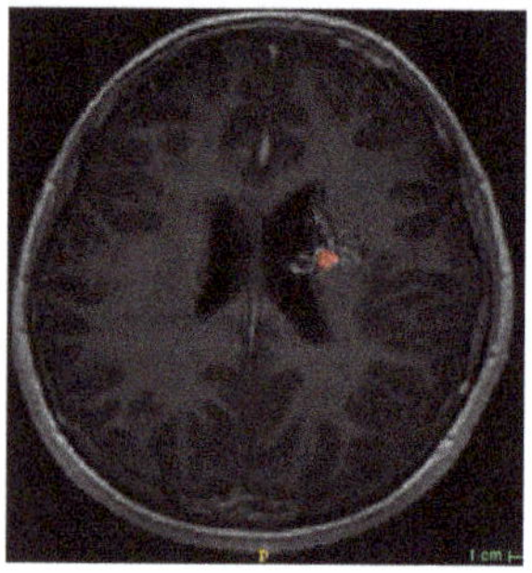	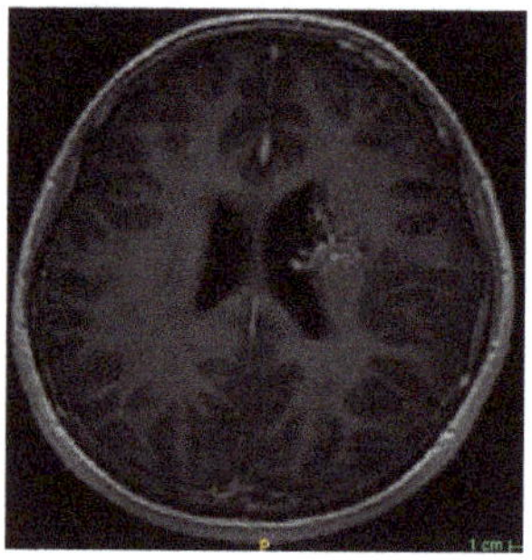	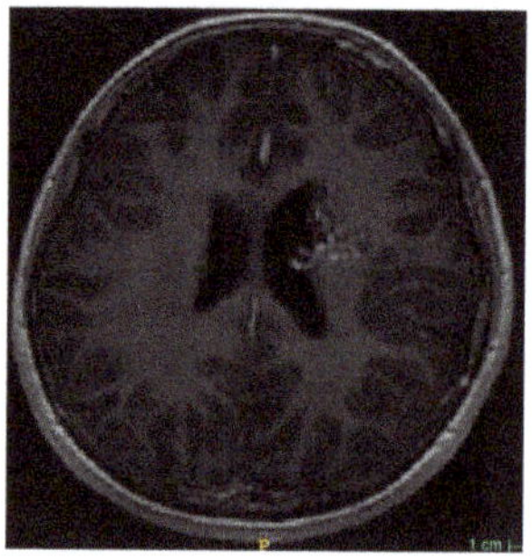

Table 3. Predictions with average dice scores.

Ground truth	DeconvNet	DeepMedic
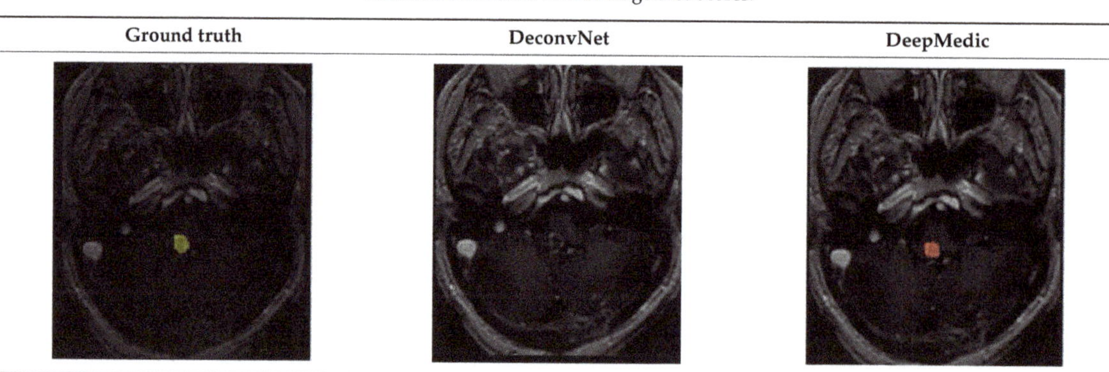		

Table 3. *Cont.*

PSPNet	U-Net	V-Net

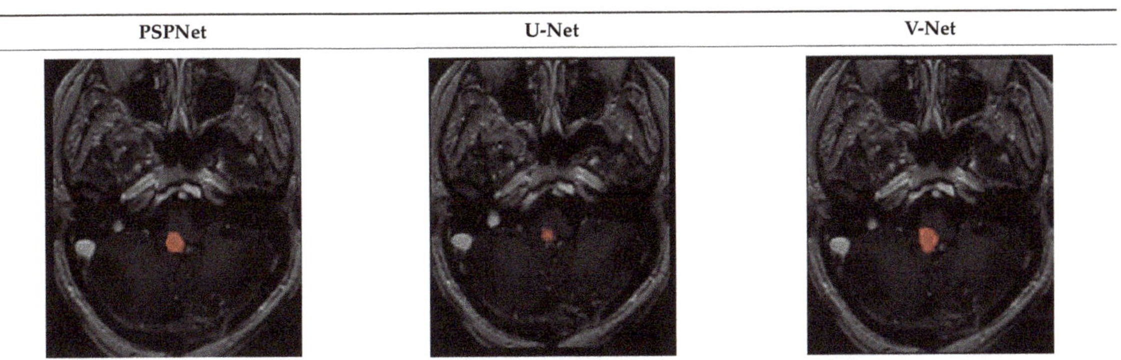

Table 4. Predictions with high dice scores.

Ground truth 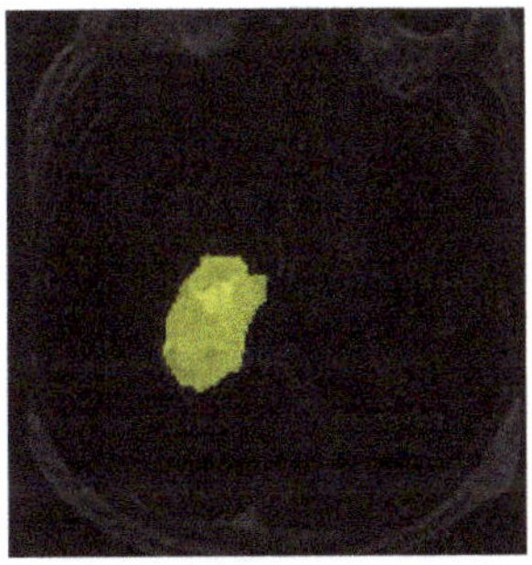	DeconvNet 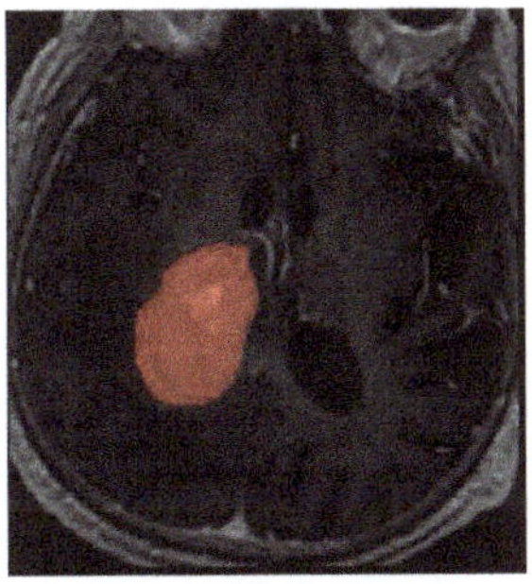	DeepMedic 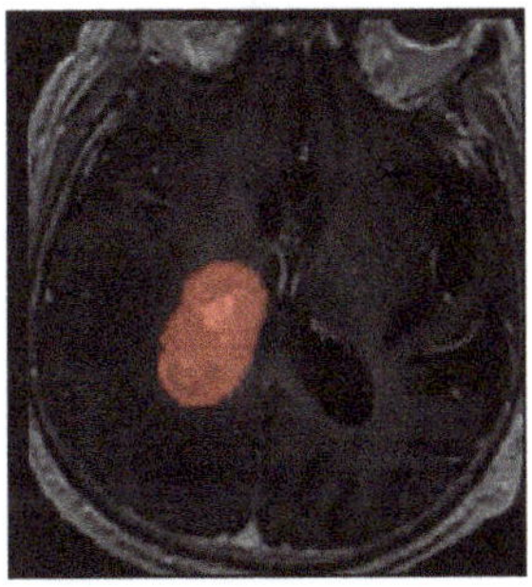
PSPNet 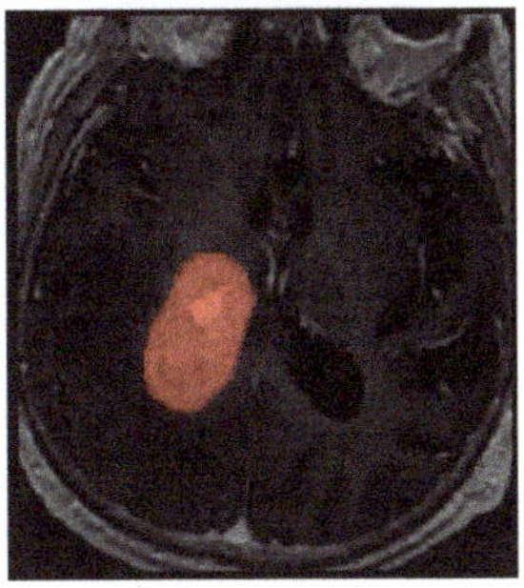	**U-Net** 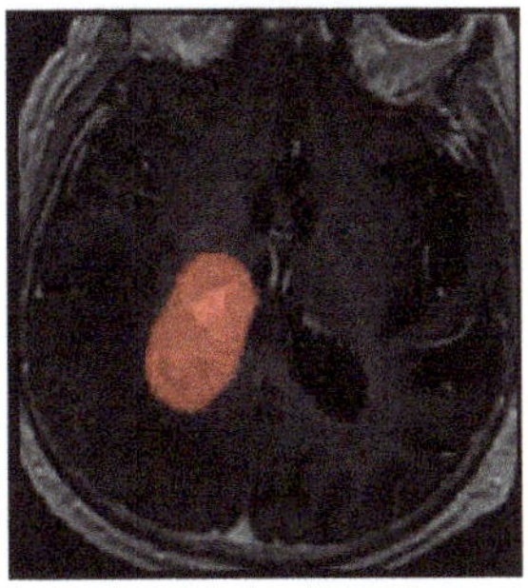	**V-Net** 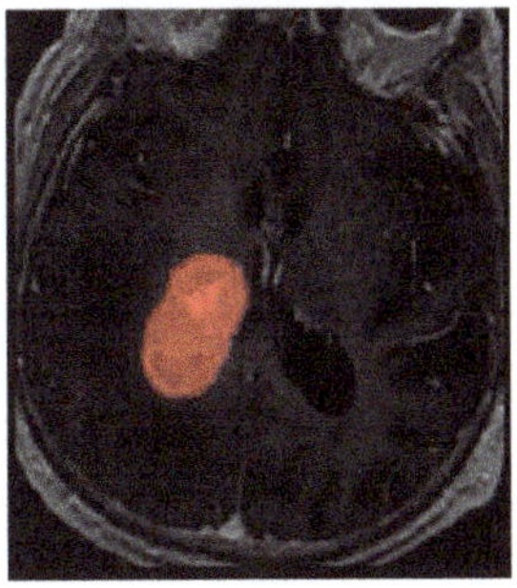

On the NTUH datasets, the performance was also affected by the types of lesions. As shown in Figure 1, we obtained better results for brain metastases, meningiomas, and schwannomas, while all models performed poorly on pituitary tumors, AVMs, and other tumor types. Detailed tables are attached as Appendix A, Appendix B, Appendix C.

Table 5. Performance of different models on the NTUH dataset.

Model	Numbers of Parameters	Batch Samplers	Loss Function	Precision	Sensitivity	Hard Dice
DeconvNet	12,544,324	two_dim	Cross-entropy minus log(soft-dice)	0.46	0.48	0.43
U-Net	34,524,034	two_dim	Cross-entropy minus log(soft-dice)	**0.48**	0.48	0.43
PSPNet	28,280,773	two_dim	Cross-entropy minus log(soft-dice)	0.47	0.48	0.43
V-Net	8,232,274	uniform_patch3d	Cross-entropy minus log(soft-dice)	0.39	0.54	0.41
V-Net	8,232,274	three_dim	Cross-entropy	0.2	0.56	0.25
V-Net	8,232,274	three_dim	Cross-entropy minus log(soft-dice)	**0.48**	0.51	0.46
V-Net dropout 0.1	8,232,274	three_dim	Cross-entropy minus log(soft-dice)	0.47	**0.66**	**0.51**
DeepMedic	1,301,478	center_patch3d	Cross-entropy minus log(soft-dice)	0.36	0.43	0.35
DeepMedic	1,301,478	center_patch3d	Cross-entropy	0.37	0.43	0.33

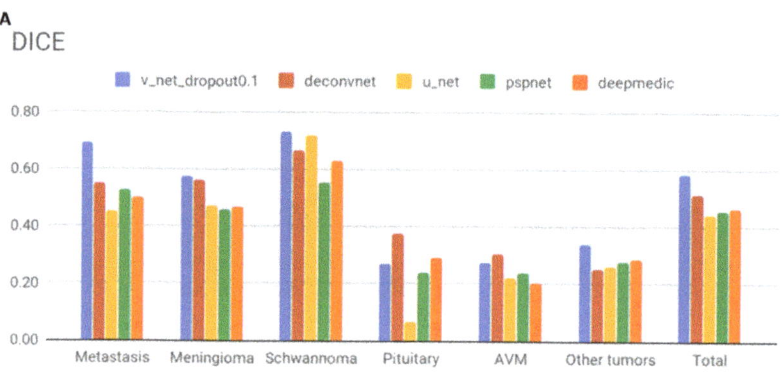

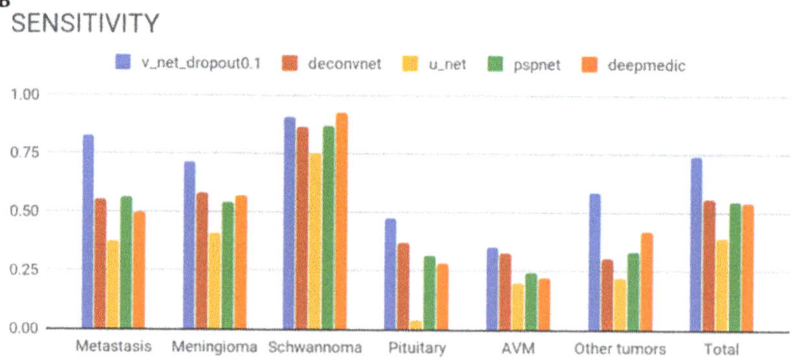

Figure 1. *Cont.*

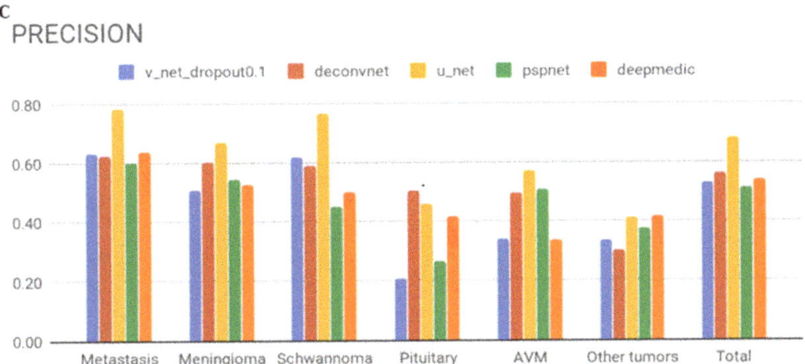

Figure 1. Bar chart results of (**A**) dice score; (**B**) sensitivity; and (**C**) precision of deep learning-based segmentation versus ground truth on different lesion types in the NTUH dataset.

As shown in Figure 2, lesions with smaller target volumes introduce lower average dice performance for each deep-learning model. V-Net, the best-performing model in the current study, obtained a fairly satisfactory dice score when lesion size exceeded the median size of all targets.

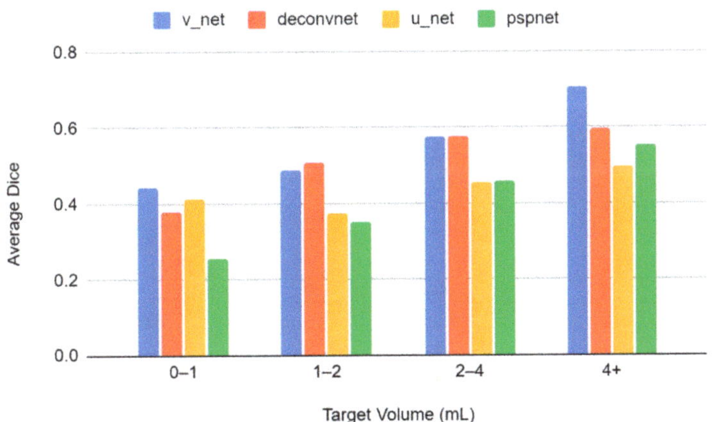

Figure 2. The performance of each model on lesions within different volume ranges.

To compare the performances of different models trained with one-channel input on the segmentation of brain lesions, we performed another experiment in which models were trained with just T1+C input with tumor core labels of the BraTS dataset. The evaluation was based on prediction of the tumor cores. As shown in Figure 3, V-Net had the highest Dice score when trained with 4-channel input with 5-class labels. Interestingly, all models performed better in this circumstance than when trained with only T1+C images. Of note, V-Net and PSPNet could not yield comparable results when trained with only T1+C images, implying that they are more sensitive to the change from multimodal to single modality inputs. While the models trained with one-channel inputs yielded lower performances in segmentation, they still performed better than their counterparts trained with the NTUH dataset.

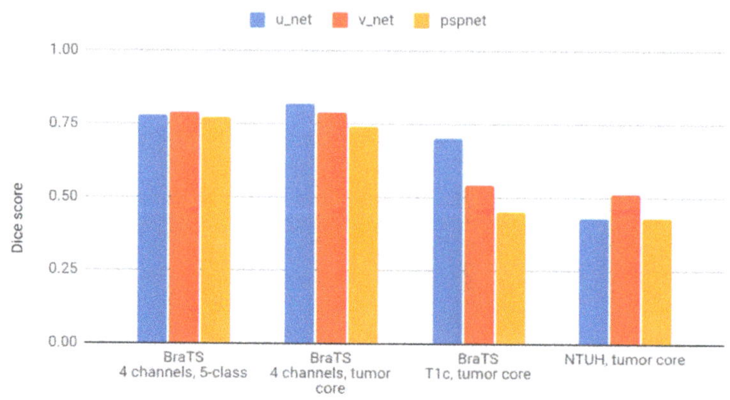

Figure 3. The performance of each model trained with four-channel or one-channel inputs.

Because of the nature of PSPNet and DeepMedic, they took a significantly longer time for inference, as shown in Table 6. V-Net had the least number of parameters and the shortest inference time. We also found that adding dropouts in V-Net further improved its performance, which we have noted in the table with 0.1 being the dropout rate.

Table 6. Inference time on our hardware and parameters of different networks.

	V-Net Dropout 0.1	DeconvNet	U-Net	PSPNet	DeepMedic
Inference time (minutes:seconds)	02:51	04:00	04:01	14:43	17:73
Number of parameters	8.23M	12.5M	34.5M	28.3M	13M

4. Discussion

4.1. Segmentation Performance: NTUH vs. BraTS Dataset

The performance on our radiosurgery dataset was inferior to that on BraTS. Many factors might lead to such a result. First of all, the tumor volumes in the NTUH dataset are typically smaller than those in BraTS 2015. On average, the tumor occupied 1.23% of the whole image volume in the BraTS dataset, but only 0.145% in ours. It should also be noted that a significant portion of our dataset contained multiple targets, which is much less likely for glioma patients (BraTS). The lesions in NTUH dataset are thus more difficult to detect.

Moreover, there is significant heterogeneity in our dataset. To evaluate whether our model could achieve similar segmentation performance under a more realistic scenario, we used the dataset containing cranial lesions of various pathology, which is different from the BraTS dataset with only glioma cases. In a strict sense, we also have some images of non-neoplastic diseases such as AVM. Additionally, some of the tumors are extra-axial (outside the brain parenchyma) and may even extend extracranially, so we cannot perform skull stripping like BraTS. Due to the heterogeneity of tumor types and sites, we may need a much larger dataset to reach similar performance.

Our results indicated that better performance was correlated with more training samples (as in metastases, meningioma, and schwannoma, Figure 2) and larger lesion dimensions (Figure 3). We also report the effect of input channels (of BraTS) in this revision.

Another reason is that we only used one image set (T1+C) to predict instead of four sequences used in the BraTS dataset. Less information might lead to deteriorated performance.

It is also worth mentioning that our dataset is quite imbalanced disease-wise. From the performance of the models we trained, we could observe that this imbalance resulted in serious bias issues for minority patients. We found it quite difficult to train a model by the traditional soft-dice loss or cross-entropy loss. Using the weighted cross-entropy loss gave us a 0.25 dice score, while our modification of subtracting a log-soft-dice term improved the dice score to 0.40. Such difference may result from tumor size since tumors in our dataset were of fewer voxels on average. In addition to the data variety, the weighted cross-entropy function could be very unstable and thus harmful to the optimization. Empirically, we found that the model will most likely fail in 10 epochs and predict nothing but the background for all inputs. By adding another term with the dice score, the new loss function provides better guidance to the model, and we could empirically observe the significant improvements.

We added images of trigeminal neuralgia in the training set as negative samples, in which there was no real space-occupying lesion. We did not expect the machine to learn how to identify trigeminal neuralgia. Instead, it can be considered that images of trigeminal neuralgia are examples of the heterogeneity of real clinical datasets. This artificial impurity was meant to mimic the systematic bias that could occur in a larger and unpurified dataset to infer the availability of deep learning models.

Although the targets in our dataset were defined and contoured by experienced clinicians, it should be noted that they were the targets we wanted to treat. Therefore, in very few cases, not every lesion detected by human experts was labeled. For example, a patient with brain metastases may also have a small meningioma, which may be stable and will not be labeled and treated by radiosurgery. If an algorithm detects that meningioma is this rare, decreased precision and dice score can be expected. However, from the clinical experience of our expert neurosurgeons and radiation oncologists, the rate of intentionally ignored meningiomas and pituitary adenomas was estimated around 1%. This estimation was in parallel with the reported prevalence of intracranial incidentaloma. On the other hand, the estimated rate for ignored brain metastases was much higher (5%), because our clinical experts might decide not to treat small lesions (less than 5–10 mm or visible only on one axial slice) in patients with multiple brain metastases [41,42]. As a result, this should not impede the training due to its rarity, and most meningiomas were labeled.

4.2. Performance on Different Types of Tumor

We can see that these models performed better for brain metastases, meningiomas, and schwannomas, where there were more than 300 cases each. They performed best for schwannomas, probably because most of these are vestibular ones, whose locations are always around internal auditory meatus.

On the other hand, these models performed poorly for pituitary tumors, AVMs, and other tumor types. Besides the relatively small number of cases for training, pituitary tumors and AVMs are not always readily visible for humans using only the T1+C series. For example, dynamic contrast-enhanced MRI may be required to visualize pituitary tumors. AVMs are sometimes not visible even using time-of-flight (TOF) MRI, so computed tomography angiography and/or digital subtraction angiography may be required for target contouring.

4.3. Comparison between Deep Learning Models

With respect to the input format, there are two classes of model architectures. The 2D model predicts tumors in just one slice and completely discards the information along the *z*-axis, while the 3D model utilizes the full information on the MRI volume. This results in a trade-off between features and overfitting. When receiving more features, it is more likely to overfit the unrelated noise, especially with such a small dataset. Patch sizes in previous works range from $16 \times 16 \times 16$ to $64 \times 64 \times 64$ mm^3 [18,43–55], of which Kamnitsas et al. outperformed the others. Thus, in the current proposed work, we restrained the receptive field and predicted on inputs patches with the size of $64 \times 64 \times 64$ mm^3. We examined this trade-off in our benchmark experiment on the BraTS dataset. Surprisingly, when experimenting with V-Net on our dataset, small patch-wise prediction became detrimental, but receiving the full brain volume guaranteed the best performance.

Overall, the 3D models seem to be more appealing. The 3D models present the full potential of convolution networks, reducing the number of parameters and becoming far more efficient due to their convolution nature. Specifically, V-Net has approximately 1/30 of the parameters compared to U-Net, shortest inference time, and the best performance on dice metric. The only shortcoming of 3D models is the requirement of GPU RAM due to the large input. In our experiments, we solve this by using a smaller batch-size. Furthermore, replacing batch normalization with dropout is quite effective in preventing overfitting because of the small batch size.

We compared the performances of the models trained with one-channel inputs of the NTUH and BraTS datasets. When the models were trained with one-channel inputs, the segmentation performances were slightly better than when they were trained with four-channel inputs. It could be inferred that the models perform better on a dataset with less heterogeneity in lesion types as well as lesion sizes.

4.4. Comparison to Previous Studies Addressing Deep Learning-Based Segmentation in SRS Treatment Planning

Efforts to identify the targets and the OARs prior to SRS treatment are crucial for dosimetry planning to protect the organs other than the lesions themselves. Several studies have benefited from deep learning methods on the classification and nomenclature standardization of the OARs [56,57]. The above-mentioned studies could advance computer-assisted radiation therapy.

To evaluate the benchmark performed in this study on the segmentation of brain lesions, previous studies addressing the segmentation of brain tumors in the treatment planning process during SRS will be reviewed. Of all types of brain lesions, asymptomatic or unresectable metastases warrant SRS without maximal surgical resection. As SRS serves as the first-line treatment for oligometastatic lesions, which denotes metastases of lesser than five lesions, contouring the lesions is of important clinical significance. The models previously used included modified DeepMedic [11,27], an ensemble of DeepMedic and 3D U-Net [28], and CNN [29].

Tumor volume tremendously affects the performance of segmentation; higher variety in tumor sizes and smaller lesions usually imply adversity in segmentation. Smaller lesions, while not affecting dice scores much, are not easily detected in methods with lower sensitivity. Liu et al. (2017) [27] proposed a modification of DeepMedic and managed to reach a dice score of 0.67. In their study, the number of brain metastases per case varied from 1 to 93 (5.679 ± 8.917), and the mean tumor size was 672 ± 1994 mm^3. Lu et al. (2019) [28] ensembled two neural networks, namely 3D U-Net and DeepMedic, yielding a good performance in segmentation with a median dice score of 0.74. The median size of the tumors in their dataset was 980 mm^3, while the smallest tumor was 3 mm^3. Fong et al. (2019) [29] trained the convoluted neural network with multiplanar slices, yielding a dice score of 0.77. Lu et al. (2021) [11] implemented an ensemble of 3D U-Net and DeepMedic and enhanced the prediction significantly, especially for non-experts. In their dataset, the median volume of the lesion was 890 mm^3. In our dataset, the lesions possessed a median size of 656 mm^3 and a mean of 2833 ± 6389 mm^3, while the smallest lesion was 13.05 mm^3. Generally speaking, with the highest dice score of 0.51, sensitivity of 0.66, and precision of 0.48, the lesions in our dataset had a higher size variety and smaller median size. The inconsistency in the lesion characteristics could cause difficulties for the deep learning models to extract features and hinder the prediction.

Ensemble models introduced higher segmentation performance than a single model in the previous studies [11,27–29]. Although in our study, V-Net with a dropout rate of 0.1 outperformed other methods in segmentation of brain lesions in the NTUH dataset, we did not perform a benchmark on the ensemble methods. It remains undetermined whether ensemble models yield better performance as well as which models ensembled could enhance segmentation the most.

As the difference in the imaging sequences used in the training process is a determinant of segmentation performance, the sequences used in previous works are discussed. Liu et al. (2017) used contrast-enhanced T1-weighted images [27] while Lu et al. (2019) used CT and T1-weighted MRI scans with contrast as the input [28]. Multiplanar slices of MPRAGE (magnetization-prepared rapid acquisition with gradient echo) images were taken as input in Fong et al. (2019) [29]. Lu et al. (2021) used contrast-enhanced CT and T1-weighted MR scans [11]. Out of the three studies, methods with MPRAGE as the input sequence yielded the highest dice score compared to the ground truth. Brain tumors on FLAIR, which is often used to contour the clinical target volume (CTV), mostly appeared as confluent hyperintense signals, introducing higher sensitivity and lower precision. On the other hand, brain tumors were mostly discrete on MPRAGE, an MRI modality taking advantage of gradient echo [58]. Despite the fact that higher precision could be achieved with MPRAGE, it is currently of lower significance in contouring before SRS. Of note, studies have shown that simultaneous use of different imaging modalities promised a better performance in segmentation compared to single modality use [38]. In our study, only contrast-enhanced T1-weighted MR images were used, and this could be a determinant of lower segmentation performance.

The required dataset size to yield high performances could not be confirmed, as we collected the data available to train the models and only draw conclusions from the current dataset. It is probably true that a larger dataset may generate better or different results, but such a dataset was not available to us.

4.5. Limitation of This Study

Compared to previous works investigating samples that underwent SRS, a relatively large dataset was implemented in the current study. However, the results suggest that the numbers of pituitary tumors, AVMs, and other tumors are probably insufficient for good results. Since the numbers of above lesions in a single institute may be insufficient, federated learning can be a potentially practical approach for better results.

Contrast-enhanced T1-weighted MR imaging was the only modality used as input in our study. Some tumors such as low-grade glioma or pituitary tumors are non-enhancing, introducing great difficulty in the detection and segmentation of these types of lesions. Simultaneous use of multiple imaging modalities could be the solution to this. Reviewing previous works, the sensitivity for detection of smaller brain lesions (<3 mm) with 3D U-Net, whether trained with black-blood or gradient echo modalities, decreased significantly compared to larger brain lesions ($\geq$10 mm, 0.981, 3–10 mm 0.829, <3 mm 0.235) [59]. The same trend could be observed in studies performed with 2-stage MetNet ($\geq$6 mm 0.99, 3–6 mm 0.87, $\leq$3 mm 0.25) [60] or GoogLeNet [61]. The 2-stage MetNet [60] and BMDS net [62] could achieve satisfactory segmentation prediction on tumors larger than 6 mm, with dice scores of 0.87 and 0.83, respectively. In our dataset counterpart, the diameters of 10.5% lesions were smaller than 6 mm, 45% lesions smaller than 10 mm, and 95.7% smaller than 3 cm. The small lesion sizes in our NTUH dataset contributed to the dice score lower than 0.6 predicted with V-Net.

The way dice score is derived could mask the effect of contouring small lesions. In our work, dice score was calculated per voxel, which favored larger lesions compared to dice score derived per lesion. Clinically, SRS is indicated and is of significant importance for patients with smaller brain lesions, whereas for patients who are surgical candidates with larger lesions, standard care remains surgery with adjuvant stereotactic radiation therapy or whole-brain radiation therapy. As for patients with diffuse lesions, whole-brain radiation therapy is the standard treatment due to the lack of level 1 evidence to support the use of SRS in the patient population [63] (p. 865). Contouring deflection on the gross tumor volume (GTV) of such small lesions could introduce a huge impact on later target contouring, compromising organs at risk (OAR). Take brain metastases, for example, current guidelines for contouring for SRS generally indicate a 1.5 cm expansion from GTV to generate CTV. In our dataset, the smallest volume of brain lesion being 20 mm^3 implies a 3.4 mm diameter, and the volume difference of CTV with GTV is about 3000 mm^3. This expansion in target volume significantly differs if a small lesion was not correctly contoured. As a consequence, a dice score per lesion provides benefit in some circumstances.

Evidence derived from trials concerning treatment response to SRS based on either deep-learning segmentation or manual segmentation is still an unmet need. Several studies implemented multiple modalities (PET/MRI) in order to train machine learning models for tumor segmentation, which suggested that biological target volume (BTV) could be promising in helping CTV definition during SRS treatment and their ability to indicate dose escalation on biologically active targets [64,65]. Despite the effort in assisting CTV definition by taking advantage of the training set of multi-modalities, whether the addition in modalities to either of the learning methods improves clinical treatment response is yet undetermined.

5. Conclusions

We benchmarked five commonly used deep learning segmentation models on our SRS dataset. We confirmed that these approaches also work on a heterogeneous dataset, but with decreased performance. We discovered that the V-Net architecture worked best for this specific task. With the top dice scores, the smallest size of the model, and the shortest inference time, V-Net may be a good choice to improve upon. We also found that when training on the dataset with such heterogeneity and class imbalance, using weighted cross-entropy loss with log-soft-dice term significantly improved the performance.

Supplementary Materials: GitHub repository, https://github.com/raywu0123/Brain-Tumor-Segmentation, (accessed on 25 September 2021).

Author Contributions: Conceptualization, S.W., Y.W., H.C., and F.X.; Methodology, S.W., H.C., and Y.W.; Software, S.W. and H.C.; Validation, S.W., Y.W., and F.X.; Formal analysis, Y.W., C.L., and F.X.; Investigation, S.W., H.C., Y.W., and F.X.; Resources, F.X.; Data curation, S.W., H.C., Y.W., and F.X.; Writing—original draft preparation, Y.W., C.L., and F.X.; Writing—review and editing, Y.W., F.T.S., H.L., W.T., C.L., F.L., F.H., and F.X.; Visualization, C.L., Y.W., and F.X.; Supervision, F.X.; Project administration, F.X.; Funding acquisition, F.X. All authors have read and agreed to the published version of the manuscript.

Funding: This work was supported by the Ministry of Science and Technology, Taiwan, ROC, grant numbers 107-2634-F-002-015, 110-2634-F-002-032, 110-2314-B-002-161.

Institutional Review Board Statement: The study was conducted according to the guidelines of the Declaration of Helsinki, and approved by Research Ethics Committee of National Taiwan University Hospital (201708071RINC, 6 October 2017).

Informed Consent Statement: Informed consent was waived by the committee.

Data Availability Statement: The datasets generated and/or analyzed during the current study are available from the corresponding author on reasonable request.

Conflicts of Interest: The funders had no role in the design of the study; in the collection, analyses, or interpretation of data; in the writing of the manuscript, or in the decision to publish the results.

Appendix A

DICE	v_net_dropout0.1	Deconvnet	u_net	Pspnet	Deepmedic
Metastasis	0.69	0.55	0.45	0.52	0.5
Meningioma	0.57	0.56	0.48	0.46	0.47
Schwannoma	0.73	0.66	0.72	0.55	0.63
Pituitary	0.27	0.38	0.07	0.24	0.29
AVM	0.27	0.31	0.22	0.24	0.2
Other tumors	0.34	0.25	0.26	0.28	0.29
Total	0.59	0.52	0.44	0.46	0.46

Appendix B

SENSITIVITY	v_net_dropout0.1	Deconvnet	u_net	Pspnet	Deepmedic
Metastasis	0.82	0.55	0.37	0.56	0.5
Meningioma	0.71	0.58	0.41	0.54	0.57
Schwannoma	0.91	0.86	0.75	0.87	0.92
Pituitary	0.48	0.37	0.04	0.31	0.29
AVM	0.36	0.33	0.2	0.24	0.22
Other tumors	0.58	0.3	0.22	0.33	0.42
Total	0.74	0.56	0.39	0.55	0.54

Appendix C

PRECISION	v_net_dropout0.1	Deconvnet	u_net	Pspnet	Deepmedic
Metastasis	0.63	0.62	0.78	0.6	0.64
Meningioma	0.51	0.6	0.67	0.54	0.53
Schwannoma	0.62	0.59	0.76	0.45	0.5
Pituitary	0.21	0.5	0.46	0.27	0.42
AVM	0.34	0.49	0.57	0.51	0.34
Other tumors	0.34	0.3	0.41	0.38	0.42
Total	0.53	0.56	0.68	0.52	0.54

References

1. Adler, J.R., Jr.; Colombo, F.; Heilbrun, M.P.; Winston, K. Toward an expanded view of radiosurgery. *Neurosurgery* **2004**, *55*, 1374–1376. [CrossRef] [PubMed]
2. Chao, S.T.; Dad, L.K.; Dawson, L.A.; Desai, N.B.; Pacella, M.; Rengan, R.; Xiao, Y.; Yenice, K.M.; Rosenthal, S.A.; Hartford, A. ACR–ASTRO Practice Parameter for the Performance of Stereotactic Body Radiation Therapy. *Am. J. Clin. Oncol.* **2020**, *43*, 545–552. [CrossRef] [PubMed]
3. Schell, M.C.; Bova, F.J.; Larson, D.A.; Leavitt, D.D.; Latz, W.R.; Podgorsak, E.B.; Wu, A. *Stereotactic Radiosurgery*; AAPM Report NO. 54; American Association of Physicists in Medicine: Alexandria, VA, USA, 1995.
4. Seung, S.K.; Larson, D.A.; Galvin, J.M.; Mehta, M.P.; Potters, L.; Schultz, C.J.; Yajnik, S.V.; Hartford, A.C.; Rosenthal, S.A. American College of Radiology (ACR) and American Society for Radiation Oncology (ASTRO) Practice Guideline for the Performance of Stereotactic Radiosurgery (SRS). *Am. J. Clin. Oncol.* **2013**, *36*, 310–315. [CrossRef] [PubMed]
5. Shin, H.-C. Hybrid clustering and logistic regression for multi-modal brain tumor segmentation. In Proceedings of the MICCAI-BRATS 2012, Nice, France, 1 October 2012.
6. Bauer, S.; Fejes, T.; Slotboom, J.; Wiest, R.; Nolte, L.-P.; Reyes, M. Segmentation of brain tumor images based on integrated hierarchical classification and regularization. In Proceedings of the MICCAI-BRATS 2012, Nice, France, 1 October 2012.
7. Zhao, L.; Wu, W.; Corso, J.J. Brain tumor segmentation based on GMM and active contour method with a model-aware edge map. In Proceedings of the MICCAI-BRATS 2012, Nice, France, 1 October 2012; pp. 19–23.
8. Xiao, Y.; Hu, J. Hierarchical random walker for multimodal brain tumor segmentation. In Proceedings of the MICCAI-BRATS 2012, Nice, France, 1 October 2012.
9. Subbanna, N.; Arbel, T. Probabilistic gabor and markov random fields segmentation of brain tumours in mri volumes. In Proceedings of the MICCAI-BRATS 2012, Nice, France, 1 October 2012; pp. 28–31.
10. Zikic, D.; Glocker, B.; Konukoglu, E.; Shotton, J.; Criminisi, A.; Ye, D.; Demiralp, C.; Thomas, O.M.; Das, T.; Jena, R. Context-sensitive classification forests for segmentation of brain tumor tissues. In Proceedings of the MICCAI-BRATS 2012, Nice, France, 1 October 2012; pp. 22–30.

11. Lu, S.-L.; Xiao, F.-R.; Cheng, J.C.-H.; Yang, W.-C.; Cheng, Y.-H.; Chang, Y.-C.; Lin, J.-Y.; Liang, C.-H.; Lu, J.-T.; Chen, Y.-F.; et al. Randomized multi-reader evaluation of automated detection and segmentation of brain tumors in stereotactic radiosurgery with deep neural networks. *Neuro-Oncology* **2021**, *23*, 1560–1568. [CrossRef]
12. Havaei, M.; Davy, A.; Warde-Farley, D.; Biard, A.; Courville, A.; Bengio, Y.; Pal, C.; Jodoin, P.-M.; Larochelle, H. Brain tumor segmentation with Deep Neural Networks. *Med. Image Anal.* **2017**, *35*, 18–31. [CrossRef]
13. Kamnitsas, K.; Ledig, C.; Newcombe, V.F.J.; Simpson, J.P.; Kane, A.D.; Menon, D.K.; Rueckert, D.; Glocker, B. Efficient multi-scale 3D CNN with fully connected CRF for accurate brain lesion segmentation. *Med. Image Anal.* **2017**, *36*, 61–78. [CrossRef] [PubMed]
14. Ronneberger, O.; Fischer, P.; Brox, T. U-Net: Convolutional Networks for Biomedical Image Segmentation. In *International Conference on Medical Image Computing and Computer-Assisted Intervention*; Springer: Cham, Switzerland, 2015; pp. 234–241. [CrossRef]
15. Dong, H.; Yang, G.; Liu, F.; Mo, Y.; Guo, Y. Automatic Brain Tumor Detection and Segmentation Using U-Net Based Fully Convolutional Networks. In *Communications in Computer and Information Science*; Springer: Cham, Switzerland, 2017; pp. 506–517.
16. Livne, M.; Rieger, J.; Aydin, O.U.; Taha, A.A.; Akay, E.M.; Kossen, T.; Sobesky, J.; Kelleher, J.D.; Hildebrand, K.; Frey, D.; et al. A U-Net Deep Learning Framework for High Performance Vessel Segmentation in Patients With Cerebrovascular Disease. *Front. Neurosci.* **2019**, *13*, 97. [CrossRef]
17. Bakas, S.; Akbari, H.; Sotiras, A.; Bilello, M.; Rozycki, M.; Kirby, J.; Freymann, J.B.; Farahani, K.; Davatzikos, C. Advancing The Cancer Genome Atlas glioma MRI collections with expert segmentation labels and radiomic features. *Sci. Data* **2017**, *4*, 170117. [CrossRef]
18. Kamnitsas, K.; Bai, W.; Ferrante, E.; McDonagh, S.; Sinclair, M.; Pawlowski, N.; Rajchl, M.; Lee, M.; Kainz, B.; Rueckert, D.; et al. Ensembles of Multiple Models and Architectures for Robust Brain Tumour Segmentation. In *Machine Learning and Knowledge Discovery in Databases*; Springer: Cham, Switzerland, 2017; pp. 450–462.
19. Shelhamer, E.; Long, J.; Darrell, T. Fully convolutional models for semantic segmentation. *IEEE Trans. Pattern Anal. Mach. Intell.* **2016**, *39*, 640–651. [CrossRef]
20. Milletari, F.; Navab, N.; Ahmadi, S.-A. V-Net: Fully Convolutional Neural Networks for Volumetric Medical Image Segmentation. In Proceedings of the 2016 Fourth International Conference on 3D Vision (3DV), Stanford, CA, USA, 25–28 October 2016; pp. 565–571. [CrossRef]
21. Militello, C.; Rundo, L.; Vitabile, S.; Russo, G.; Pisciotta, P.; Marletta, F.; Ippolito, M.; D'Arrigo, C.; Midiri, M.; Gilardi, M.C. Gamma Knife treatment planning: MR brain tumor segmentation and volume measurement based on unsupervised Fuzzy C-Means clustering. *Int. J. Imaging Syst. Technol.* **2015**, *25*, 213–225. [CrossRef]
22. Hamamci, A.; Kucuk, N.; Karaman, K.; Engin, K.; Unal, G. Tumor-Cut: Segmentation of Brain Tumors on Contrast Enhanced MR Images for Radiosurgery Applications. *IEEE Trans. Med. Imaging* **2011**, *31*, 790–804. [CrossRef]
23. Hu, M.; Zhong, Y.; Xie, S.; Lv, H.; Lv, Z. Fuzzy System Based Medical Image Processing for Brain Disease Prediction. *Front. Neurosci.* **2021**, *15*, 965. [CrossRef]
24. Rundo, L.; Militello, C.; Russo, G.; Vitabile, S.; Gilardi, M.C.; Mauri, G. GTV cut for neuro-radiosurgery treatment planning: An MRI brain cancer seeded image segmentation method based on a cellular automata model. *Nat. Comput.* **2018**, *17*, 521–536. [CrossRef]
25. Wu, X.; Bi, L.; Fulham, M.; Feng, D.D.; Zhou, L.; Kim, J. Unsupervised brain tumor segmentation using a symmetric-driven adversarial network. *Neurocomputing* **2021**, *455*, 242–254. [CrossRef]
26. Rundo, L.; Militello, C.; Tangherloni, A.; Russo, G.; Vitabile, S.; Gilardi, M.C.; Mauri, G. NeXt for neuro-radiosurgery: A fully automatic approach for necrosis extraction in brain tumor MRI using an unsupervised machine learning technique. *Int. J. Imaging Syst. Technol.* **2018**, *28*, 21–37. [CrossRef]
27. Liu, Y.; Stojadinovic, S.; Hrycushko, B.; Wardak, Z.; Lau, S.; Lu, W.; Yan, Y.; Jiang, S.B.; Zhen, X.; Timmerman, R.; et al. A deep convolutional neural network-based automatic delineation strategy for multiple brain metastases stereotactic radiosurgery. *PLoS ONE* **2017**, *12*, e0185844. [CrossRef]
28. Lu, S.; Hu, S.; Weng, W.; Chen, Y.; Lu, J.; Xiao, F.; Hsu, F. Automated Detection and Segmentation of Brain Metastases in Stereotactic Radiosurgery Using Three-Dimensional Deep Neural Networks. *Int. J. Radiat. Oncol.* **2019**, *105*, S69–S70. [CrossRef]
29. Fong, A.; Swift, C.; Wong, J.; McVicar, N.; Giambattista, J.; Kolbeck, C.; Nichol, A. Automatic Deep Learning-based Segmentation of Brain Metastasis on MPRAGE MR Images for Stereotactic Radiotherapy Planning. *Int. J. Radiat. Oncol.* **2019**, *105*, E134. [CrossRef]
30. Sachdeva, J.; Kumar, V.; Gupta, I.; Khandelwal, N.; Ahuja, C.K. Segmentation, Feature Extraction, and Multiclass Brain Tumor Classification. *J. Digit. Imaging* **2013**, *26*, 1141–1150. [CrossRef]
31. Gros, C.; Lemay, A.; Cohen-Adad, J. SoftSeg: Advantages of soft versus binary training for image segmentation. *Med. Image Anal.* **2021**, *71*, 102038. [CrossRef]
32. Wong, J.; Huang, V.; Wells, D.; Giambattista, J.; Giambattista, J.; Kolbeck, C.; Otto, K.; Saibishkumar, E.P.; Alexander, A. Implementation of deep learning-based auto-segmentation for radiotherapy planning structures: A workflow study at two cancer centers. *Radiat. Oncol.* **2021**, *16*, 101. [CrossRef] [PubMed]
33. Shattuck, D.W.; Leahy, R.M. BrainSuite: An automated cortical surface identification tool. *Med. Image Anal.* **2002**, *6*, 129–142. [CrossRef]

34. Wu, S.-R.; Wu, P.Y.; Chang, H.Y. Brain-Tumor-Segmentation/Models at Master • raywu0123/Brain-Tumor-Segmentation. Available online: https://github.com/raywu0123/Brain-Tumor-Segmentation/tree/master/models (accessed on 25 September 2021).
35. Noh, H.; Hong, S.; Han, B. Learning deconvolution network for semantic segmentation. In Proceedings of the IEEE International Conference on Computer Vision, Santiago, Chile, 7–13 December 2015; pp. 1520–1528.
36. Pennig, L.; Shahzad, R.; Caldeira, L.; Lennartz, S.; Thiele, F.; Goertz, L.; Zopfs, D.; Meißner, A.-K.; Fürtjes, G.; Perkuhn, M.; et al. Automated Detection and Segmentation of Brain Metastases in Malignant Melanoma: Evaluation of a Dedicated Deep Learning Model. *Am. J. Neuroradiol.* **2021**, *42*, 655–662. [CrossRef] [PubMed]
37. Jünger, S.T.; Hoyer, U.C.I.; Schaufler, D.; Laukamp, K.R.; Goertz, L.; Thiele, F.; Grunz, J.; Schlamann, M.; Perkuhn, M.; Kabbasch, C.; et al. Fully Automated MR Detection and Segmentation of Brain Metastases in Non-small Cell Lung Cancer Using Deep Learning. *J. Magn. Reson. Imaging* **2021**. [CrossRef]
38. Charron, O.; Lallement, A.; Jarnet, D.; Noblet, V.; Clavier, J.-B.; Meyer, P. Automatic detection and segmentation of brain metastases on multimodal MR images with a deep convolutional neural network. *Comput. Biol. Med.* **2018**, *95*, 43–54. [CrossRef] [PubMed]
39. Zhao, H.; Shi, J.; Qi, X.; Wang, X.; Jia, J. Pyramid Scene Parsing Network. In Proceedings of the 2017 IEEE Conference on Computer Vision and Pattern Recognition (CVPR), Honolulu, HI, USA, 21–26 July 2017; pp. 2881–2890.
40. Wu, S.R.; Wu, P.Y.; Chang, H.Y. Brain-Tumor-Segmentation/Models/Batch_Samplers at Master • raywu0123/Brain-Tumor-Segmentation. Available online: https://github.com/raywu0123/Brain-Tumor-Segmentation/tree/master/models/batch_samplers (accessed on 25 September 2021).
41. Neugut, A.I.; Sackstein, P.; Hillyer, G.C.; Jacobson, J.S.; Bruce, J.; Lassman, A.B.; Stieg, P.A. Magnetic Resonance Imaging-Based Screening for Asymptomatic Brain Tumors: A Review. *Oncologist* **2019**, *24*, 375–384. [CrossRef]
42. Nakasu, S.; Notsu, A.; Nakasu, Y. Prevalence of incidental meningiomas and gliomas on MRI: A meta-analysis and meta-regression analysis. *Acta Neurochir.* **2021**, 1–15. [CrossRef]
43. Andermatt, S.; Pezold, S.; Cattin, P. Multi-dimensional gated recurrent units for brain tumor segmentation. In Proceedings of the International MICCAI BraTS Challenge 2017, Quebec City, QC, Canada, 14 September 2017; pp. 15–19.
44. Amorim, P.H.A.; Chagas, V.S.; Escudero, G.G.; Oliveira, D.D.C.; Pereira, S.M.; Santos, H.M.; Scussel, A.A. 3D u-nets for brain tumor segmentation in miccai 2017 brats challenge. In Proceedings of the International MICCAI BraTS Challenge 2017, Quebec City, QC, Canada, 14 September 2017.
45. Castillo, L.S.; Daza, L.A.; Rivera, L.C.; Arbeláez, P. Volumetric multimodality neural network for brain tumor segmentation. In Proceedings of the 13th International Conference on Medical Information Processing and Analysis, San Andres Island, Colombia, 5–7 October 2017; p. 105720E.
46. Feng, X.; Meyer, C. Patch-based 3d u-net for brain tumor segmentation. In Proceedings of the International MICCAI BraTS Challenge 2017, Quebec City, QC, Canada, 14 September 2017.
47. Zhou, C.; Ding, C.; Lu, Z.; Zhang, T. Brain tumor segmentation with cascaded convolutional neural networks. In Proceedings of the International MICCAI BraTS Challenge 2017, Quebec City, QC, Canada, 14 September 2017; pp. 328–333.
48. Isensee, F.; Kickingereder, P.; Wick, W.; Bendszus, M.; Maier-Hein, K.H. Brain Tumor Segmentation and Radiomics Survival Prediction: Contribution to the BRATS 2017 Challenge. In *International MICCAI Brainlesion Workshop*; Springer: Cham, Switzerland, 2018; pp. 287–297. [CrossRef]
49. Li, Y.; Shen, L. MvNet: Multi-view deep learning framework for multimodal brain tumor segmentation. In Proceedings of the International MICCAI BraTS Challenge 2017, Quebec City, QC, Canada, 14 September 2017.
50. Pourreza, R.; Zhuge, Y.; Ning, H.; Miller, R. Brain Tumor Segmentation in MRI Scans Using Deeply-Supervised Neural Networks. In *International MICCAI Brainlesion Workshop*; Springer: Cham, Switzerland, 2018; pp. 320–331.
51. Zhou, F.; Li, T.; Li, H.; Yu, K.; Wang, Y.; Zhu, H. TP-CNN: A two-phase convolution neural network based model to do automatic brain tumor segmentation by using BRATS 2017 data. In Proceedings of the International MICCAI BraTS Challenge 2017, Quebec City, QC, Canada, 14 September 2017; pp. 334–341.
52. Zhu, J.; Wang, D.; Teng, Z.; Lio, P. A multi-pathway 3d dilated convolutional neural network for brain tumor segmentation. In Proceedings of the International MICCAI BraTS Challenge 2017, Quebec City, QC, Canada, 14 September 2017; pp. 342–347.
53. Hu, Y.; Xia, Y. Automated brain tumor segmentation using a 3D deep detection-classification model. In Proceedings of the International MICCAI BraTS Challenge 2017, Quebec City, QC, Canada, 14 September 2017.
54. Chen, S.; Ding, C.; Zhou, C. Brain tumor segmentation with label distribution learning and multi-level feature representation. In Proceedings of the International MICCAI BraTS Challenge 2017, Quebec City, QC, Canada, 14 September 2017.
55. Beers, A.; Chang, K.; Brown, J.; Sartor, E.; Mammen, C.; Gerstner, E.; Rosen, B.; Kalpathy-Cramer, J. Sequential 3d u-nets for brain tumor segmentation. In Proceedings of the International MICCAI BraTS Challenge 2017, Quebec City, QC, Canada, 14 September 2017; pp. 20–23.
56. Yang, Q.; Chao, H.; Nguyen, D.; Jiang, S. A Novel Deep Learning Framework for Standardizing the Label of OARs in CT. In *Workshop on Artificial Intelligence in Radiation Therapy*; Springer: Cham, Switzerland, 2019; pp. 52–60.
57. Yang, Q.; Chao, H.; Nguyen, D.; Jiang, S. Mining Domain Knowledge: Improved Framework Towards Automatically Standardizing Anatomical Structure Nomenclature in Radiotherapy. *IEEE Access* **2020**, *8*, 105286–105300. [CrossRef]
58. Brant-Zawadzki, M.; Gillan, G.D.; Nitz, W.R. MP RAGE: A three-dimensional, T1-weighted, gradient-echo sequence—Initial experience in the brain. *Radiology* **1992**, *182*, 769–775. [CrossRef]

59. Park, Y.W.; Jun, Y.; Lee, Y.; Han, K.; An, C.; Ahn, S.S.; Hwang, D.; Lee, S.-K. Robust performance of deep learning for automatic detection and segmentation of brain metastases using three-dimensional black-blood and three-dimensional gradient echo imaging. *Eur. Radiol.* **2021**, *31*, 6686–6695. [CrossRef]
60. Zhou, Z.; Sanders, J.W.; Johnson, J.M.; Gule-Monroe, M.; Chen, M.; Briere, T.M.; Wang, Y.; Son, J.B.; Pagel, M.D.; Ma, J.; et al. MetNet: Computer-aided segmentation of brain metastases in post-contrast T1-weighted magnetic resonance imaging. *Radiother. Oncol.* **2020**, *153*, 189–196. [CrossRef] [PubMed]
61. Grøvik, E.; Yi, D.; Iv, M.; Tong, E.; Rubin, D.; Zaharchuk, G. Deep learning enables automatic detection and segmentation of brain metastases on multisequence MRI. *J. Magn. Reson. Imaging* **2020**, *51*, 175–182. [CrossRef] [PubMed]
62. Xue, J.; Wang, B.; Ming, Y.; Liu, X.; Jiang, Z.; Wang, C.; Liu, X.; Chen, L.; Qu, J.; Xu, S.; et al. Deep learning–based detection and segmentation-assisted management of brain metastases. *Neuro-Oncology* **2020**, *22*, 505–514. [CrossRef]
63. Hansen, E.K.; Roach, M., III. *Handbook of Evidence-Based Radiation Oncology*; Springer: Berlin/Heidelberg, Germany, 2018.
64. Wang, X.; Cui, H.; Gong, G.; Fu, Z.; Zhou, J.; Gu, J.; Yin, Y.; Feng, D. Computational delineation and quantitative heterogeneity analysis of lung tumor on 18F-FDG PET for radiation dose-escalation. *Sci. Rep.* **2018**, *8*, 10649. [CrossRef]
65. Rundo, L.; Stefano, A.; Militello, C.; Russo, G.; Sabini, M.G.; D'Arrigo, C.; Marletta, F.; Ippolito, M.; Mauri, G.; Vitabile, S.; et al. A fully automatic approach for multimodal PET and MR image segmentation in gamma knife treatment planning. *Comput. Methods Programs Biomed.* **2017**, *144*, 77–96. [CrossRef] [PubMed]

Article

Multichannel Multiscale Two-Stage Convolutional Neural Network for the Detection and Localization of Myocardial Infarction Using Vectorcardiogram Signal

Jay Karhade [1], Samit Kumar Ghosh [1], Pranjali Gajbhiye [1], Rajesh Kumar Tripathy [1] and U. Rajendra Acharya [2,3,4,*]

1 Department of Electrical and Electronics Engineering, BITS-Pilani, Hyderabad Campus, Hyderabad 500078, India; f20180852@hyderabad.bits-pilani.ac.in (J.K.); samitnitrkl@gmail.com (S.K.G.); gajbhiyepranjali@gmail.com (P.G.); rajeshiitg13@gmail.com (R.K.T.)
2 School of Engineering, Ngee Ann Polytechnic, Singapore 599489, Singapore
3 Department of Bioinformatics and Medical Engineering, Asia University, Taichung 41354, Taiwan
4 School of Management and Enterprise, University of Southern Queensland, Springfield, Ipswich, QLD 4300, Australia
* Correspondence: Rajendra_Udyavara_ACHARYA@np.edu.sg

Abstract: Myocardial infarction (MI) occurs due to the decrease in the blood flow into one part of the heart, and it further causes damage to the heart muscle. The 12-channel electrocardiogram (ECG) has been widely used to detect and localize MI pathology in clinical studies. The vectorcardiogram (VCG) is a 3-channel recording system used to measure the heart's electrical activity in sagittal, transverse, and frontal planes. The VCG signals have advantages over the 12-channel ECG to localize posterior MI pathology. Detection and localization of MI using VCG signals are vital in clinical practice. This paper proposes a multi-channel multi-scale two-stage deep-learning-based approach to detect and localize MI using VCG signals. In the first stage, the multivariate variational mode decomposition (MVMD) decomposes the three-channel-based VCG signal beat into five components along each channel. The multi-channel multi-scale VCG tensor is formulated using the modes of each channel of VCG data, and it is used as the input to the deep convolutional neural network (CNN) to classify MI and normal sinus rhythm (NSR) classes. In the second stage, the multi-class deep CNN is used for the categorization of anterior MI (AMI), anterior-lateral MI (ALMI), anterior-septal MI (ASMI), inferior MI (IMI), inferior-lateral MI (ILMI), inferior-posterior-lateral (IPLMI) classes using MI detected multi-channel multi-scale VCG instances from the first stage. The proposed approach is developed using the VCG data obtained from a public database. The results reveal that the approach has obtained the accuracy, sensitivity, and specificity values of 99.58%, 99.18%, and 99.87%, respectively, for MI detection. Moreover, for MI localization, we have obtained the overall accuracy value of 99.86% in the second stage for our proposed network. The proposed approach has demonstrated superior classification performance compared to the existing VCG signal-based MI detection and localization techniques.

Keywords: myocardial infarction; vectorcardiogram; multivariate VMD; deep CNN; accuracy

Citation: Karhade, J.; Ghosh, S.K.; Gajbhiye, P.; Tripathy, R.K.; Acharya, U.R. Multichannel Multiscale Two-Stage Convolutional Neural Network for the Detection and Localization of Myocardial Infarction Using Vectorcardiogram Signal. *Appl. Sci.* **2021**, *11*, 7920. https://doi.org/10.3390/app11177920

Academic Editors: Leonardo Rundo, Carmelo Militello and Andrea Tangherloni

Received: 26 July 2021
Accepted: 25 August 2021
Published: 27 August 2021

1. Introduction

The obstruction in one of the coronary arteries of the heart causes the myocardial infarction (MI) disease [1,2]. Typically, the MI is progressed in three phases [3]. These three phases are (a) ischemic phase, (b) acute phase, and (c) myocardial necrosis phase. The 12-lead ECG signal is used in the clinical study for the early detection and localization of MI pathology [4]. The ST-segment elevations, inverted T-waves, and pathological Q-waves are the morphological changes observed in the ECG signals of different leads in MI pathology [5]. The morphological changes in the ECG signals of the channels or leads, such as V1, V2, V3, and V4, are used to diagnose anterior MI (AMI) [6]. Similarly, inferior MI is diagnosed based on the variations in the morphologies of ECG signals for II, III, and aVF channels. Moreover, the morphological variations in the ECG signals of I, aVL, V5, and V6 channels are used to diagnose left lateral MI pathology [6]. In 12-lead ECG, no ECG lead capture the information about the diagnosis of posterior MI [6,7]. However, the reciprocal changes in the V1 and V2 channel ECG signals are used in the clinical study to diagnose posterior MI [8]. Vectorcardiogram (VCG) is an orthogonal three lead system which measures the heart's electrical activity along transverse, sagittal, and frontal planes, and it has been used for the detection of MI pathology [3,9]. The 12-lead ECG can be derived from the VCG signal using various transformation techniques [10]. In VCG, one of the orthogonal leads reveals the posterior view of the heart [11]. Hence, the method based on the analysis of VCG signal information is helpful to detect and localize MI pathology. The continuous recording and monitoring of VCG signal information for MI disease diagnosis is cumbersome, and hence automated approaches have been used for the accurate detection and localization of MI using VCG signals [3]. The development of novel approaches to detect and localize MI pathology using the VCG signals is challenging in clinical study.

In recent years, various approaches have been developed to detect MI using VCG signals [12–15]. The methods based on the evaluation of various VCG signal morphological features, such as difference in ST-T vector magnitude, area of ST-segment vector, and other T-wave features, have been used to detect MI disease [14,16–18]. Similarly, in [13], authors have applied independent component analysis (ICA) and principal component analysis (PCA) for projecting VCG signal feature vector into a lower-dimensional space. They have extracted various morphological features from the VCG signal to formulate the feature vector. The neural network-based classifier has been used for the detection of MI using reduced dimension feature vector of VCG signal [13]. In [14], authors have computed octant and vector-based features from VCG signals and used a decision tree model to detect MI pathology. These methods require the detection of P, Q, R, S, T-onset points manually in the VCG signal to compute the morphological features [3]. In literature, various wavelet-based techniques, such as multi-scale recurrent quantification analysis (MRQA) [15], and complex wavelet sub-band features [3] have been used to detect MI using VCG signals. In [15], each channel of the VCG signal is decomposed into sub-band signals using discrete wavelet transform (DWT). From each sub-band signal, the recurrent quantification analysis (RQA) based non-linear features have been extracted, and Gaussian discriminant analysis (GDA) classifier is used for the detection of MI [15]. Moreover, in [3], the dual-tree complex wavelet transform (DT-CWT) has been used to decompose the VCG signal into sub-band signals along each channel. The entropy and L1-norm features have been extracted from each sub-band signal. The relevance vector machine (RVM) classifier has been used to detect MI from these VCG signals features [3]. Along with cardiac signal processing, cardiac imaging today represents an important area of clinical research that has achieved excellent results in recent years, such as deep-learning approaches [19]; especially, this led to the development of computer-assisted tools capable of segmenting the whole heart [20,21], as well as identifying specific regions of interest [22]. In the wavelet-based approach, the pre-defined basis functions and the number of decomposition levels are used to compute sub-band signals from VCG signal [23]. Additionally, the mentioned VCG signal-based approaches have considered only for MI detection. The automated classification of various types of MI pathologies has not been considered using VCG signals. The existing VCG-based approaches have considered various feature extraction and machine learning methods to detect MI. In recent years, various deep learning-based approaches have been used to detect and localize MI using 12-lead ECG signals [2,24,25]. The deep learning-based MI detection and localization methods do not require extracting features from 12-lead ECG signals [2]. The deep learning-based methods have not been explored for the detection and localization of MI using VCG signals. Therefore, a deep learning-based approach can be developed to detect and localize MI using VCG signals.

The multivariate variational mode decomposition (MVMD) is a recently proposed signal processing technique to decompose the multi-channel signals into components or modes [26]. This method is fully signal-driven and does not consider any basis functions and decomposition levels like DWT to obtain components of non-stationary signals. The univariate version of VMD has been used for the analysis of ECG signals for the detection of ventricular tachycardia and atrial fibrillation episodes [27,28]. The VCG is a multi-channel signal, and, therefore, the MVMD can be used to decompose the signal into modes. Moreover, deep learning-based methods have been used in the multi-scale or modal domain of ECG signals to detect cardiac ailments [29]. For VCG signal, the deep learning method has not been explored in the multi-scale domain to detect and localization of MI. The novelty of this work is to develop a multi-channel multi-scale deep learning-based framework to detect and localize MI using VCG signals. The important contributions of this work are given as follows:

1. The MVMD is introduced to decompose the VCG signals into sub-band signals or modes;
2. A multi-channel multi-scale two-stage deep convolutional neural network (CNN) framework is proposed for the detection and localization of MI;
3. The MI types, such as AMI, IMI, ILMI, ALMI, ASMI, and IPLMI, are classified in the second stage of the proposed multi-channel multi-scale deep CNN (MMD-CNN) model;
4. The multi-channel multi-scale two-stage deep CNN performance is evaluated using hold-out and 10-fold cross-validation (CV) schemes.

The remaining sections of this paper are written as follows. The explanation regarding the VCG signal database is written in Section 2. In Section 3, the proposed approach for MI detection and localization is described. Section 4 presents the results and discussion of the proposed approach. In Section 5, conclusions of this paper is summarized.

2. VCG Signal Database

In this work, the VCG signals from the PTB diagnostic database (https://www.physionet.org/content/ptbdb/1.0.0/ (accessed on 20 June 2021)). were used to develop the proposed multi-channel multi-scale two-stage deep CNN approach [30,31]. The PTB database from Physionet comprises both 12-lead ECG and 3-lead VCG recordings of normal sinus rhythm (NSR) and various heart diseases, such as MI, hypertrophy, cardiomyopathy, bundle branch block, and dysrhythmia, respectively [30]. Each VCG signal has been sampled at 1000 samples per second in the PTB database, and the amplitude value of each lead VCG varied between −16.384 mV to 16.384 mV. In this study, we have used 73 VCG recordings from 52 healthy controls (HC) subjects of PTB diagnostic database. Similarly, 99 VCG recordings from 148 subjects with MI pathology are used. For MI localization, 13, 20, 11, 21, 21, and 13 VCG recordings from AMI, IMI, ALMI, ASMI, ILMI, and IPLMI classes, respectively, are considered. In the PTB diagnostic database [30,31], the number of VCG recordings for MI class is higher than the healthy class. A higher difference in the number of VCG instances between MI and healthy classes may cause the over-fitting problem during the training of the proposed MMDCNN model. Due to this reason, we have considered only 99 VCG recordings from the MI class in this work. Each VCG recording in the PTB diagnostic database contains three orthogonal leads (V_x, V_y, V_z), which represent the electrical activity of heart in three different planes [28].

3. Method

The proposed MI detection and localization stages are shown in a flow-chart form in Figure 1a,b, respectively. The MI detection stage comprises the filtering of VCG signal, segmentation of VCG recordings into beats, decomposition of VCG beat into multi-scale VCG tensors using MVMD, and deep CNN to detect MI pathology. Similarly, the localization stage consists of the classification of AMI, IMI, ALMI, ASMI, ILMI, and IPLMI beats using MI detected multi scale VCG tensor data. The following section briefly discuss each part of the flow-chart, as shown in Figure 1.

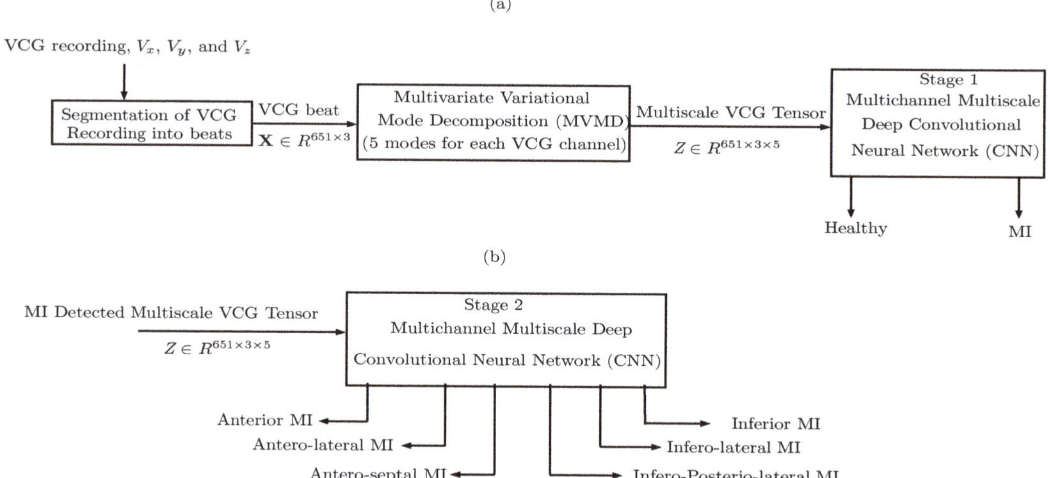

Figure 1. (**a**) Stagel 1: MI detection block using VCG signals. (**b**) Stage 2: MI localization using MI detected multi-scale VCG tensor.

3.1. Segmentation of VCG Data

In this work, we have performed the amplitude normalization for each lead of VCG recording [28]. The samples of raw VCG signal corresponding to each lead is divided by the maximum absolute value the signal to obtain the normalized VCG signal [6]. After normalization of each lead VCG recording, we have detected the R-peak in the V_x lead of VCG signals. The beat by beat segmentation of each VCG recording is performed using a window of size 651 samples [2]. The 251 samples before each R-peak of the V_x lead VCG signal and 400 samples after R-peak are considered for the beat by beat segmentation of VCG signal [2]. The number of MI and NSR VCG beats used for the proposed MI detection work are shown in Table 1. Similarly, the number of VCG beats evaluated for each type of MI are also shown in Table 1.

Table 1. Number of VCG beats used for MI detection and localization.

		Stage 1: MI detection				
Class		NSR			MI	
VCG beats		9874			13982	
		Stage 2: MI localization				
Class	AMI	ALMI	ASMI	IMI	ILMI	IPLMI
VCG beats	1664	1778	3057	2807	3049	1981

3.2. Multivariate VMD for VCG Signal Analysis

In this study, we have used MVMD to evaluate the modes of VCG beat along each orthogonal lead. The MVMD is the extension of VMD algorithm used to decompose multichannel signals into modes [26]. The VCG beat is given as $v_m(n)$, with $n = 1, 2, ...N$. N is the number of samples in the VCG beat. The parameter m is denoted as the mth orthogonal lead of VCG beat. The VCG beat synthesized from its modes is given as follows:

$$v_m(n) = \sum_{i=1}^{k} u_m^i(n) \tag{1}$$

where, $u_m^i(n)$ is the ith mode of mth lead VCG signal $u_m^i(n) = \left[u_1^i(n), u_2^i(n), u_3^i(n)\right]$ is also interpreted as the multivariate modulated oscillations of VCG signal with $i = 1, 2, ...k$, and k is the total number of modes [26]. The vector analytic representation of ith mode of mth lead VCG is written as follows [26]:

$$\tilde{u}_m^i(n) = u_m^i(n) + j\mathrm{H}\left(u_m^i(n)\right) \tag{2}$$

where, $\mathrm{H}(u_m^i(n))$ is the Hilbert transform of ith mode of mth lead VCG signal [26]. In MVMD, the objective is to evaluate the modes of VCG signal based on the criteria as (a) the sum of bandwidth of components or modes of VCG should be minimum and (b) sum of all modes should recover the VCG signal along each lead [26]. The optimization problem of MVMD for the decomposition of VCG signal is formulated as follows [26]:

$$\begin{aligned} &\min_{u_m^i(n), w^i} \left\{ \sum_{i=1}^{k} \sum_{m=1}^{M} \left\| \frac{\partial}{\partial n} \left[\tilde{u}_m^i(n) e^{-jw^i n} \right] \right\|_F^2 \right\} \\ &\text{s.t.} \quad \sum_{i=1}^{k} u_m^i(n) = v_m(n), \quad m = 1, 2, \text{and } 3 \end{aligned} \tag{3}$$

where, $\|\bullet\|_F$ is the representation of Frobenious norm [26]. The optimization problem in Equation (3) can be reformulated using augmented Lagrangian and it is given as follows:

$$L\left\{u_m^i(n), w^i, \eta_m(n)\right\} = \beta \sum_{i=1}^{k} \sum_{m=1}^{M} \left\| \frac{\partial}{\partial n} \left[u_m^i(n) e^{-jw^i n} \right] \right\|_F^2 + \sum_{m=1}^{M} \left\| v_m(n) - \sum_{i=1}^{k} u_m^i(n) \right\|_F^2 + \sum_{m=1}^{M} \left\langle \eta_m(n), v_m(n) - \sum_{i=1}^{k} u_m^i(n) \right\rangle \tag{4}$$

where, $\eta_m(n)$ is the Lagrangian multiplier for mth lead VCG beat, and β is interpreted as the penalty factor for MVMD. The modes of VCG beat along each lead is iteratively evaluated based on the solution of Equation (4) using alternating direction method of multipliers (ADMM) [26]. The complete algorithm of MVMD for the extraction of modes from the non-stationary signals has been given in [26]. In this study, we have evaluated five modes from the VCG beat along each orthogonal lead. The multi-scale VCG tensor is formulated using the modes of VCG beat and the size of multi-scale VCG tensor is $651 \times 3 \times 5$.

For NSR class, the V_x, V_y, and V_z lead VCG beat are shown in Figure 2a,g,m, respectively. The modes of V_x, V_y, and V_z lead VCG beats evaluated using MVMD are shown Figure 2b–f,h–l,n–r, respectively. Similarly, the V_x, V_y, and V_z channel VCG beats for IPLMI class are shown in Figure 3a,g,m, respectively. For IPLMI class, the modes of V_x, V_y, and V_z lead VCG beats are depicted in Figure 3b–f,h–l,n–r, respectively. It can be observed from these plots that the modes of each lead VCG beat have different shape and amplitude values for IPLMI and NSR classes. In VCG signal, the clinical parameters, such as QRS-complex shape, special QRS-T angle, T-wave shape are different for healthy and MI cases [32]. The study in [33] has reported the physiological parameters of VCG signal for MI class, such as QRS-loop maximum vector magnitude, QRS-area perimeter ratio, and ST-vector magnitude, have higher mean values than those of healthy class. Similarly, the VCG parameters, such as QRS-loop volume, QRS-loop planar area, maximum of the distance between QRS-centroid and QRS-loop, and QRS-perimeter have the lowest mean values for MI class as compared to healthy class [33]. For the AMI case, there is abnormal posterior deviation in the QRS-vector of VCG signal [34]. Similarly, for the posterior-lateral MI case, the pathological changes, such as oriented T-loop and maximal leftward deviation of frontal plane QRS-vector are observed [35]. The transverse plane QRS-vector maximum value greater than 1.5 mV is also used as the criteria for the detection of inferior and posterior MI using VCG signals [11]. These differences in the morphological parameters of VCG signal for NSR and various types of MI cases can be captured in the modes which are evaluated using MVMD. Therefore, the deep CNN model designed using the modes of the VCG beat can be used to detect and localize MI.

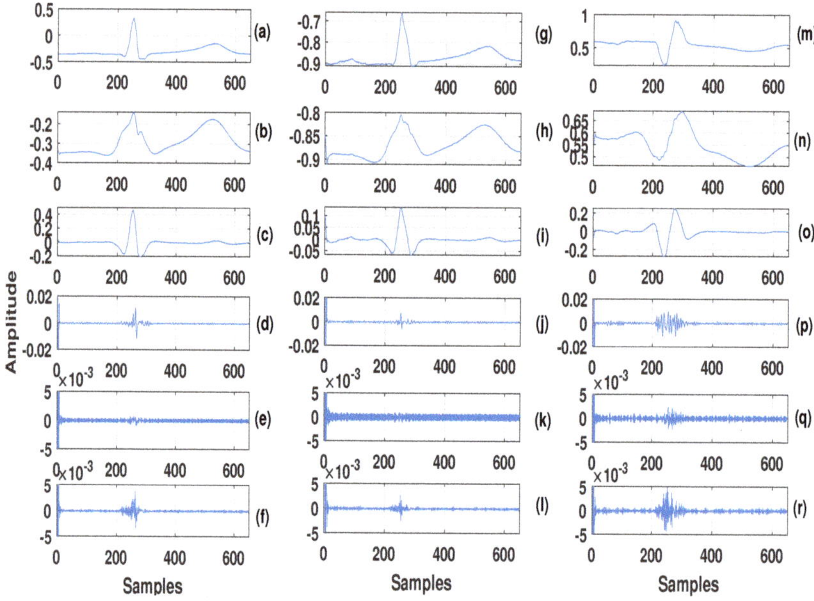

Figure 2. (**a**) V_x lead VCG signal for NSR class. (**b–f**) mode 1 to 5 of V_x lead VCG signal for NSR class. (**g**) V_y lead VCG signal for NSR class. (**h–l**) mode 1 to 5 of V_y lead VCG signal for NSR class. (**m**) V_z lead VCG signal for NSR class. (**n–r**) mode 1 to 5 of V_z lead VCG signal for NSR class.

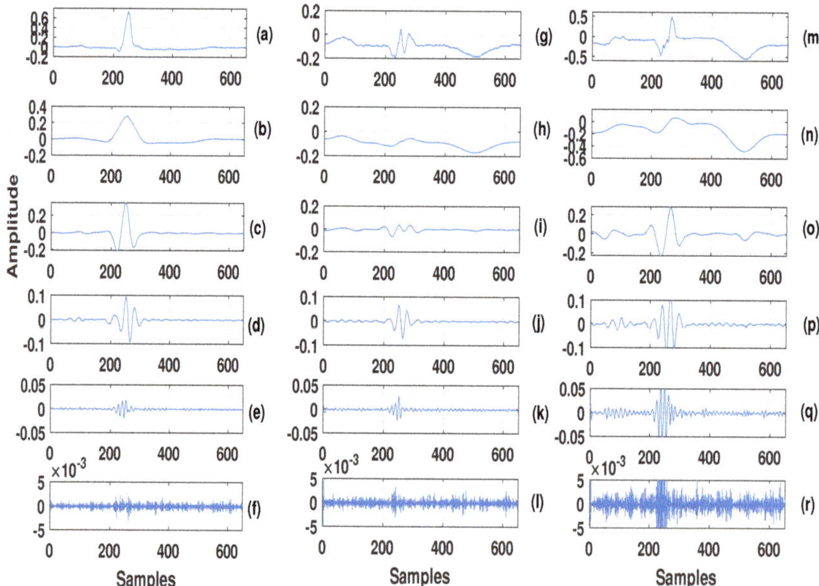

Figure 3. (**a**) V_x lead VCG signal for IPLMI class. (**b–f**) mode 1 to 5 of V_x lead VCG signal for IPLMI class. (**g**) V_y lead VCG signal for IPLMI class. (**h–l**) mode 1 to 5 of V_y lead VCG signal for IPLMI class. (**m**) V_z lead VCG signal for IPLMI class. (**n–r**) mode 1 to 5 of V_z lead VCG signal for IPLMI class.

3.3. Multi-Channel Multi-Scale Deep Convolutional Neural Network

In this work, a novel MMDCNN model is proposed to detect and localize MI. The Python codes for the MMDCNN model is available at (https://github.com/JayKarhade/MI_VCG_DL (accessed on 20 August 2021)). The MMDCNN architecture shown in Figure 4 comprises 12 layers. The first and last layers are interpreted as input and output layers of MMDCNN model. The input layer contains the multi-scale VCG tensor. The output layer consists of two neurons for MI detection stage, one for NSR class and the other for MI class. Similarly, for the MI localization stage, the output layer contains six neurons corresponding to six types of MI classes as AMI, IMI, ALMI, ASMI, ILMI, and IPLMI, respectively. The MMDCNN contains four convolutions, two max-pooling, and four dense layers for both MI detection and localization stages. The mathematical expression to compute the tth feature map for first convolution layer is given as follows [29,36]:

$$\mathbf{X}_t^{(l)}(\widetilde{n}) = h(\sum_{n=1}^{N}\sum_{m=1}^{M}\sum_{i=1}^{I}\mathbf{X}(n,m,i)\mathbf{K}_t(\widetilde{n}-n+\frac{N}{2},m,i)+b_t) \tag{5}$$

where $\mathbf{X}(n,m,i)$ is the input to the MMDCNN and $i = 1,2\ldots\ldots I$ and $m = 1,2\ldots M$, respectively. The parameters I and M are total number of modes and channels, respectively. Similarly, the mathematical expression for the evaluation of feature maps in other convolution layers are evaluated as follows [29,36]:

$$\mathbf{X}_t^{(l)}(\widetilde{n}) = h(\sum_{n=1}^{N}\sum_{c=1}^{C}\mathbf{X}_{\widetilde{t}}^{(l-1)}(n,c)\widetilde{\mathbf{K}}_t(\widetilde{n}-n+\frac{N}{2},c)+\widetilde{b}_t) \tag{6}$$

$\mathbf{X}_{\widetilde{t}}^{(l-1)}(n,c)$ is the $\widetilde{t}$th feature map at $(l-1)$th convolution layer. Similarly, the feature maps for second, third and fourth convolution layers are evaluated using Equation (6). The $\mathbf{X}_t^{(l)}$ is denoted as the tth feature map for lth convolution layer. Moreover, the mathematical expression to evaluate the pooling layer feature map is given as follows [29,36]:

$$\mathbf{X}_t^{(l)}(\widetilde{n}) = \text{max-pooling}(\mathbf{X}_t^{(l-1)}(\widetilde{n})) \tag{7}$$

For dense layers, the feature vector is evaluated as follows [37]:

$$\mathbf{a}^{(l)} = h(\mathbf{a}^{(l-1)}\overline{\mathbf{W}}^{(l)} + \overline{b}^{(l)}) \tag{8}$$

where $\mathbf{a}^{(l)}$ is the feature vector for lth dense layer. $\overline{\mathbf{W}}^{(l)}$ is the weight matrix between $(l-1)$th dense and lth dense layers [37]. $\overline{b}^{(l)}$ is the bias for lth dense layer. The categorical cross-entropy-based cost function is used for MMDCNN for both detection and localization stages [38]. The hyper-parameters used for MMDCNN in detection and localization stages are shown in Table 2. In this study, for both MI detection and localization stages, the hold-out validation and 10-fold cross-validation (CV) methods [37] are used to select the training and test VCG beats. For hold-out validation 78.75%, 11.25%, and 10% VCG beats are used as training, validation, and testing, respectively, for MMDCNN model during detection and localization phases. We have used the performance measures such as accuracy, sensitivity, specificity, and Kappa scores for the MI detection using MMDCNN classifier [37,39]. Similarly, for MI localization, the overall accuracy (OA), individual accuracy (IA), and Kappa score are used to evaluate the performance in the second stage MMDCNN [6].

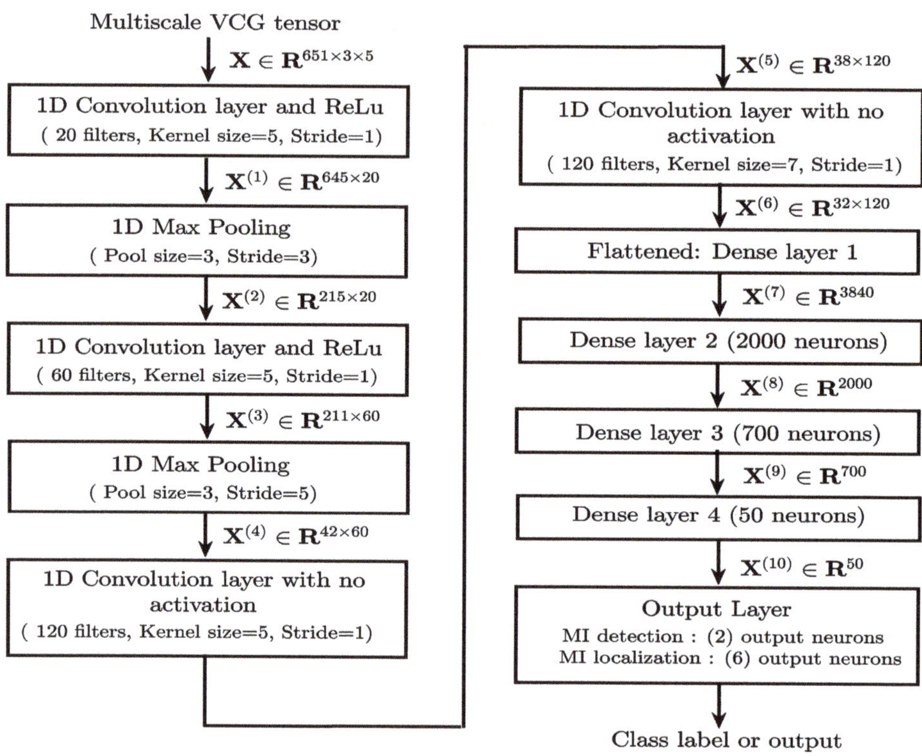

Figure 4. Proposed MMDCNN model to detect and localize MI using VCG beats.

Table 2. Hyper-parameters used MI detection and localization using our proposed MMDCNN model.

		Hold-out (MI Detection)			10-fold (MI Detection)		
Parameters	Optimizer	Batch size	Epochs	Learning rate	Batch size	Epochs	Learning rate
Values	Adam	1024	15	0.0001	1024	15	0.0001
		Hold-out (MI Localization)			10-fold (MI Localization)		
Parameters	Optimizer	Batch size	Epochs	Learning rate	Batch size	Epochs	Learning rate
Values	Adam	1024	15	0.00004	256	15	0.0001

4. Results and Discussions

The results evaluated using the proposed MMDCNN for MI detection and localization using VCG signals are shown in this section. In Table 3, we have shown the accuracy, sensitivity, specificity, and kappa score values for our proposed MMDCNN model with hold-out CV. Similarly, for MI detection, the accuracy vs. epoch plots for training and validation VCG instances obtained using MMDCNN are illustrated in Figure 5. It is evident from this plot that both training and validation accuracy values are 100% after 10th epoch. Similarly, we have shown the confusion matrix obtained using the proposed MMDCNN for MI detection using VCG signals for one random hold-out trial in Table 4. The number of false-positive and false-negative values are 1 in the confusion matrix table. The accuracy, sensitivity, specificity, and kappa values for this random hold-out validation are 99.9%, 99.89%, 99.92%, and 0.998, respectively. The average values of accuracy, sensitivity, specificity, and kappa scores over five trial-based random validation are more than 99% (as seen from Table 3).

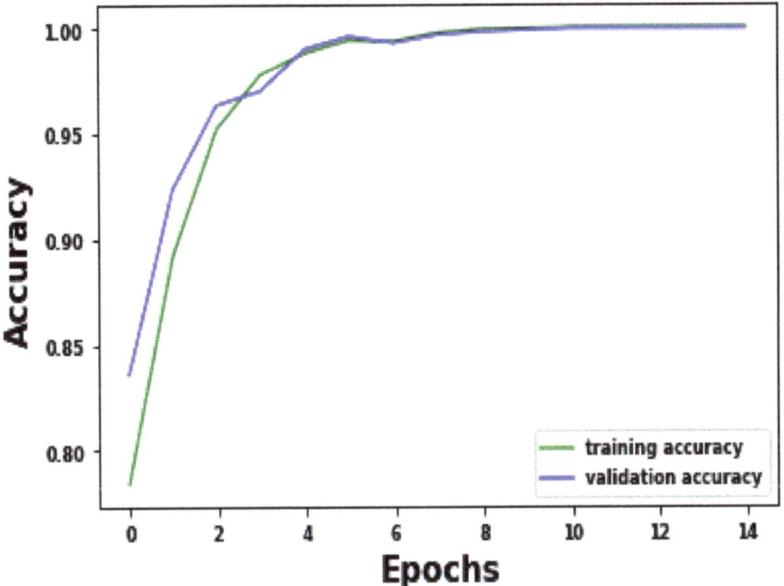

Figure 5. Graphs of accuracy vs. epochs of multi-channel multi-scale deep CNN obtained using training and validation VCG instances for MI detection.

Table 3. Classification results of two class multi-channel multi-scale deep CNN obtained for the detection of MI using hold-out validation.

Accuracy (%)	Sensitivity (%)	Specificity (%)	Kappa
99.58 ± 0.38	99.18 ± 0.90	99.87 ± 0.07	0.990 ± 0.01

Table 4. Confusion matrix for one-trail of hold-out CV for MI detection.

		Predicted	
		Healthy	**MI**
Actual	Healthy	978	1
	MI	1	1390

For MI detection, the classification results obtained for the proposed first stage MMDCNN using 10-fold CV are shown in Table 5. It can be observed from this table that, the accuracy values are more than 99.50% for each fold. Similar high percentages in the sensitivity and specificity are seen in each fold using the first stage MMDCNN method for MI detection. It can also be observed that the Cohen kappa score is more than 0.99 for each fold. From these 10-fold CV results, It can be noted that the proposed first stage deep CNN successfully detected MI using the modes of VCG beats.

Table 5. Results obtained using multi-channel multi-scale deep CNN with 10-fold CV.

Folds	1	2	3	4	5	6	7	8	9	10	Value ($\mu \pm \sigma$)
Accuracy (%)	99.7	100	99.95	100	99.74	99.95	100	100	100	100	99.93 ± 0.11
Sensitivity (%)	100	100	100	100	99.38	100	100	100	100	100	99.93 ± 0.19
Specificity (%)	99.49	100	99.99	100	100	99.92	100	100	100	100	99.94 ± 0.16
Kappa	0.993	1	0.999	1	0.994	0.999	1	1	1	1	0.998 ± 0.002

The confusion matrix obtained using one random trial-based hold-out validation for MI localization with second stage MMDCNN is shown in Table 6. Similarly, we have shown the accuracy vs. epoch plots for training and validation of multi-scale VCG tensor instances in Figure 6. It can be observed from these plots that both training and validation accuracy values obtained are more than 99% after 10th epoch using the second stage MMDCNN model. It can be seen from Table 6 that the number of true positives for AMI, IMI, ALMI, ASMI, ILMI, and IPLMI classes are obtained as 162, 284, 185, 287, 301, and 201, respectively. Three multi-scale VCG tensor instances, which belong to IMI, are classified as ALMI class. Similarly, the classification results of the proposed second-stage MMDCNN obtained for MI localization using hold-out validation are shown in Table 7. It can be noted that the average IA values are more than 99% for AMI, IMLI, ALMI, ILMI, and IPLMI classes. For ASMI class, the IA value is 94.38%. The OA and kappa values obtained are 98.77% and 0.982, respectively, using the proposed second-stage MMDCNN model.

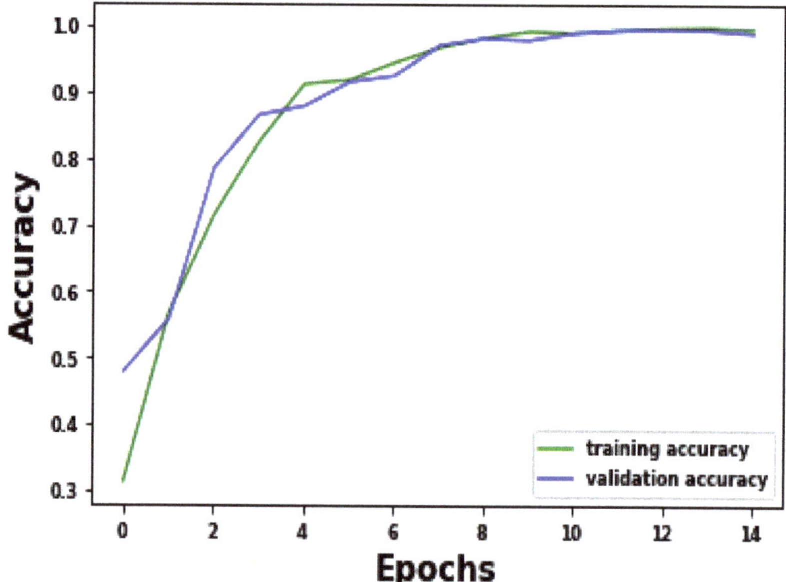

Figure 6. Plots of accuracy vs. epochs of multi-channel multi-scale deep CNN obtained using training and validation VCG instances for MI localization.

Table 6. Confusion matrix obtained using the proposed second stage multi-channel multi-scale deep CNN classifier for MI localization.

		Predicted Classes					
		AMI	ALMI	ASMI	IMI	ILMI	IPLMI
Actual Classes	AMI	162	0	0	0	0	0
	ALMI	0	185	0	0	1	0
	ASMI	0	0	287	0	0	0
	IMI	0	3	0	284	0	0
	ILMI	0	0	0	0	301	1
	IPLMI	0	0	0	0	0	200

Table 7. Classification results obtained for MI localization using proposed MMDCNN model with hold-out validation.

Parameters	Value ($\mu \pm \sigma$)
IA_{AMI} (%)	99.79 ± 0.35
IA_{ALMI} (%)	99.64 ± 0.31
IA_{ASMI} (%)	94.38 ± 5.56
IA_{IMI} (%)	99.53 ± 0.53
IA_{ILMI} (%)	99.66 ± 0.00
IA_{IPLMI} (%)	99.66 ± 0.57
OA (%)	98.77 ± 0.96
Kappa	0.982 ± 0.014

Moreover, we have shown the classification results of second-stage MMDCNN for MI localization using a 10-fold CV and these results are shown in Table 8. It can be observed from these results that for ILMI class, the accuracy value of each fold is more than 99%. Similarly, for IMI class, apart from 5th fold, more than 99% accuracy values are observed for other folds. For IPLMI and AMI classes, more than 98% accuracy values are obtained in each fold using second stage MMDCNN model. Similarly, more than 97% accuracy values are obtained using MMDCNN classifier for ASMI and and ALMI classes. The overall accuracy (OA) values are obtained as more than 99% at each fold. The kappa value of more than 0.97 is observed for each fold using MMDCNN classifier.

Table 8. Classification results obtained for MI localization using proposed MMDCNN model with 10-fold CV.

Folds	1	2	3	4	5	6	7	8	9	10	$\mu \pm \sigma$
IA_{AMI} (%)	100	98.79	100	99.4	100	100	100	100	100	100	99.81 ± 0.40
IA_{ALMI} (%)	99.43	97.75	97.75	96.57	99.43	98.31	97.19	99.43	97.75	98.3	98.19 ± 0.99
IA_{ASMI} (%)	97.38	95.09	100	97.37	100	100	99.67	98.69	99.01	100	98.72 ± 1.64
IA_{IMI} (%)	100	99.64	100	99.28	98.93	99.64	98.57	100	99.64	99.28	99.49 ± 0.48
IA_{ILMI} (%)	99.67	99.66	99.34	99.67	99.01	99.67	99.01	99.67	100	100	99.57 ± 0.34
IA_{IPLMI} (%)	99.49	100	98.98	98.48	99.49	99.49	98.48	100	98.98	98.48	99.18 ± 0.59
OA(%)	99.72	99.09	99.44	99.23	99.37	99.65	98.88	99.58	99.37	99.44	99.37 ± 0.25
Kappa	0.990	0.978	0.993	0.979	0.993	0.994	0.986	0.994	0.991	0.993	0.989 ± 0.006

The classification results of MMDCNN models evaluated using the selected modes of each lead VCG signal, and all modes of high-pass filtered VCG signals for MI detection with hold-out validation are shown in Table 9. It is observed that the average accuracy value of MMDCNN is 99.58% using mode 1 and mode 2 of each lead VCG signal. The average accuracy value remains the same as the accuracy of MMDCNN model using all modes of VCG signals for MI detection. Mode 1 and mode 2 capture the significant information of the VCG signal after decomposition using MVMD. Henceforth, the accuracy value remains the same for MI detection using selected modes and the MMDCNN classifier. Moreover, we have also evaluated the classification performance of the MMDCNN model using all modes of high-pass filtered VCG signal for MI detection. A high-pass Butterworth filter with a cut-off frequency of 0.5 Hz is applied to each lead VCG signal to remove baseline wondering artifacts [6,28]. It is observed from Table 9 that average accuracy, average kappa score, average sensitivity, and average specificity values are improved after the filtering of baseline wandering artifact from VCG signals. In Table 10, we have shown the individual accuracy value for each MI class, OA, and kappa scores of MMDCNN classifier for MI localization using mode 1 and mode 2 of each lead VCG signal and all modes of high-pass filtered VCG signals, respectively. It is observed that the OA value obtained using the MMDCNN model is less using mode 1 and mode 2 of VCG signals as compared to all modes of VCG signals. Similarly, the OA and kappa values are improved using the modes of high-pass filtered VCG signals with the MMDCNN classifier. For MI localization, the IA values for ASMI, IMI, and ILMI classes are also improved using the modes of high-pass filtered VCG signals composed with the MMDCNN classifier. Moreover, we have also evaluated the classification results of the MMDCNN classifier using all modes of VCG signal with leave one out (LOO) CV strategy. The VCG beats of one recording are considered during testing of the MMDCNN model, whereas the VCG beats of all other VCG recordings are used to train the MMDCNN classifier. The same procedure is applied to all VCG recordings, and it can also be interpreted as a 171-fold CV strategy. The LOO CV or pre-recording-based MI detection results are shown in Figure 7. It is observed that out of 172 VCG recordings, 114 recordings are correctly classified with 100% accuracy. The OA value obtained using MMDCNN classifier with LOO CV strategy is 87.65%.

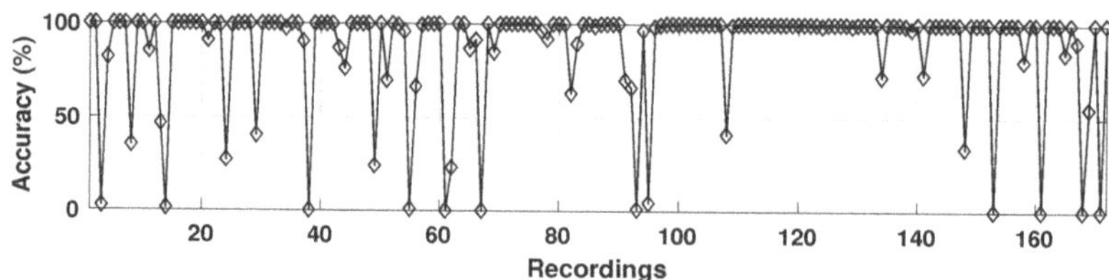

Figure 7. Accuracy values obtained using the proposed MMDCNN classifier for MI detection with leave one out CV strategy.

Table 9. Classification results of MMDCNN obtained using selected modes and all modes for the detection of MI.

Mode Selection	Accuracy (%)	Sensitivity (%)	Specificity (%)	Kappa
Mode 1 and mode 2 of each lead VCG signals	99.58 ± 0.37	99.17 ± 0.89	99.86 ± 0.06	0.991 ± 0.007
All modes from high-pass filtered VCG signals	99.92 ± 0.02	99.84 ± 0.05	100 ± 0	0.998 ± 0.0005

Table 10. Classification results obtained for MI localization using proposed MMDCNN model with mode selection.

Parameters	Mode 1 and Mode 2 from Each Lead VCG Signals	All Modes from High-Pass Filtered VCG Signals
IA_{AMI} (%)	100 ± 0	100 ± 0
IA_{ALMI} (%)	99.17 ± 0.29	98.55 ± 2.22
IA_{ASMI} (%)	93.09 ± 5.16	99.26 ± 0.27
IA_{IMI} (%)	99.30 ± 0.28	100 ± 0
IA_{ILMI} (%)	99.50 ± 0.19	99.86 ± 0.14
IA_{IPLMI} (%)	99.25 ± 0.50	98.45 ± 2.45
OA (%)	98.37 ± 1.43	99.44 ± 0.56
Kappa	0.976 ± 0.014	0.992 ± 0.006

We have also formulated the seven-class classification scheme as (Healthy vs. AMI vs. ALMI vs. ASMI vs. IMI vs. ILMI vs. IPLMI) using MMDCNN classifier with all modes of VCG signals. The seven class classification results obtained using the MMDCNN model are shown in Table 11. It can be observed that for healthy, ALMI, IMI, and IPLMI classes, the IA values are 69.87%, 83.22%, 82.61%, and 41.03%, respectively. The OA value of MMDCNN classifier obtained is 81.48%, which is less than the proposed two-stage MMDCNN model for MI detection and localization.

Table 11. Classification results obtained using MMDCNN classifier for seven class classification scheme with hold-out validation.

Parameters	Value ($\mu \pm \sigma$)
Healthy (%)	69.87 ± 28.87
IA_{AMI} (%)	99.46 ± 0.38
IA_{ALMI} (%)	83.22 ± 10.14
IA_{ASMI} (%)	99.63 ± 0.25
IA_{IMI} (%)	82.61 ± 7.93
IA_{ILMI} (%)	98.44 ± 2.28
IA_{IPLMI} (%)	41.03 ± 18.18
OA (%)	81.48 ± 0.63
Kappa	0.777 ± 0.008

The classification performance of the proposed first stage MMDCNN classifier is compared with the existing techniques for MI detection using VCG signals with a 10-fold CV-based technique. The comparison results are shown in Table 12. The work reported in [15] has computed features from each lead of VCG signal using multi-scale recurrent quantification analysis (MRQA). The Gaussian discriminant analysis (GDA) based classification model has been used to detect MI using MRQA based VCG features. The sensitivity and specificity values of 96.50% and 75% have been obtained in their work. Similarly, in [14], the combination of octant and vector-based features have been obtained using VCG signal. The classification and regression tree (CART) based model has been used for the detection of MI. The classification performance, such as the sensitivity and specificity values of 97.28% and 96%, respectively, are reported. The complex wavelet sub-band features of VCG coupled with the RVM classifier have obtained the sensitivity and specificity values of 98.40%, and 98.66%, respectively, for MI detection [28]. The proposed MMDCNN model has obtained better classification performance than the existing machine learning-based methods for MI detection using VCG signals. The advantages of our proposed MMDCNN based approach are given as follows:

- A novel two-stage based MMDCNN model is proposed to detect and localize MI using VCG beats;
- The multi-scale analysis of VCG signal is performed using MVMD based multi-variate signal driven approach;
- The approach has demonstrated more than 99% accuracy for MI detection;
- The extraction of raw features from VCG signals are not required using the proposed approach for both detection and localization stages;
- The second stage MMDCNN model successfully classified six types of MI with an accuracy of more than 99%.

Table 12. Comparison of proposed MI detection approach with existing methods obtained using VCG signals (with 10-fold CV).

Authors	Features Extracted	Classifiers Used	Sensitivity (%)	Specificity (%)
Yang et al., 2012 [14]	Octant and vector features evaluated from VCG signal	CART	97.28	95
Yang, 2011 [15]	DWT domain RQA features from VCG Signal	GDA	96.50	75
Tripathy et al., 2017 [28]	Complex wavelet sub-band features from VCG signal	RVM	98.40	98.66
Proposed work	Multi-channel and multi-scale domain learnable features	CNN	99.93	99.94

In this work, the proposed approach has considered only 99 VCG recordings from different MI classes in the second stage for MI localization. The approach can be tested using VCG recordings from a huge database containing more subjects. The MVMD based multi-scale approach is used in this study to decompose the VCG signal. The other multi-scale analysis methods, such as multivariate empirical mode decomposition (MEMD) [40], multivariate projection based empirical wavelet transform (MPEWT) [41], and fast and adaptive based MEMD [42] can be used for the decomposition of VCG signals.

5. Conclusions

The multi-channel multi-scale two-stage deep CNN model is proposed to detect and localize MI using VCG signals. The MVMD is used to decompose the VCG beat into modes along with each orthogonal lead. The multi-channel multi-scale VCG tensor has been formulated and used as input to the deep CNN model to detect and localize MI. For MI detection, the proposed first-stage MMDCNN model obtained an average accuracy value of 99.93% with 10-fold CV. The second-stage MMDCNN model produced an average overall accuracy (OA) value of 99.37% for MI localization. The average OA values are more than 99% for AMI, IMI, ILMI, and ILMI classes. The proposed first-stage MMDCNN classifier obtained a higher accuracy value than the existing VCG based approaches for MI detection. The MMDCNN model can also be explored to detect other cardiac ailments, such as atrial fibrillation, hypertrophy, cardiomyopathy, ventricular arrhythmia, and bundle branch block using VCG signals.

Author Contributions: Conceptualization, J.K.; Data curation, R.K.T. and S.K.G.; Formal analysis, J.K. and U.R.A.; Methodology, P.G. and R.K.T.; Project administration, U.R.A.; Resources, R.K.T.; Supervision, R.K.T.; Writing—original draft, R.K.T., P.G.; Writing—review and editing, U.R.A. All authors have read and agreed to the published version of the manuscript.

Funding: This research received no external funding.

Institutional Review Board Statement: Not applicable.

Informed Consent Statement: Not applicable.

Data Availability Statement: The codes of the work is available at (https://github.com/JayKarhade/MI_VCG_DL (accessed on 20 August 2021)).

Conflicts of Interest: The authors declare no conflict of interest.

References

1. Antman, E.M.; Anbe, D.T.; Armstrong, P.W.; Bates, E.R.; Green, L.A.; Hand, M.; Hochman, J.S.; Krumholz, H.M.; Kushner, F.G.; Lamas, G.A.; et al. ACC/AHA guidelines for the management of patients with ST-elevation myocardial infarction: A report of the American College of Cardiology/American Heart Association Task Force on Practice Guidelines (Committee to Revise the 1999 Guidelines for the Management of Patients with Acute Myocardial Infarction). *J. Am. Coll. Cardiol.* **2004**, *44*, E1–E211.
2. Tripathy, R.K.; Bhattacharyya, A.; Pachori, R.B. Localization of myocardial infarction from multi-lead ECG signals using multiscale analysis and convolutional neural network. *IEEE Sens. J.* **2019**, *19*, 11437–11448. [CrossRef]
3. Tripathy, R.; Dandapat, S. Detection of myocardial infarction from vectorcardiogram using relevance vector machine. *Signal Image Video Process.* **2017**, *11*, 1139–1146. [CrossRef]
4. Hedén, B.; Ohlin, H.; Rittner, R.; Edenbrandt, L. Acute myocardial infarction detected in the 12-lead ECG by artificial neural networks. *Circulation* **1997**, *96*, 1798–1802. [CrossRef]
5. Boateng, S.; Sanborn, T. Acute myocardial infarction. *Disease-a-Month* **2013**, *59*, 83–96. [CrossRef]
6. Sharma, L.; Tripathy, R.; Dandapat, S. Multiscale energy and eigenspace approach to detection and localization of myocardial infarction. *IEEE Trans. Biomed. Eng.* **2015**, *62*, 1827–1837. [CrossRef]
7. Khan, J.N.; Chauhan, A.; Mozdiak, E.; Khan, J.M.; Varma, C. Posterior myocardial infarction: Are we failing to diagnose this? *Emerg. Med. J.* **2012**, *29*, 15–18. [CrossRef]
8. Parale, G.; Kulkarni, P.; Khade, S.; Athawale, S.; Vora, A. Importance of reciprocal leads in acute myocardial infarction. *JAPI* **2004**, *52*, 376–379.
9. Prabhakararao, E.; Dandapat, S. Automated detection of posterior myocardial infarction from VCG signals using stationary wavelet transform based features. *IEEE Sens. Lett.* **2020**, *4*, 1–4. [CrossRef]
10. Schreck, D.M.; Fishberg, R.D. Derivation of the 12-lead electrocardiogram and 3-lead vectorcardiogram. *Am. J. Emerg. Med.* **2013**, *31*, 1183–1190. [CrossRef]

11. Loperfido, F.; Digaetano, A.; Guccione, P.; Desantis, F.; Vigna, C.; Laurenzi, F.; Solfanelli, N.; Ferrazza, A.; Pennestri, F.; Manzoli, U. Assessment of left ventricular hypertrophy by ECG and VCG in patients with inferior and posterior myocardial infarction. A comparison with echocardiographic data. *J. Electrocardiol.* **1986**, *19*, 247–256. [CrossRef]
12. Bortolan, G.; Christov, I. Myocardial infarction and ischemia characterization from T-loop morphology in VCG. In Proceedings of the Computers in Cardiology 2001, Rotterdam, The Netherlands, 23–26 September 2001; Volume 28, pp. 633–636
13. Dehnavi, A.R.M.; Farahabadi, I.; Rabbani, H.; Farahabadi, A.; Mahjoob, M.P.; Dehnavi, N.R. Detection and classification of cardiac ischemia using vectorcardiogram signal via neural network. *J. Res. Med. Sci. Off. J. Isfahan Univ. Med Sci.* **2011**, *16*, 136.
14. Yang, H.; Bukkapatnam, S.T.; Le, T.; Komanduri, R. Identification of myocardial infarction (MI) using spatio-temporal heart dynamics. *Med. Eng. Phys.* **2012**, *34*, 485–497. [CrossRef] [PubMed]
15. Yang, H. Multiscale recurrence quantification analysis of spatial cardiac vectorcardiogram signals. *IEEE Trans. Biomed. Eng.* **2010**, *58*, 339–347. [CrossRef] [PubMed]
16. Correa, R.; Arini, P.D.; Correa, L.; Valentinuzzi, M.E.; Laciar, E. Acute myocardial ischemia monitoring before and during angioplasty by a novel vectorcardiographic parameter set. *J. Electrocardiol.* **2013**, *46*, 635–643. [CrossRef] [PubMed]
17. Correa, R.; Arini, P.D.; Valentinuzzi, M.E.; Laciar, E. Novel set of vectorcardiographic parameters for the identification of ischemic patients. *Med. Eng. Phys.* **2013**, *35*, 16–22. [CrossRef] [PubMed]
18. Correa, R.; Arini, P.D.; Correa, L.S.; Valentinuzzi, M.; Laciar, E. Novel technique for ST-T interval characterization in patients with acute myocardial ischemia. *Comput. Biol. Med.* **2014**, *50*, 49–55. [CrossRef]
19. Chen, C.; Qin, C.; Qiu, H.; Tarroni, G.; Duan, J.; Bai, W.; Rueckert, D. Deep learning for cardiac image segmentation: A review. *Front. Cardiovasc. Med.* **2020**, *7*, 25. [CrossRef]
20. Oktay, O.; Ferrante, E.; Kamnitsas, K.; Heinrich, M.; Bai, W.; Caballero, J.; Cook, S.A.; De Marvao, A.; Dawes, T.; O'Regan, D.P.; et al. Anatomically constrained neural networks (ACNNs): Application to cardiac image enhancement and segmentation. *IEEE Trans. Med. Imaging* **2017**, *37*, 384–395. [CrossRef]
21. Dahiya, N.; Yezzi, A.; Piccinelli, M.; Garcia, E. Integrated 3D anatomical model for automatic myocardial segmentation in cardiac CT imagery. *Comput. Methods Biomech. Biomed. Eng. Imaging Vis.* **2019**, *7*, 690–706. [CrossRef]
22. Militello, C.; Rundo, L.; Toia, P.; Conti, V.; Russo, G.; Filorizzo, C.; Maffei, E.; Cademartiri, F.; La Grutta, L.; Midiri, M.; et al. A semi-automatic approach for epicardial adipose tissue segmentation and quantification on cardiac CT scans. *Comput. Biol. Med.* **2019**, *114*, 103424. [CrossRef]
23. Tripathy, R.K.; Bhattacharyya, A.; Pachori, R.B. A novel approach for detection of myocardial infarction from ECG signals of multiple electrodes. *IEEE Sens. J.* **2019**, *19*, 4509–4517. [CrossRef]
24. Acharya, U.R.; Fujita, H.; Oh, S.L.; Hagiwara, Y.; Tan, J.H.; Adam, M. Application of deep convolutional neural network for automated detection of myocardial infarction using ECG signals. *Inf. Sci.* **2017**, *415*, 190–198. [CrossRef]
25. Liu, W.; Zhang, M.; Zhang, Y.; Liao, Y.; Huang, Q.; Chang, S.; Wang, H.; He, J. Real-time multilead convolutional neural network for myocardial infarction detection. *IEEE J. Biomed. Health Inform.* **2017**, *22*, 1434–1444. [CrossRef]
26. ur Rehman, N.; Aftab, H. Multivariate variational mode decomposition. *IEEE Trans. Signal Process.* **2019**, *67*, 6039–6052. [CrossRef]
27. Tripathy, R.; Sharma, L.; Dandapat, S. Detection of shockable ventricular arrhythmia using variational mode decomposition. *J. Med. Syst.* **2016**, *40*, 79. [CrossRef] [PubMed]
28. Tripathy, R.; Paternina, M.R.A.; Arrieta, J.G.; Pattanaik, P. Automated detection of atrial fibrillation ECG signals using two stage VMD and atrial fibrillation diagnosis index. *J. Mech. Med. Biol.* **2017**, *17*, 1740044. [CrossRef]
29. Panda, R.; Jain, S.; Tripathy, R.; Acharya, U.R. Detection of shockable ventricular cardiac arrhythmias from ECG signals using FFREWT filter-bank and deep convolutional neural network. *Comput. Biol. Med.* **2020**, *124*, 103939. [CrossRef] [PubMed]
30. Bousseljot, R.; Kreiseler, D.; Schnabel, A. Nutzung der EKG-Signaldatenbank CARDIODAT der PTB über das Internet. 1995. Available online: https://www.degruyter.com/document/doi/10.1515/bmte.1995.40.s1.317/html (accessed on 25 August 2021).
31. Goldberger, A.L.; Amaral, L.A.; Glass, L.; Hausdorff, J.M.; Ivanov, P.C.; Mark, R.G.; Mietus, J.E.; Moody, G.B.; Peng, C.K.; Stanley, H.E. PhysioBank, PhysioToolkit, and PhysioNet: Components of a new research resource for complex physiologic signals. *Circulation* **2000**, *101*, e215–e220. [CrossRef]
32. Hasan, M.A.; Abbott, D. A review of beat-to-beat vectorcardiographic (VCG) parameters for analyzing repolarization variability in ECG signals. *Biomed. Eng. (Biomedizinische Technik)* **2016**, *61*, 3–17. [CrossRef]
33. Correa, R.; Arini, P.; Correa, L.; Valentinuzzi, M.; Laciar, E. New VCG and ECG indexes for early identification of acute myocardial infarction patients. In Proceedings of the VI Latin American Congress on Biomedical Engineering CLAIB 2014, Paraná, Argentina, 29–31 October 2014; pp. 369–372.
34. Yamauchi, K.; Segal, M.; Tatematsu, H.; Simonson, E. Analysis of discrepancies between VCG and ECG interpreation of anterior wall myocardial infarction. *J. Electrocardiol.* **1977**, *10*, 171–178. [CrossRef]
35. Zema, M.J. Electrocardiographic tall R waves in the right precordial leads: Comparison of recently proposed ECG and VCG criteria for distinguishing posterolateral myocardial infarction from prominent anterior forces in normal subjects. *J. Electrocardiol.* **1990**, *23*, 147–156. [CrossRef]
36. Jain, P.; Gajbhiye, P.; Tripathy, R.; Acharya, U.R. A two-stage Deep CNN Architecture for the Classification of Low-risk and High-risk Hypertension Classes using Multi-lead ECG Signals. *Inform. Med. Unlocked* **2020**, *21*, 100479. [CrossRef]

37. Maheshwari, D.; Ghosh, S.; Tripathy, R.; Sharma, M.; Acharya, U.R. Automated accurate emotion recognition system using rhythm-specific deep convolutional neural network technique with multi-channel EEG signals. *Comput. Biol. Med.* **2021**, *134*, 104428. [CrossRef] [PubMed]
38. Jani, V.P.; Ostovaneh, M.R.; Chamera, E.; Lima, J.A.; Ambale-Venkatesh, B. Automatic segmentation of left ventricular myocardium and scar from LGE-CMR images utilizing deep learning with weighted categorical cross entropy loss function weight initialization. *Circulation* **2019**, *140*, A15934–A15934.
39. Becker, G. Creating comparability among reliability coefficients: The case of Cronbach alpha and Cohen kappa. *Psychol. Rep.* **2000**, *87*, 1171E–1182E. [CrossRef] [PubMed]
40. Rehman, N.; Mandic, D.P. Multivariate empirical mode decomposition. *Proc. R. Soc. A Math. Phys. Eng. Sci.* **2010**, *466*, 1291–1302. [CrossRef]
41. Tripathy, R.K.; Ghosh, S.K.; Gajbhiye, P.; Acharya, U.R. Development of automated sleep stage classification system using multivariate projection-based fixed boundary empirical wavelet transform and entropy features extracted from multichannel EEG signals. *Entropy* **2020**, *22*, 1141. [CrossRef]
42. Tripathy, R.; Gajbhiye, P.; Acharya, U.R. Automated sleep apnea detection from cardio-pulmonary signal using bivariate fast and adaptive EMD coupled with cross time–frequency analysis. *Comput. Biol. Med.* **2020**, *120*, 103769. [CrossRef]

Article

Quantitative and Qualitative Image Analysis of *In Vitro* Co-Culture 3D Tumor Spheroid Model by Employing Image-Processing Techniques

Mukta Sharma [1,*], Venkanagouda S. Goudar [2], Manohar Prasad Koduri [3,4], Fan Gang Tseng [2,5] and Mahua Bhattacharya [1]

1 Department of Information Technology, ABV-Indian Institute of Information Technology and Management, Gwalior, Madhya Pradesh 474015, India; mb@iiitm.ac.in
2 Department of Engineering and System Sciences, National Tsing Hua University, Hsinchu 30013, Taiwan; vsg@gapp.nthu.edu.tw (V.S.G.); fangang@ess.nthu.edu.tw (F.G.T.)
3 Department of Mechanical, Materials and Aerospace, School of Engineering, University of Liverpool, Harrison Hughes Building, Liverpool L693GH, UK; mfgtjcjh@liverpool.ac.uk
4 International Intercollegiate PhD Program, National Tsing Hua University, Hsinchu 30013, Taiwan
5 Research Center for Applied Sciences, Academia Sinica, Taipei, Frontier Research Center on Fundamental and Applied Sciences of Matters, National Tsing Hua University, Hsinchu 30013, Taiwan
* Correspondence: mukta@iiitm.ac.in

Abstract: This work proposes a novel region-estimation (RE) algorithm using the quantification of colon-cancer (HCT-8) and fibroblasts (NIH3T3) cells to estimate the densest region of colon-cancer cells in *in vitro* 3D co-cultured spheroids. Cells were labelled with different cell tracker dyes to track the cells. The technique involves staining cells with cell trackers The quantification of HCT-8 and NIH3T3 cells by the RE algorithm leads to distribution pattern analysis of cells from the core to the periphery, which ultimately estimates the densest region of HCT-8 cells in an in vitro 3D cell spheroid. Cell quantification by the RE algorithm was compared with the results of cell quantification by ImageJ software. Results demonstrated the distribution patterns of cells from the core to the peripheral region of the *in vitro* 3D cell spheroid. The overall experimentation showed that the proposed methodology outperformed state-of-the-art approaches in terms of segmentation, quantification, and reducing biasing error.

Keywords: distribution patterns; fibroblast cells; HCT-8 colon-cancer cells; nature-inspired techniques; quantification; segmentation

Citation: Sharma, M.; Goudar, V.S.; Koduri, M.P.; Tseng, F.G.; Bhattacharya, M. Quantitative and Qualitative Image Analysis of *In-Vitro* Co-Culture 3D Tumor Spheroid Model by Employing Image-Processing Techniques. *Appl. Sci.* **2021**, *11*, 4636. https://doi.org/10.3390/app11104636

Academic Editors: Leonardo Rundo, Carmelo Militello and Andrea Tangherloni

Received: 24 February 2021
Accepted: 31 March 2021
Published: 19 May 2021

1. Introduction

In mimicking the structural and natural complexity of living tissue, current technology such as *in vitro* 3D spheroid cell culture models is evolving compared to the 2D cell culture model. *In vivo* cell growth and cell signaling are highly dependent on the extracellular matrix (ECM) and the interaction produced by different kinds of cells. *In vitro* 3D cell culture models aid in the study of molecular level tissue function by employing co-culture models and developing drugs for the cancer model in mitigating animal usage for drug testing. In oxygen and nutrients, the gradient can be closely mimicked by 3D cell culture techniques compared to 2D cell culture techniques [1]. However, 3D *in vitro* spheroid models seriously suffer from image acquisition and standalone image processing algorithms. As a result, user intervention during analysis might lead to heavier biases, thereby leading to erroneous results.

Such issues can be overcome by different automated computer-aided design (CAD) tasks such as segmentation, i.e., the extraction of the region of interest from the images. In the literature, several studies showed multiple usages of automatic segmentation approaches for variable cellular types [2,3]. Al-Kofahi et al. [4] stated that the segmentation of cells through automation is an essential step in image cytometry and histometry. M. Sharma et al. [5] proposed a novel non-linear segmentation model to discriminate and quantify living or dead cells. Xing and Yang [6] focused on digital pathology and microscopy image analysis, and extensively reviewed the techniques. Considerable progress was achieved in the past, but algorithms still suffer from biasing and need to boost their accuracy and robustness, consume less time, and self act against the upcoming applications. Color-based segmentation using traditional clustering algorithms, on the other hand, is relatively easy, and complexity is lesser when compared to that of segmentation techniques [7]. It is likewise more relevant for biomedical image segmentation, as the count of clusters is known beforehand. However, traditional clustering techniques suffer from various issues such as being trapped in local optima, and having sensitivity to initial cluster centers and boundary-level constraints [8]. In recent advances, many nature-inspired algorithms arose to solve these clustering problems. The hybridization of nature-inspired algorithms with each other and with traditional clustering techniques was described by Krisna et al. [9], Rana et al. [10], and Chowdhury et al. [11] to resolve clustering issues. However, all of these clustering algorithms need much parameter initialization (Table S1), increasing their complexity and manual intervention. Moreover, incorrect parameter initialization affects the end outcome. To skip the overhead of parameter settings, a nature-inspired algorithm called the teacher learning-based optimization (TLBO) algorithm [12] was proposed.

2. Material Preparations

In this study, a poly-di-methyl-siloxane (PDMS) based microwell array chip was utilized to co-culture NIH3T3 and HCT-8 cells in *in vitro*. A PDMS based microwell array chip was used to construct the tumor spheroids. The well known soft lithography process was used to fabricate the microwells, and fabrication steps were followed as given by Patra et al. [13]. Images were obtained by scanning a horizontal cross-sectional view using fluorescence based confocal microscopy. To make the cells visible under confocal microscopy, NIH3T3 cells were labeled with CellTracker™ Green CMFDA (5-chloromethylfluorescein diacetate) dye (Thermo Fisher Scientific, China). HCT-8 cells were labeled with CellTracker™ Blue CMHC (4-chloromethyl-7-hydroxycoumarin) dye (Thermo Fisher Scientific, China). Dyes were functionalized as per the manufacturer's instructions. Briefly, both cell types were incubated with the respective cell tracker dye (50 µL) for 30 min at 37 °C. Cells were further washed 3 times with PBS, mixed, and seeded on the microwell array chip for spheroid formation.

HCT-8 and NIH3T3 cells (blue and green stained cells) were co-cultured in different ratios of 2:0.5; 2:1; 2:2; and 2:4, as shown in Figure 1; While preparing the *in vitro* 3D cell spheroid, the spheroid shaping capabilities expanded and became quicker by including more NIH3T3 cells with the HCT-8 cells. To analyze the cellular distribution in the spheroids, 3D images were captured by using confocal microscopy. Z-direction images were captured (using 10x objective) from the bottom of each spheroid with 6 µm step sizes of 50 slices, i.e., a total of 300 µm, which was equal to the spheroid size, as shown in Figure 2.

This study compares the different intra and inter-domain clustering techniques in the clustering of colored cells. The outcome of the best performing algorithms was used to quantify the cells and estimate the densest region of the colon-cancer cells over an *in vitro* 3D cell spheroid using a novel region estimation algorithm based on a distance transform (DT) technique.

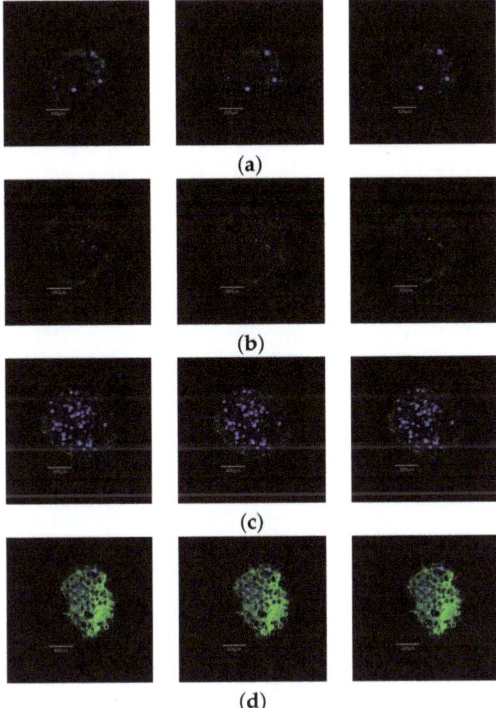

Figure 1. Sample images of each ratio for horizontal cross-section of 3D cell spheroid: (**a**) 2:0.5; (**b**) 2:1; (**c**) 2:2; (**d**) 2:4.

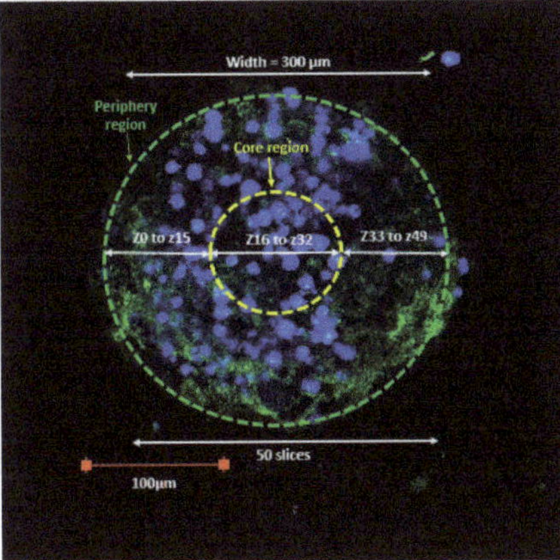

Figure 2. Partitioning of *in vitro* 3D cell spheroid into regions.

3. Problem Formulation

The spheroid was divided into 50 slices (z_{0--49}) for each ratio (Figure 2). The width of the whole spheroid was 300 µm, which means that each slice had 6 µm thickness. Therefore, the whole *in vitro* 3D cell spheroid was partitioned into two regions: (1) the core region (from approx. slice nos. z_{16--32}) and (2) the peripheral region (from approx. slice nos. $z_{0--15}and_{33--49}$). The densest region was where the concentrations of the HCT-8 cells in the *in vitro* 3D cell spheroid were the maximum between the two regions. The estimation of the densest region of the colon-cancer cells in the *in vitro* 3D cell spheroid was required for further analysis and treatment planning. With this, the technology of ultralow attachment surfaces of *in vitro* techniques is of the recent trend [14,15] recently became popular [16,17] in reducing the burden animal model. These microwell technologies have been used for the past few decades [18], and several cancer tumor models were constructed *in vitro*, which may provide a better environment to build *in vitro* 3D culturing of patient derived xenografts (PDX). Further improving image processing techniques may advance our understanding of stromal cell distribution in real tumor tissue *in vitro*, which helps in personalized medicine. Cells have different proliferation rates, but the environment depends on initial cell seeding density in the spheroid. We are currently experimenting on the aspect of the proliferation rate and monitoring them at different time points.

In order to understand the regional distribution of *in vitro* 3D cell spheroids, the total number N_b of HCT-8 cells and N_g NIH3T3 cells present in the *in vitro* 3D cell spheroid was considered. The cells were considered to be circular. The counts of HCT-8 and NIH3T3 cells were evaluated for horizontal cross sectional images of the *in vitro* 3D cell spheroid for each slice (slice nos. z_{0--49}). The identification of the densest region depends on the maximal concentrations of the HCT-8 cells in two different regions i.e., core (B_c) and periphery (B_p). Therefore, it was formulated as

$$R_{dense} = \begin{cases} CR, ifCount(B_c) > Count(G_p) \\ ||\, Count(B_c) > Count(B_p) \\ PR, ifCount(B_p) > Count(G_c) \\ ||\, Count(B_p) > Count(B_c)) \end{cases} \tag{1}$$

where G_c and G_p are the count of NIH3T3 cells for the core and peripheral regions, respectively; B_c and B_p are the count of HCT-8 cells for the core and peripheral regions, respectively. The proposed region estimation algorithm finds the densest region R_{dense} of the HCT-8 cells and gives the distribution patterns of the HCT-8 and NIH3T3 cells from the core to the peripheral region over the *in vitro* 3D cell spheroid.

4. Proposed Methodology

The overall proposed methodology comprises two main steps: (1) foreground cell clustering, and (2) region estimation and quantification, as shown in Figure 3. The original images of the *in vitro* 3D cell spheroid were first converted from the RGB color space to the $L^*a^*b^*$ color space. Unlike the RGB color model, the $L^*a^*b^*$ color model is approximately close to human vision. This color model provides uniformity in the range of perception [19]. Extracted $a * b*$ components from the RGB images were given as input to the TLBO clustering algorithm. The approach has two resultant phases, the teaching phase and the learner phase, as described in [12]. Initially, a K number of clusters were taken, each cluster datum or pixel datum was defined as a learner, and centroids that were selected randomly for each cluster are called teachers. After that, each learner's Euclidean distance with the centroid for all k clusters and fitness value was evaluated. Using the learner-phase steps given in [12], each learner was modified. Likewise, the centroids or the best learners and the existing solution are modified. After reaching maximal iterations I_{max}, the foreground cells were separately clustered. The HCT-8 and NIH3T3 cluster cell images were then converted into a binary image using Ostu's global thresholding technique [20].

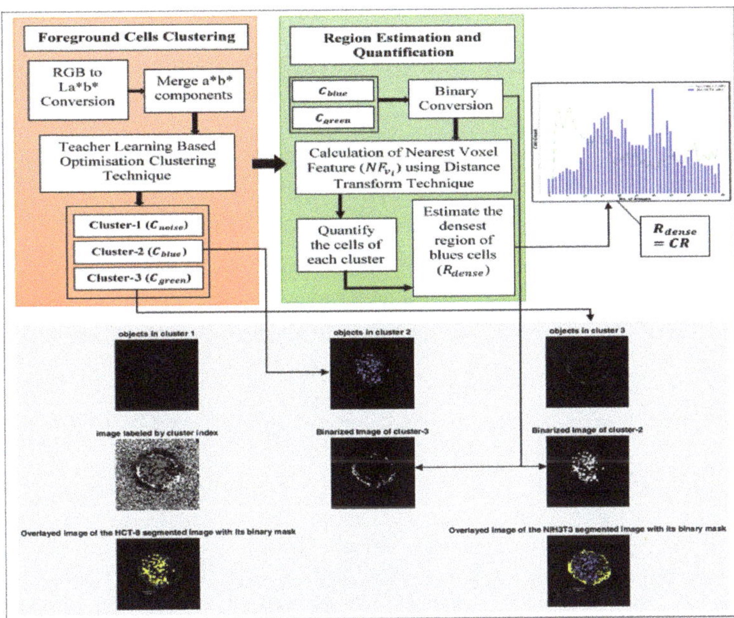

Figure 3. Resultant view of proposed methodology.

Binary images $B_b and B_r$ were then given as input in Algorithm 1 to estimate the densest HCT-8 cell region by finding the cell count. The nearest feature voxel NF_{v_i} of the B_b and B_r were calculated as follows:

$$NF_{v_i} = \begin{cases} 0 & v_i \in B_{v_i} \\ \min(\|v_i, v_j\|) & (v_i, v_j) \in F_{v_i} \end{cases} \tag{2}$$

where, $\|v_i, v_j\| = \|v_i - v_{i_0}, v_j - v_{j0}\|, \forall(v_{i_0}, v_{i_0}) \in B_{v_i}$ is any distance metric.

The outcome of DT depends on the selection of distance metrics, which varies from application to application. However, among all other metrics, the Euclidean distance metric was the most generalized because its measurement corresponds to the way in which objects are measured in the real world, and is rotation-invariant. The metric uses the L_2 norm and is defined as $\|v_i, v_j\|_2 = \sqrt{v_i^2 + v_j^2}$. After obtaining the nearest feature voxel matrix, maxima propagation is applied using the inverted mask matrix of NF_{v_i} and by limiting the propagation using some fraction ($c_1 = 0.5$) of the maximal distance of the nearest feature voxel matrix NF_{v_i}. To control the extent of propagation of MP_i, MP_i is dilated with the 3×3 matrix of all ones. This operation transforms the finer distance image into coarser segments. After that, the different numbers of segments or levels as Lvl_i are found using the "unique" function of MATLAB. Then, background voxels were removed from the Lvl_i matrix. The area threshold value was calculated to impose the minimal area constraints for segments. After that, the 8-connected component (or blobs or cells) area and pixel indices were found using the "region props()" inbuilt function of MATLAB, which simply sums the pixels of a particular region (area) by using their assigned labels. Then, the area is thresholded by using the extracted areas and finding the count of blobs or cells encountered within that area region. This gave the count of HCT-8 and NIH3T3 cells in different regions of the *in vitro* 3D cell spheroid. On the basis of these counts, i.e., $B_c, B_p, G_c and G_p$ the densest HCT-8 cell region (R_{dense}) using Equation (1) was estimated.

Algorithm 1 Region-estimation algorithm.

Input: Binary images (B_b and B_r) of HCT-8 cluster and NIH3T3 cluster images, respectively.
Output: Count of cells ($B_c, B_p, G_c and G_p$) and region of drug delivery (R_{dense})

$Count = 0, [m, n] = size(B_b)$, where m belongs to number of rows, and n belongs to number of columns of B_b.
Calculate nearest feature voxel NF_{v_i} of the B_b and B_r using Equation (2).
Calculate mask as:
for i=1 to n **do**
    $mask = (1 - NF_{v_i})$
end for
$max_d = \lceil \sum_{i=1}^{n} max(NF_{v_i}) \rceil \times c_1$, where c_1 is the constant.
Apply the maxima propagation on the distance-transformed image as:
for r=1 to max_d **do**
    $MP_i = MP_i \bigoplus [1]_{3\times3}$
    $MP_i = MP_i \times mask$
end for
Extract unique values from the MP_i matrix and save it as a Lvl_i matrix in sorted order
Removes B_{v_i} from Lvl_i
Evaluate minimal area threshold as:
$TH_a = \pi \times (max_d)^2 \times c_2$, where c_2 is the constant.
for k=1 to length(Lvl_i) **do**
    $L_i = MP_i == Lvl_i, \forall i = 1, 2, ..., n$
    Calculate area and pixel indices for each 8-connected component (object) in the binary image L_i as:
    $P_r = regionprops(L_i,' Area',' PixelIdxList')$, where P_r is the structure containing the specified properties values.
    Extract areas of the objects or cells as:
    $A_{cells_j} = [P_r.Area]$
    Threshold area as:
    $A'_{cells_j} = A_{cells} > TH_a$
    Calculate count of cells that passes imposed area A'_{cells} as:
    $Count = Count + \sum_{j=1}^{n} A'_{cells_j}$, where n is the dimension of matrix A'_{cells}
    Assign unique IDs to mark segments after area thresholding to create final segmented image as:
    for p=1 to length(A'_{cells}) **do**
        if 1 == A'_{cells_p} **then**
            $idx = P_r(p).PixelIdxList$
            $seg(idx) = random(m * n)$
        end if
    end for
end for

$$R_{dense} = \begin{cases} CR, ifCount(B_c) > Count(G_p) \\ \quad || Count(B_c) > Count(B_p) \\ PR, ifCount(B_p) > Count(G_c) \\ \quad || Count(B_p) > Count(B_c)) \end{cases}, \text{ where}$$

$CRandPR$ denote the core and peripheral region of the *in vitro* 3D cell spheroid, respectively (refer to Equation (1)).

5. Experiment Analysis

The proposed methodology was simulated using MATLAB R2017a on a system with Intel 7th generation 4770 @3.40 GHz, and validated using the dataset as mentioned in Section 1. The dataset was divided into 4 types of ratio images of NIH3T3 and HCT-8 cells. In the experiment analysis to evaluate the proposed methodology's performance, the capability of the foreground cell clustering algorithm (TLBO) was evaluated on the basis of qualitative and quantitative results as described in Section 5.1. Likewise, the region estimation algorithm's ability for cell quantification and in reducing the biasing error was evaluated by comparing it with ImageJ software as described in Section 5.2. All results are shown for 2:0.5 ratio images for reference, and the rest of the ratio images figures, graphs, and tables are provided as supplementary material.

5.1. Comparative Analysis: Qualitative and Quantitative

In this section, the reason is provided for selecting the TLBO nature-inspired clustering algorithm to extract the foreground cells of the *in vitro* 3D cell spheroid. However, comparing the ground truth images of colon-cancer cells was not possible due to the unavailability of segmentation masks to compare them with the TLBO clustering segmentation results. Therefore, the TLBO clustering algorithm results were compared with other nature-inspired clustering and traditional clustering approaches: Particle Swarm Optimization (PSO), Genetic Algorithm (GA), Invasive Weed Optimization (IWO), k-means (KM), k-medoids (KMed), and Fuzzy C-means (FCM). Results were compared on the basis of qualitative and quantitative results. On 200 iterations, the value of the fitness metrics (clustering cost as shown in Equation (3)) converged, as shown in Figure 4. Thus, all clustering experiments were performed by running the algorithms on 200 iterations. The parameter values used for each algorithm are shown in Table S1 (provided in the supplementary material).

$$Cost_j = \frac{\sum_{k=1}^{n_{max}} (d_k)}{n_{max}} \tag{3}$$

$$d_k = \sqrt{(x_s - y_t)^2 + (x_s - y_t)^2}$$

$\forall s, t \in 1, 2, ..., C_n$, where n_{max} is the number of distances d_k calculated within the clusters.

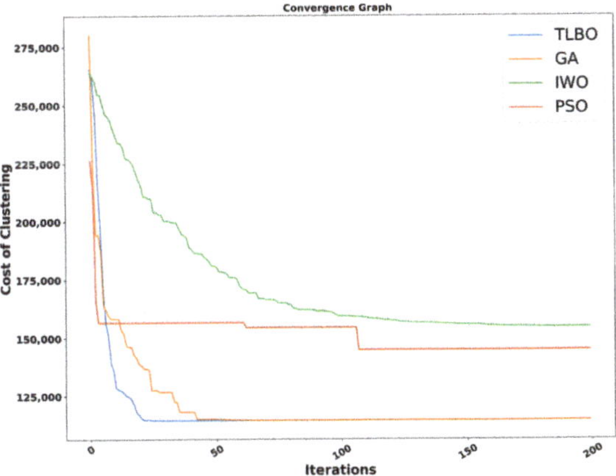

Figure 4. Cost of clustering graph of each algorithm to show convergence up to 200 iterations.

A qualitative comparison among clustering algorithms is shown in Tables 1–3 for the cellular ratio of 2:0.5 using three different images. Different cellular ratios of 2:1, 2:2, and 2:4 are shown in Tables S2–S4, respectively. The tables' information contains labeled images, the three clusters, and the graphical representation of clusters and their centroids. In the clustering of HCT-8 cells, the TLBO algorithm had greater potential than that of other algorithms. A similar phenomenon was observed for other cellular ratios, as shown in Tables S2–S4. In the quantitative comparison, all clustering algorithms were compared on the basis of quantization error (QE) [21] and best cost (BC) (Equation (3)). The quantization error of 2:0.5 ratio images is shown in Figure 5a,b, and the best cost is shown in Figure 5c,d. The different cellular ratios of quantization error and best cost are shown in Figures S5–S10.

Qualitative and quantitative analysis suggested that TLBO clustering performance was better than that of other nature-inspired clustering approaches (PSO, GA, IWO) and traditional clustering approaches (k-means, k-medoids, FCM). Therefore, the TLBO clustering algorithm was selected for foreground cell clustering in the proposed methodology.

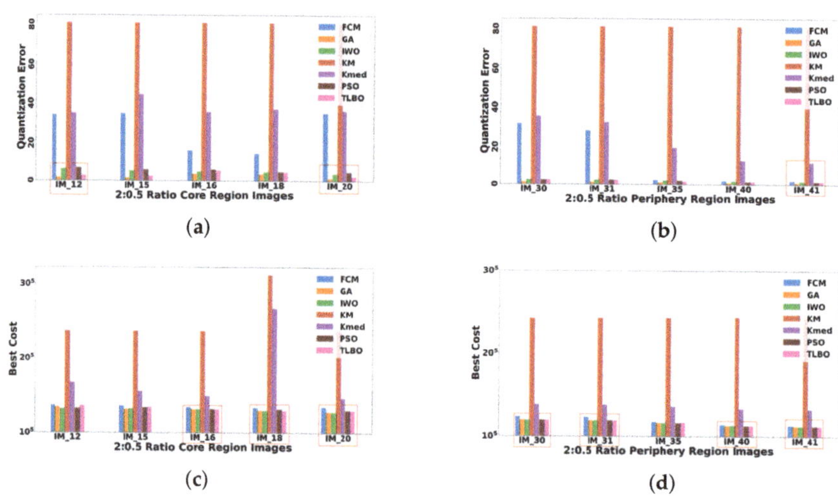

Figure 5. Comparative analysis between teacher learning based optimization (TLBO) and other existing nature-inspired clustering techniques using 2:0.5 ratio slices based on (**a**,**b**) quantization error (QE) and (**c**,**d**) best cost (BC).

Table 1. Comparative Analysis of visual results of TLBO with other clustering algorithms for 2:0.5 ratio image of Figure 1a.

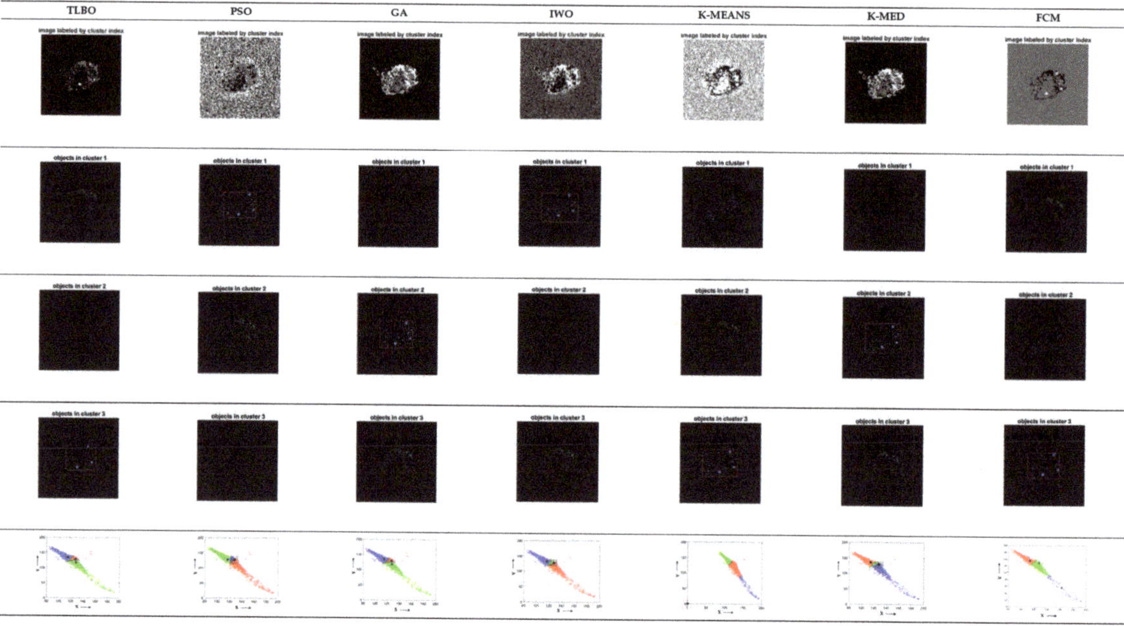

Table 2. Comparative analysis of visual TLBO results with other clustering algorithms for 2:0.5 ratio image of Figure 1a.

TLBO	PSO	GA	IWO	K-MEANS	K-MED	FCM
image labeled by cluster index	image labeled by cluster index	image labeled by cluster index	image labeled by cluster index	image labeled by cluster index	image labeled by cluster index	image labeled by cluster index
objects in cluster 1	objects in cluster 1	objects in cluster 1	objects in cluster 1	objects in cluster 1	objects in cluster 1	objects in cluster 1
objects in cluster 2	objects in cluster 2	objects in cluster 2	objects in cluster 2	objects in cluster 2	objects in cluster 2	objects in cluster 2
objects in cluster 3	objects in cluster 3	objects in cluster 3	objects in cluster 3	objects in cluster 3	objects in cluster 3	objects in cluster 3

Table 3. Comparative analysis of visual TLBO results with other clustering algorithms for 2:0.5 ratio image of Figure 1a.

TLBO	PSO	GA	IWO	K-MEANS	K-MED	FCM
image labeled by cluster index	image labeled by cluster index	image labeled by cluster index	image labeled by cluster index	image labeled by cluster index	image labeled by cluster index	image labeled by cluster index
objects in cluster 1	objects in cluster 1	objects in cluster 1	objects in cluster 1	objects in cluster 1	objects in cluster 1	objects in cluster 1
objects in cluster 2	objects in cluster 2	objects in cluster 2	objects in cluster 2	objects in cluster 2	objects in cluster 2	objects in cluster 2
objects in cluster 3	objects in cluster 3	objects in cluster 3	objects in cluster 3	objects in cluster 3	objects in cluster 3	objects in cluster 3

5.2. Quantitative Results of Region-Estimation Algorithm

In cell quantification, a thresholding process was employed. During the start of processing or in the optimization threshold, images were inspected or visualized manually, and the best set of images (in this case, 2:1) that were visualized were taken in optimizing the threshold value. This might be because of variability in laser power excitation during the acquisition of images, and intensity based image variations had minimal or no effect on the quantification of cellular analysis. Employing one ratio threshold to other cellular ratios did not affect the results, as cellular sizes were uniform in all cellular ratios. Fibroblast (NIH3T3) cell-to-cell interactions are tight and they form tight junctions; therefore, cell boundaries may not be distinguishable. NIH3T3 cells were clustered, and various thresholding values were applied to find the optimal threshold values, as shown in Figure 6. Figure 6 indicates the segmented NIH3T3 cells overlain with binary masks to evaluate the most relevant threshold value. Values were obtained by multiplying a scalar quantity with maximal intensity value in the image, and passing the binary thresholding data. It was important to choose an optimal threshold value because, by choosing a higher threshold value, intensities start merging into another, and for small threshold values, tiny cells seem to appear that are oversegmented. By visual inspection, and from Figure 6 and 7, the value of T = 0.125 corresponded to the best possible results for all image slices.

In this way, clustered NIH3T3 cells were converted into their binary masks using the optimal threshold. The binarized mask was then fed into the RE algorithm 1 to count the NIH3T3 cells. The results of the overlain images of HCT-8 and NIH3T3 binary masks on its original images are shown in Figure 8.

The distribution of NIH3T3 and HCT-8 cells was as shown in Figure 9a,b for horizontal cross-sectional views. Figure 9a shows the blue (HCT-8) cell count (BCC), and green (NIH3T3) cell count (GCC); and Figure 9b shows the blue (HCT-8) cell area (BCA) and green (NIH3T3) cell area (GCA) of the *in vitro* 3D cell spheroid for 2:0.5 ratio images. For other cellular ratios, the distribution pattern is shown in Figures S11–S13. Cell-count distribution gave better observation for HCT-8 cells, whereas the area plot gave better distribution pattern analysis for NIH3T3 cells. The limitation in NIH3T3 cell count can be attributed to the high overlap and the staining procedure of the green cell tracker. After evaluating the count for HCT-8 and NIH3T3 cells (as shown in Figure 9a,b), further evaluation for the region-estimation algorithm (Algorithm 1) was performed for each ratio image. The concentration of HCT-8 cells was the maximum in the core region compared to the whole *in vitro* 3D cell spheroid. The densest region of HCT-8 cells in the *in vitro* 3D cell spheroid estimated by the RE algorithm 1 (R_{dense}) was the core region (CR), and the width of the core region was estimated to approximately be 100 μm.

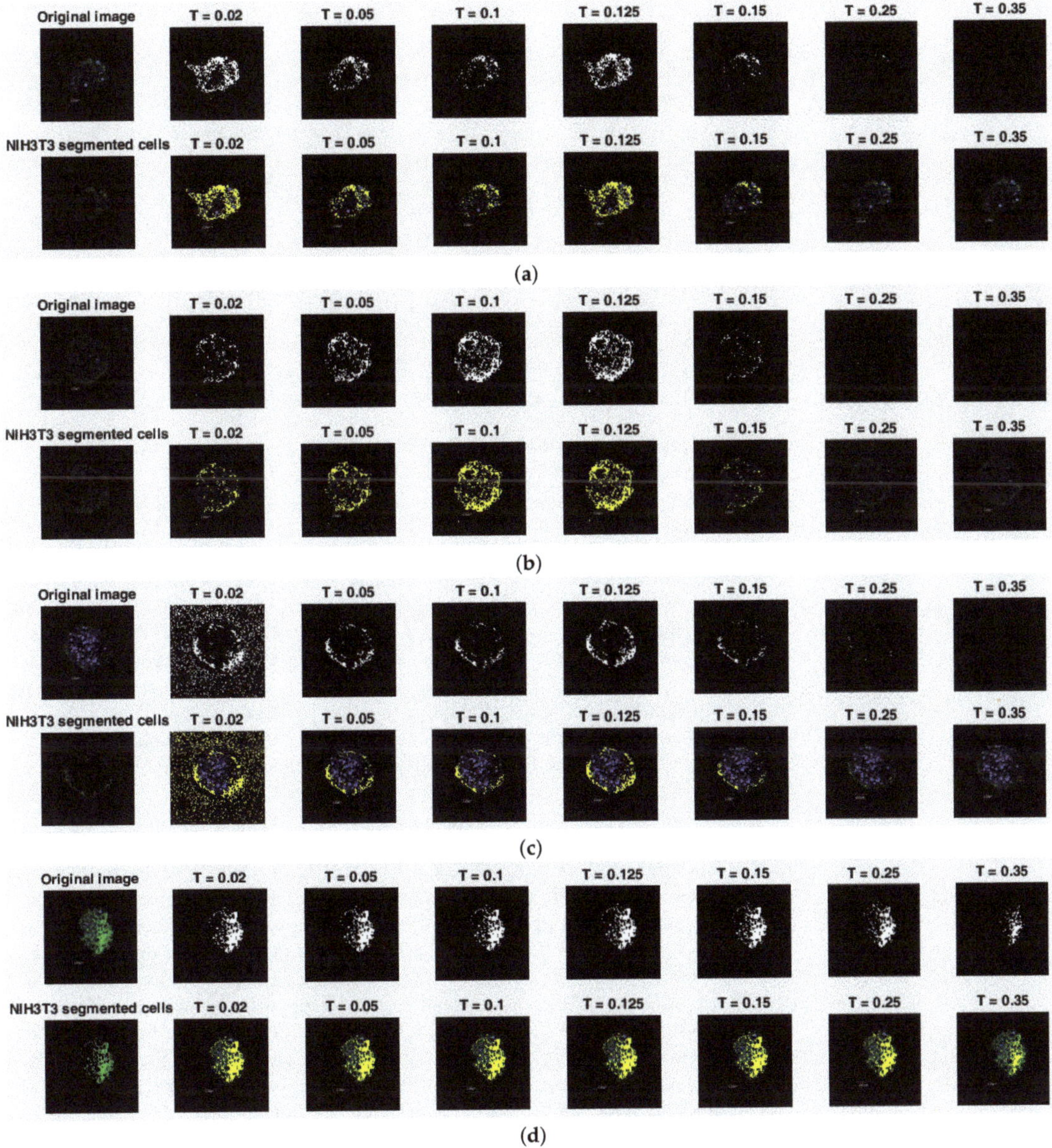

Figure 6. Optimal threshold evaluation for NIH3T3 cells using various threshold segmentation results. (**a**) 2:0.5 (2_05_z12), (**b**) 2:1 (2_1_z14), (**c**) 2:2 (2_2_z17), and (**d**) 2:4 (2_4_z10) ratio images.

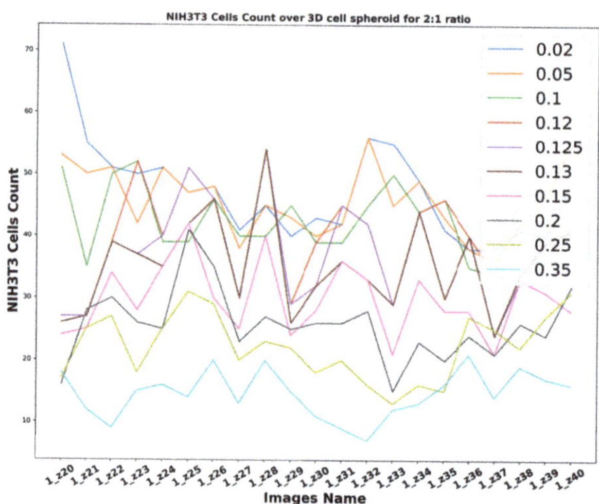

Figure 7. NIH3T3 cells count for 2:1 ratio of horizontal cross-sectional images for various threshold values.

Standalone image analysis software (Image J) was used in comparing the cellular quantification of the results. The procedure to segment and quantify HCT-8 and NIH3T3 cells for all cellular ratio images by ImageJ software was as shown in Figure 10, and the corresponding quantification data was as shown in Figure 11. The process steps during quantification using ImageJ introduce manual biasing: segmentation, thresholding, pixel size, and circularity bias. Briefly, the quantification process was as follows. Slices were input to the ImageJ software, and images were split into three channels: red, green, and blue. After splitting the channels, thresholding was adjusted. A watershed algorithm was then used for segmentation. After segmentation, images were analyzed by providing the pixel size (30–infinity) and circularity (0.30–1.00). Most biases were eliminated during the TLBO clustering and quantification approach except for thresholding bias. The ImageJ software separately clustered the HCT-8 and NIH3T3 cells along with background noise. Though it was required to perform the segmentation algorithm in both approaches, the watershed algorithm was applied in the ImageJ software, whereas TLBO clustering was applied in the proposed methodology. However, the watershed segmentation technique had the drawback of having excessive oversegmentation [22]. In contrast, TLBO separately clustered the HCT-8 and NIH3T3 cells without any background pixels.

Comparative analysis of GCC and BCC using the proposed methodology with the counts of both types of cells obtained from ImageJ software [14] from the peripheral to the core region (slice no. z_{0--49}) for 2:0.5 ratio images is shown in Figure 12. Other cellular ratios are presented in Figures S14–S16. Because of circularity and pixel-size bias, there was much observable difference between the BCC and GCC of ImageJ software, and the BCC and GCC obtained after applying the proposed methodology. The counting of cells by the proposed methodology showed that most HCT-8 cells were concentrated towards the core region, whereas NIH3T3 cells were more concentrated towards the peripheral region. The difference between manual cell counting and the proposed methodology was approximately 35%, 40%, 60%, and 80% for 2:05, 2:01, 2:2, and 2:4, respectively. Moreover, the proposed methodology accurately measured the physiological approximation of the cell count for all cell-ratio processes [15]. The difference percentage of the counting results of each ratio was variate, from 35% to 80%. This may be due to two reasons:

1. Figure 11 shows that the segmentation results of the ImageJ software for the HCT-8 and NIH3T3 cell clusters also included background pixels (noise). This happened for

each ratio image, which were further processed and detected (or counted) as blobs (or cells) by the software.

2. Moreover, the ImageJ software needed some parameter adjustment (threshold value, circularity, and size) for segmenting and counting the cells. Biasing error affected the final results.

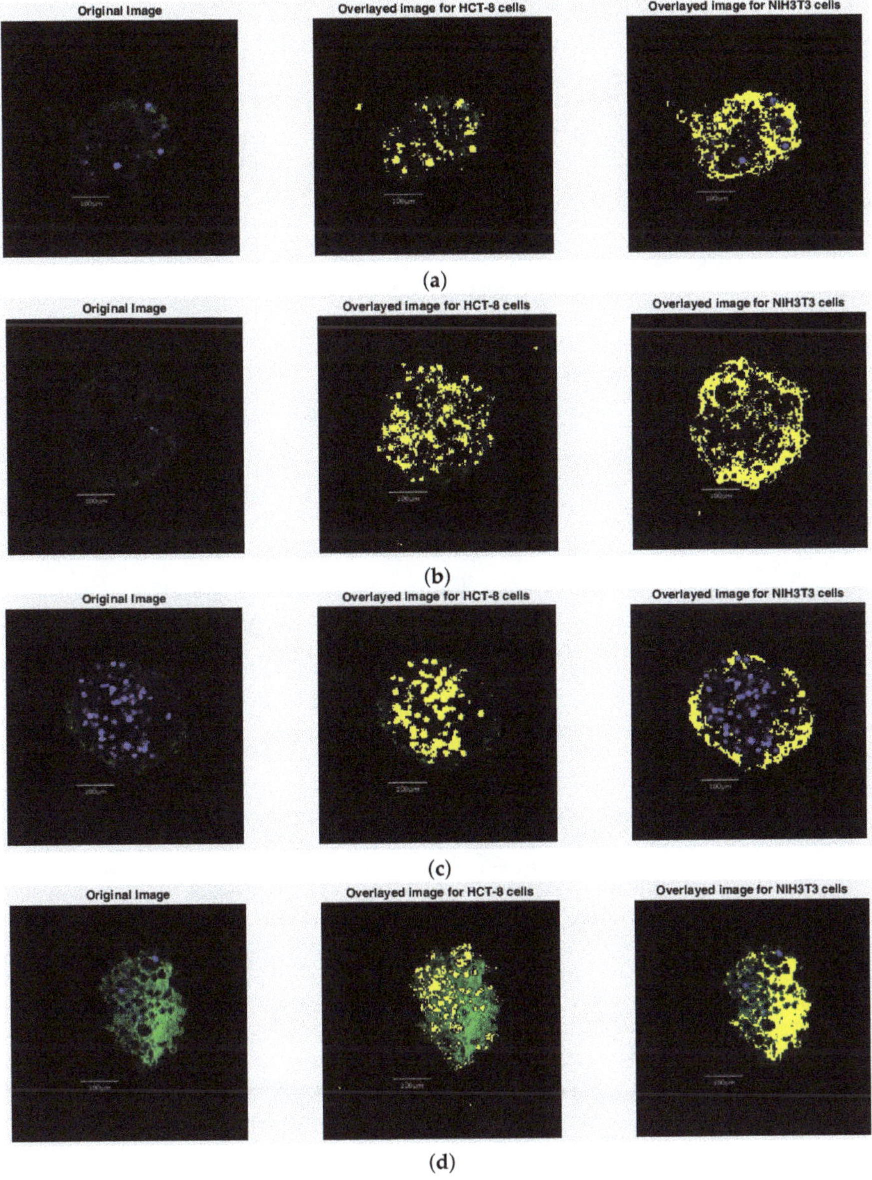

Figure 8. Overlain cluster and original images for HCT-8 and NIH3T3 cells: (**a**) 2:0.5 (2_05_z12), (**b**) 2:1 (2_1_z14), (**c**) 2:2 (2_2_z17), and (**d**) 2:4 (2_4_z10) ratio images.

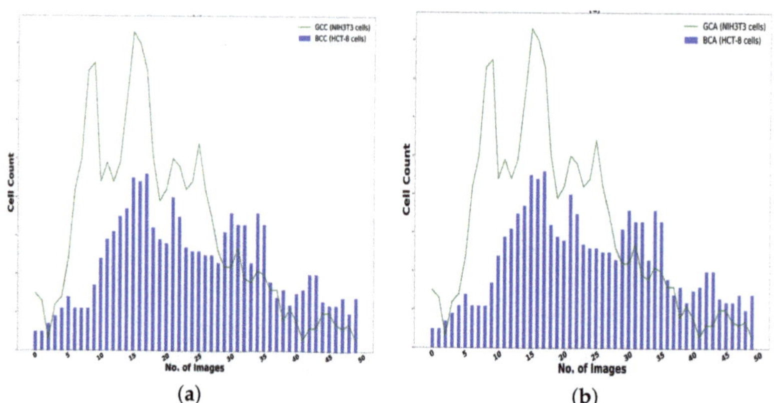

Figure 9. (**a**) Blue cell count (BCC) and green cell count (GCC) plots. (**b**) Blue cell area (BCA) and green cell area (GCA) from periphery to core of *in vitro* 3D cell spheroid of 2:0.5 ratio slices.

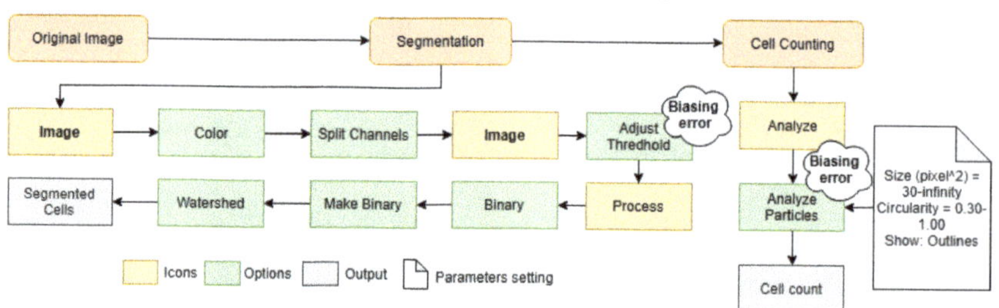

Figure 10. Schematic diagram of procedure followed by ImageJ software for segmentation and quantification of HCT-8 and NIH3T3 cells.

The equation used to evaluate the percentage difference between manual cell counting and proposed-methodology counting is as follows:

$$Diff(\%) = \frac{\sum_{i=1}^{N} M_i - \sum_{i=1}^{N} P_i}{T_c} \times 100 \quad (4)$$

where $\sum_{i=1}^{N} M_i$ is the sum of the manual cell count $\forall i = 1 \text{ to } N$, and N is the number of slices. Similarly, $\sum_{i=1}^{N} P_i$ is the sum of the proposed methodology cell count $\forall i = 1 \text{ to } N$. T_c is the total number of cells present in the *in vitro* 3D cell spheroid; $N = 50$.

Preliminary data derived from the current nature-inspired clustering algorithm (TLBO) help to understand 3D *in vitro* systems by the spectrometric location of the extracellular matrix generating protein. Thus, it aids biological scientists in further targeted molecular studies such as polymerase chain reaction (PCR) and Western blot techniques, which are highly selected for treatment planning and diagnostic procedures [23]. Hence, the proposed methodology offers better distribution analysis of HCT-8 and NIH3T3 cells compared to the ImageJ software.

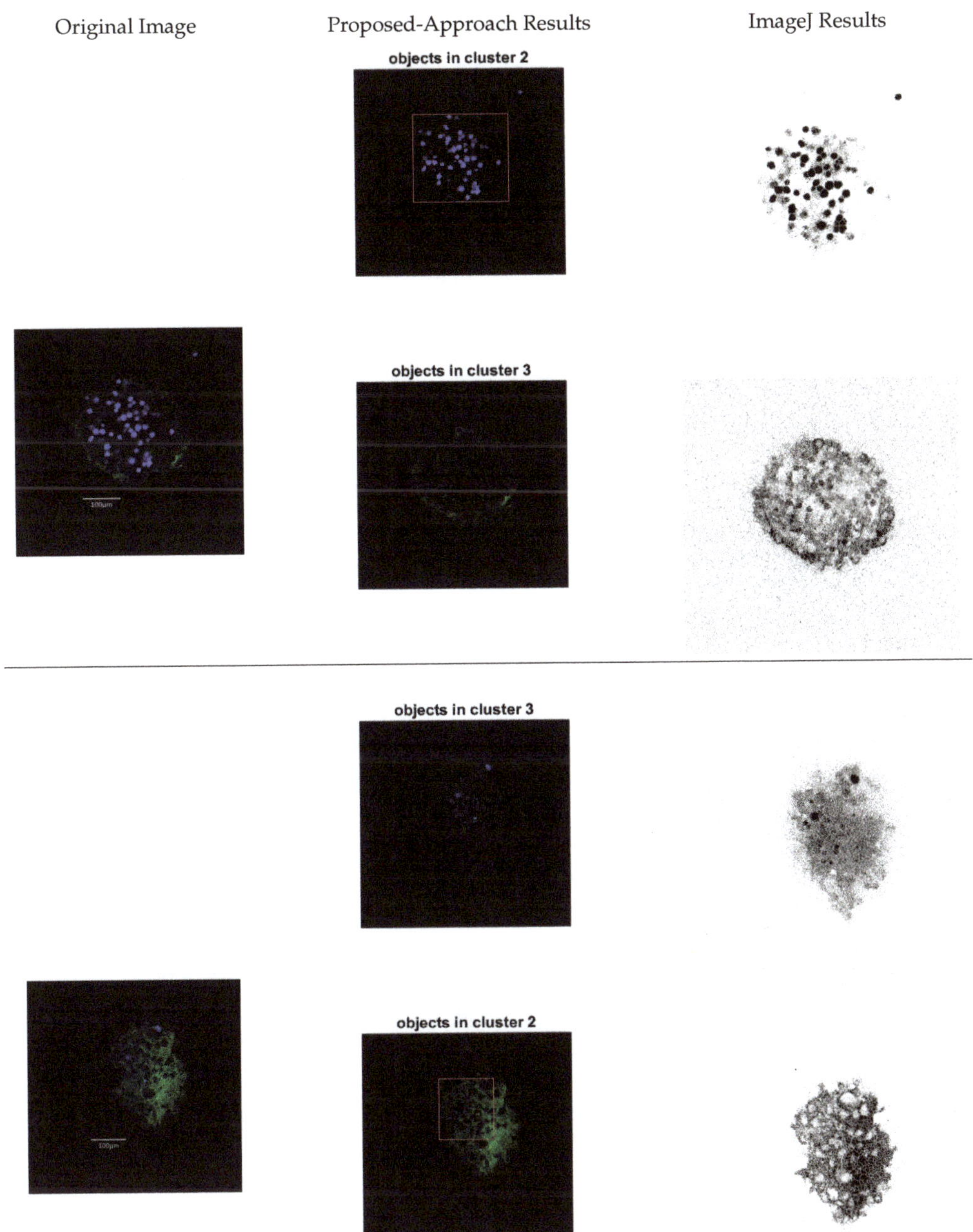

Figure 11. Comparison of TLBO algorithm and ImageJ software based on segmentation of HCT-8 and NIH3T3 cell results.

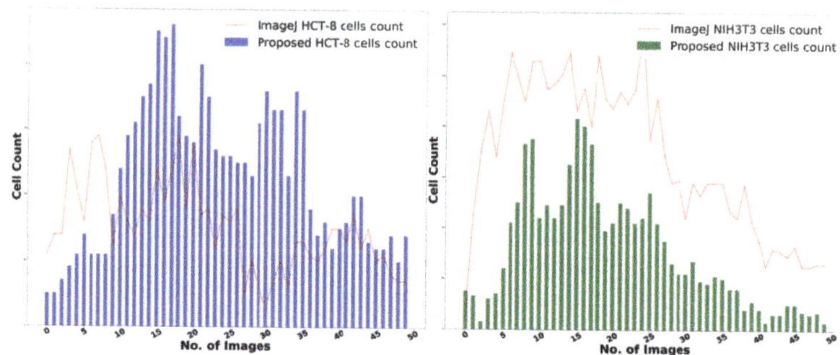

Figure 12. BCC and GCC comparison between ImageJ software and proposed methodology from periphery to core of 3D cell spheroid for 2:0.5 ratio images.

6. Conclusions

The overall proposed methodology analyzes colon-cancer cells' distribution patterns and fibroblast cells in *in vitro* 3D cell spheroids. On the basis of comparative analysis, the TLBO clustering algorithm was best suited for the cells' color-based segmentation. The region estimation algorithm estimates the densest region of HCT-8 cells in *in vitro* 3D cell spheroids on the basis of cell quantification. Compared to manual segmentation and quantification by ImageJ software, the proposed methodology reduced the biasing error for cell quantification. The current acquisition methodology was based on and majorly limited by fluorescent trackers and confocal characterization techniques. The possibility of findings can be improved by employing transfecting cells using fluorescent proteins rather than mere cell trackers, and advanced microscopy techniques such as light-sheet microscopy. So, this results in a fast acquisition process and preserves cellular 3D co-cultural spheroids' dynamic nature. This enhances the efficient utilization of the distance transform technique and the nature-inspired clustering algorithm. Therefore, one can extend the work by removing the biasing problem of the proposed methodology in quantifying NIH3T3 cells and thresholding the biasing effect.

Supplementary Materials: The following are available online at https://www.mdpi.com/2076-3417/11/10/4636/s1.

Author Contributions: Conceptualization, methodology, validation, writing—original-draft preparation: M.S.; data curation, conceptualization, formal analysis, visualization, writing—review and editing: V.S.G. and M.P.K.; funding acquisition, supervision: F.G.T.; and supervision and resources: M.B. All authors have read and agreed to the published version of the manuscript.

Funding: This research was funded by the Ministry of Science and Technology (MOST) 108-2638-E-007-001-MY2.

Institutional Review Board Statement: Not applicable.

Informed Consent Statement: Not applicable.

Data Availability Statement: Not applicable.

Acknowledgments: The authors would like to thank the ABV-Indian Institute of Information Technology and Management, India for providing the opportunity for the coordination of research work with researchers of National Tsing Hua University, Taiwan, ROC. The authors also acknowledge with thanks ABV-IIITM and NTHU for providing all facilities and expenditures related to the experiment setup, laboratory equipment, computational platform, and other requirements for the entire study and documentation of the work.

Conflicts of Interest: The authors declare no conflict of interest.

References

1. Koduri, M.P.S.; Goudar, V.; Shao, Y.W.; Hunt, J.A.; Henstock, J.R.; Curran, J.; Tseng, F.G. Fluorescence-based nano-oxygen particles for spatiometric monitoring of cell physiological conditions. *ACS Appl. Mater. Interfaces* **2018**, *10*, 30163–30171. [CrossRef]
2. Yang, H.; Ahuja, N. Automatic segmentation of granular objects in images: Combining local density clustering and gradient-barrier watershed. *Pattern Recognit.* **2014**, *47*, 2266–2279. [CrossRef]
3. Sharma, M.; Bhattacharya, M. Segmentation of CA3 Hippocampal Region of Rat Brain Cells Images Based on Bio-inspired Clustering Technique. In Proceedings of the 2019 IEEE International Conference on Bioinformatics and Biomedicine (BIBM), San Diego, CA, USA, 18–21 November 2019; pp. 2438–2445.
4. Al-Kofahi, Y.; Lassoued, W.; Lee, W.; Roysam, B. Improved automatic detection and segmentation of cell nuclei in histopathology images. *IEEE Trans. Biomed. Eng.* **2009**, *57*, 841–852. [CrossRef]
5. Sharma, M.; Bhattacharya, M. Discrimination and quantification of live/dead rat brain cells using a non-linear segmentation model. *Med. Biol. Eng. Comput.* **2020**, 1–20. [CrossRef] [PubMed]
6. Xing, F.; Yang, L. Robust nucleus/cell detection and segmentation in digital pathology and microscopy images: A comprehensive review. *IEEE Rev. Biomed. Eng.* **2016**, *9*, 234–263. [CrossRef] [PubMed]
7. Mignotte, M. Segmentation by fusion of histogram-based *k*-means clusters in different color spaces. *IEEE Trans. Image Process.* **2008**, *17*, 780–787. [CrossRef] [PubMed]
8. Hathaway, R.J.; Bezdek, J.C. Local convergence of the fuzzy c-means algorithms. *Pattern Recognit.* **1986**, *19*, 477–480. [CrossRef]
9. Krishna, K.; Murty, M.N. Genetic K-means algorithm. *IEEE Trans. Syst. Man Cybern. Part B (Cybern.)* **1999**, *29*, 433–439. [CrossRef] [PubMed]
10. Rana, S.; Jasola, S.; Kumar, R. A review on particle swarm optimization algorithms and their applications to data clustering. *Artif. Intell. Rev.* **2011**, *35*, 211–222. [CrossRef]
11. Chowdhury, A.; Bose, S.; Das, S. Automatic clustering based on invasive weed optimization algorithm. In *International Conference on Swarm, Evolutionary, and Memetic Computing*; Springer: Berlin/Heidelberg, Germany, 2011; pp. 105–112.
12. Rao, R.V.; Savsani, V.J.; Vakharia, D.P. Teaching–learning-based optimization: An optimization method for continuous non-linear large scale problems. *Inf. Sci.* **2012**, *183*, 1–5. [CrossRef]
13. Patra, B.; Chen, Y.H.; Peng, C.C.; Lin, S.C.; Lee, C.H.; Tung, Y.C. A microfluidic device for Funiform-sized cell spheroids formation, culture, harvesting and flow cytometry analysis. *Biomicrofluidics* **2013**, *7*. [CrossRef]
14. Abràmoff, M.D.; Magalhães, P.J.; Ram, S.J. Image processing with ImageJ. *Biophotonics Int.* **2004**, *11*, 36–42.
15. Alkasalias, T.; Moyano-Galceran, L.; Arsenian-Henriksson, M.; Lehti, K. Fibroblasts in the tumor microenvironment: Shield or spear? *Int. J. Mol. Sci.* **2018**, *19*, 1532. [CrossRef]
16. Fang, Y.; Eglen, R.M. Three-Dimensional Cell Cultures in Drug Discovery and Development. *SLAS Discov.* **2017**, *22*, 456–472. [CrossRef]
17. Lv, D.; Hu, Z.; Lu, L.; Lu, H.; Xu, X. Three-dimensional cell culture: A powerful tool in tumor research and drug discovery. *Oncol. Lett.* **2017**, *14*, 6999–7010. [CrossRef] [PubMed]
18. Kim, S.H.; Lee, G.H.; Park, J.Y. Microwell fabrication methods and applications for cellular studies. *Biomed. Eng. Lett.* **2013**, *3*, 131–137. [CrossRef]
19. Schwarz, M.W.; Cowan, W.B.; Beatty, J.C. An experimental comparison of RGB, YIQ, LAB, HSV, and opponent color models. *ACM Trans. Graph. (TOG)* **1987**, *6*, 123–158. [CrossRef]
20. Vala, H.J.; Baxi, A. A review on Otsu image segmentation algorithm. *Int. J. Adv. Res. Comput. Eng. Technol. (IJARCET)* **2013**, *2*, 387–389.
21. Naik, A.; Satapathy, S.C.; Parvathi, K. Improvement of initial cluster center of c-means using teaching learning based optimization. *Procedia Technol.* **2012**, *6*, 428–435. [CrossRef]
22. Grau, V.; Mewes, A.U.; Alcaniz, M.; Kikinis, R.; Warfield, S.K. Improved watershed transform for medical image segmentation using prior information. *IEEE Trans. Med. Imaging* **2004**, *23*, 447–458. [CrossRef] [PubMed]
23. Zanoni, M.; Piccinini, F.; Arienti, C.; Zamagni, A.; Santi, S.; Polico, R.; Bevilacqua, A.; Tesei, A. 3D tumor spheroid models for *in vitro* therapeutic screening: A systematic approach to enhance the biological relevance of data obtained. *Sci. Rep.* **2016**, *6*, 1. [CrossRef] [PubMed]

Short Biography of Authors

Mukta Sharma received a bachelor's degree in computer science from Uttar Pradesh Technical University, India. She completed her master's degree from GLA University, India. She is currently pursuing her doctorate from the Department of Information Technology at Atal Bihari Vajpayee Indian Institute of Information Technology and Management, Gwalior, India. Under her Ph.D. studies, Mukta Sharma is actively working on the segmentation and classification of histopathological images of cells in biomedical imaging. She has published papers in several conferences and refereed journals such as MBEC, Journal of Supercomputing, Springer. Her areas of interest are image processing, computer vision, machine learning for biomedical applications, cell morphological analysis, deep learning, and medical imaging. She has a currently active IEEE membership. Contact her at mukta.24sharma@gmail.com.

Venkanagouda S. Goudar is presently a Ph.D. student under the guidance of Prof. Fan-Gang Tseng in engineering and system sciences, National Tsing Hua University, Hsinchu, Taiwan. He completed his M.Sc. degree in Microbiology from Karnataka University, Dharwad, India in 2009. He further worked as a project assistant at the Indian Institute of Science, Bangalore, India for 4 years. He later worked as a research trainee for 7 months at the Institute of Photonic Sciences (ICFO), Barcelona, Spain. During this period, he was involved in different research areas such as nanotechnology, biosensors, lab-on-a-chip, and antibacterial and antiviral surface modifications on transparent surfaces. Currently, he is working on rare cell 3D culturing and drug delivery. Overall, during his research, he was able to produce eight peer-reviewed journal publications (Sensors and Actuators, AMI, IEEE sensors, etc.) and seven conference abstracts, and to file three patents (Indian, Taiwan, and the U.S.) in the field of biosensors and drug delivery. Contact him at venkatesh.gdr23@gmail.com.

Manohar Prasad Koduri is a dual Ph.D. student of the National Tsing Hua University, Taiwan and the University of Liverpool, UK. Manohar completed his masters in integrated circuit technologies at the University of Hyderabad, India, and finished his bachelor's degree in electronics and communication engineering at Gayatri Vidya Parishad College of Engineering, Visakhapatnam, India. Manohar is currently working on developing nanosensors for 3D tissue engineering applications. His research interests include biosensors, drug delivery, electromagnetic theory, image processing, and tissue engineering. Contact him at manimanohar92@gmail.com.

Prof. Tseng Fan-Gang is presently the vice president for R&D, NTHU, leading a multidisciplinary team at national Tsing Hua University, Hsinchu, Taiwan. For the past 20 years, his research interest included BioMEMS, nano/microfluidics, biosensors, Microfuel cells, and hydrogen storage. His recent research in BioMEMS involves cancer marker diagnosis and drug delivery, single-cell diagnosis, CTC diagnosis, the study of single-cell protein and cell dynamics, and microbubble generation and the study of its applications in tissue engineering and cancer research. He was elected as a fellow of ASME in 2014. He has received 69 patents, written 9 book chapters, published more than 220 SCI/EI journal papers with an H index 41, and has more than 6774 citations and 360 conference technical papers in the related fields. Prof. Tseng co-organized or cochaired many conferences, including Micro TAS, ISMM, IEEE MEMS, IEEE NEMS, IEEE Transducers, IEEE Nano, and IEEE Nanomed. Contact him at fangangtseng@gmail.com.

Prof. Mahua Bhattacharya is currently working as a Full Professor of the ABV Indian Institute of Information Technology and Management, Gwalior. The research area of Prof. Bhattacharya is related to biomedical-image processing, the classification of tumor or cancer growth in the human brain using multimodality medical imaging, cell-image analysis under various types of environmental exposure, and the development of AI techniques for digital and smart farming. She is executing various Government of India-funded projects in collaboration with reputed institutes. She is the President of the International Neural Network Society, India Chapter. Prof. Bhattacharya has published more than 150 papers in refereed journals and international flagship conferences as book chapters. She is the reviewer of IEEE EMBC, Elsevier, Springer, and Taylor and Francis journals. She is an organizing committee member of international IEEE conferences in India and abroad. She has delivered expert lectures in different national and international academic forums. She is an editorial board member of Neural Computing and Applications, Springer. She was Indian liaison for IJCNN'19 in Budapest, Hungary. Contact her at mahuabhatta@gmail.com.

applied sciences

Review

Deep Learning for Orthopedic Disease Based on Medical Image Analysis: Present and Future

JiHwan Lee and Seok Won Chung *

Department of Orthopaedic Surgery, School of Medicine, Konkuk University, 120-1 Neungdong-ro (Hwayang-dong), Seoul 143-729, Korea; metaphoricusjh@gmail.com
* Correspondence: smilecsw@gmail.com

Abstract: Since its development, deep learning has been quickly incorporated into the field of medicine and has had a profound impact. Since 2017, many studies applying deep learning-based diagnostics in the field of orthopedics have demonstrated outstanding performance. However, most published papers have focused on disease detection or classification, leaving some unsatisfactory reports in areas such as segmentation and prediction. This review introduces research published in the field of orthopedics classified according to disease from the perspective of orthopedic surgeons, and areas of future research are discussed. This paper provides orthopedic surgeons with an overall understanding of artificial intelligence-based image analysis and the information that medical data should be treated with low prejudice, providing developers and researchers with insight into the real-world context in which clinicians are embracing medical artificial intelligence.

Keywords: artificial intelligence; orthopedics; neural network; deep learning

Citation: Lee, J.; Chung, S.W. Deep Learning for Orthopedic Disease Based on Medical Image Analysis: Present and Future. *Appl. Sci.* **2022**, *12*, 681. https://doi.org/10.3390/app12020681

Academic Editor: Fabio La Foresta

Received: 12 November 2021
Accepted: 4 January 2022
Published: 11 January 2022

1. Introduction

A convolutional neural network (CNN) is a deep learning algorithm architecture created based on a 1962 study investigating the visual process of feline brains, and it has been applied in a wide range of areas, from autonomous vehicles to medical diagnoses [1].

A traditional CNN consists of an input layer that transmits input information, a hidden layer that modifies information (filtering) received from the input layer and amplifies the features (pooling) and an output layer that finally synthesizes and outputs the information.

According to the universal approximation theorem, it has been confirmed that various linear classifications are possible even if the neural network has a shallow hidden layer, and some pioneering studies have shown that classification and detection are improved as the layers constituting the neural network become deeper (deep neural network) [2]. Since 2012, the performance of deep learning has rapidly increased in medical image analysis with the use of deep neural networks, and this has led to a decrease in the classification error rate from approximately 25% in 2011 to 3.6% in 2015.

The CNN model was developed using a pipeline in terms of classification and detection [3], and the improved CNN shows excellent judgment, essentially giving the computer a new visual organ. A CNN has thus been expected to be used for medical diagnoses. However, a CNN does not provide any information on the basis of the decision. Therefore, even if a CNN shows an excellent diagnostic ability, it can only be discussed within a limited scope in medicine, where the basis for a judgment is important [4].

This has been pointed out as a technical limitation that reduces the effectiveness of a CNN in various fields other than medicine [5]. Researchers have dubbed this limitation "black box issues" and worked to develop "explainable artificial intelligence (XAI)" to look inside the problem [6]. The term "explainable" can be expressed as "understandability", "comprehensibility" or "interpretability" and has the same meaning. XAI should not degrade the classification or prediction performance of the model in any way and should improve the explainability. Various strategies and suitable CNN architectures have been proposed to implement an appropriate XAI [7]. Unfortunately, the black box nature of deep learning has not been completely resolved, but there are some notable achievements [8]. As one of these achievements, in 2016, Zhou et al. introduced a method explaining how a CNN makes a decision through class activation mapping [9], and this method is widely used in the field of medical artificial intelligence (Figure 1) [10].

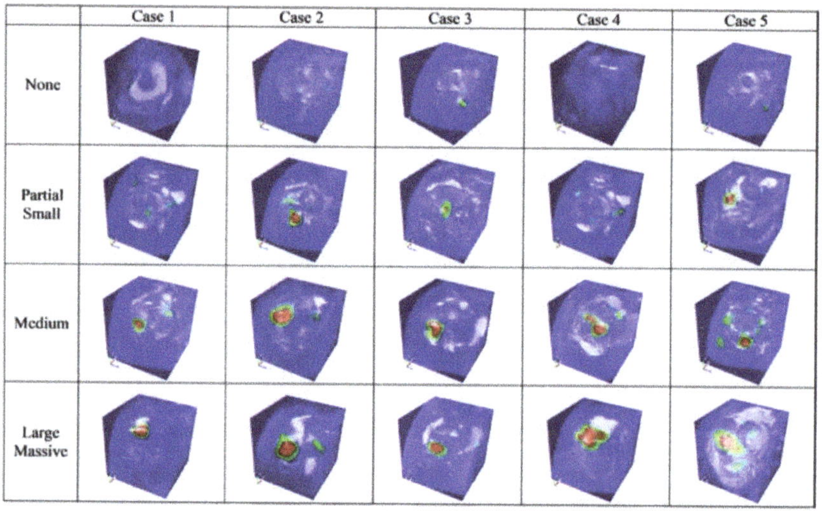

Figure 1. Image highlighting the location and size of a rotator cuff tear through a class activation map (CAM). Figure obtained from a study performed by Chung et al. [10].

In a similar context, there are attempts to improve the explainability by improving the existing CNN architecture [11]. Kim et al. modified U-Net, a CNN architecture that has strength in image segmentation, to appropriately increase the explainability. They presented an interpretable version of U-Net (SAU-Net) using an attention module for the decoder part [12].

Hence, studies introducing CNN models for diagnosing and classifying diseases using deep learning have been published in various fields of medicine, including ophthalmology and dermatology [13,14].

This trend is spreading rapidly in the field of orthopedics. Since 2017, when orthopedic disease research using deep learning was first introduced, the number of related papers has increased rapidly, and more than 300 papers in this area have been published. The search was conducted using Pubmed, MEDLINE and Embase, and papers were screened from 1 January 2017 to 2 November 2021. The search query was (orthopedic OR orthopedic) AND (deep learning). Among these studies, two orthopedic surgeons (S.W.C. and J.H.L.) independently reviewed the full text of the retrieved papers. Among these studies, 48 studies which both authors judged to be interesting and practical within the clinical context of orthopedic surgery are introduced and classified according to disease. This paper aims to provide insight into how medical artificial intelligence can help orthopedic surgeons treat patients vividly and in what context clinicians are accepting medical artificial intelligence from developers and researchers.

The authors introduce the selected papers by classifying them into the following sections: (1) Deep Learning for Fractures, (2) Deep Learning for Osteoarthritis and the Prediction of Arthroplasty Implants, (3) Deep Learning for Joint-Specific Soft Tissue Disease, (4) Miscellaneous and (5) Discussion.

2. Deep Learning for Fractures

Fractures are the most familiar ailments to orthopedists and the medical area in which deep learning methods were first applied. In 2018, Chung et al. published a CNN model for diagnosing and classifying proximal humerus fractures. Three specialists labeled 1891 anteroposterior shoulder radiographs as normal shoulders (n = 515) and 4 proximal humerus fracture types (greater tuberosity: 346; surgical neck: 514; 3-part: 269; and 4-part: 247) [15]. After labeling, a CNN model (ResNet-152) was trained with a training dataset created through augmentation of the labeled data. The CNN model recorded 96% accuracy for the normal shoulders and proximal humerus fractures, showing a higher accuracy than a general orthopedist (92.8% accuracy). This model showed a top-1 accuracy of 65–86% and an area under the curve (AUC) of 0.90–0.98 for classifying the fracture types. A recently published paper introduced a model with improved classification accuracy. In 2020, Demir et al. introduced a deep learning model to diagnose and classify humerus fractures using the exemplar pyramid method, a novel, stable feature extraction approach which showed a high classification accuracy of 99.12% [16].

Urakawa et al. trained the VGG-16 CNN model using hip plain radiographs (1773 intertrochanteric hip fracture images and 1573 normal hip images) and showed an accuracy of 95.5% [17]. Yamada et al. trained the CNN model (Xception architectural) based on 3123 hip plain and lateral radiography images, and the trained model classified fractures with 98% accuracy, which is better than orthopedists (92.2% accuracy) [18].

For the hip, as with the shoulder, there has been an attempt to classify fractures by training the CNN model. Lee et al. introduced a CNN model for training 786 anteroposterior pelvic plan radiographs using GoogLeNet-inception v3 [19]. The model classified a proximal femur fracture into type A (trochanteric region), type B (femur neck) and type C (femoral head) according to AO/OTA classification with an overall accuracy of 86.8%, showing a reasonable result. Lind et al. trained a ResNet-based CNN with anteroposterior and lateral knee radiographs, amounting to 6768 images [20]. The trained CNN model classified knee radiographic images according to the AO/OTA classification system and classified proximal tibia fractures, patellar fractures and distal femur fractures with AUCs of 0.87, 0.89 and 0.89, respectively.

The trained CNN diagnosed and classified fractures at a relatively high level in the large appendices of the shoulder, knee and hip. By contrast, a CNN model trained to diagnose and classify fractures in small joints or axial joints showed a relatively low AUC and accuracy. Farda et al. trained a PCANet-based CNN model that classified calcaneal fractures according to Sanders classification using computer tomography with 5534 datasets [21]. The trained CNN model showed 72% accuracy. In addition, Ozkaya et al. trained a CNN model based on ResNet50 with 390 anteroposterior wrist radiographic images [22]. The AUC of the learned CNN was 0.84, showing a relatively satisfactory result, but it was lower than that of experienced orthopedists.

Langerhuizen et al. compared the scaphoid fracture diagnostic accuracy between a deep learning algorithm and an orthopedist [23]. They trained the VGG16 CNN model with 150 radiographic images of scaphoid fractures and 150 images of normal wrist radiography without a fracture. Of the 150 images with scaphoid fractures, 23 could not be judged by the radiographic images and could only be confirmed through magnetic resonance imaging (MRI). The accuracy of the trained CNN model was 72%, which was lower than that of an orthopedic surgeon (84%). However, five of six occult scaphoid fractures were missed by all human observers.

An attempt was also made to diagnose the compression fractures in the spine using a trained CNN. The results showed a significant difference depending on the type of data used for learning. Chen et al. trained a ResNet-based CNN model using plain spine X-rays, and the trained CNN showed an accuracy of 73.59% [24]. By contrast, Yabu et al. presented a CNN model using MRI images as the training data. This model showed a higher accuracy (88%) than that of the surgeons [25].

In summary, fracture diagnosis using artificial intelligence showed a high level of accuracy. The trained CNN model conducted fracture diagnosis (binary classification) with a higher accuracy than fracture classification (multiclass classification), and this gap is expected to decrease as more advanced CNN models are developed.

In classifying fractures, small and axial joints showed a lower accuracy than large joints (Table 1). This may be a limitation of a CNN-based approach, which makes judgments by recognizing the contrast information (e.g., normal margin of the cortical bone and the fracture line or normal joint line) and spatial information of the images. The authors believe that this limitation can be overcome using more powerful CNN models.

Table 1. Summary of diagnostic performance for detecting/classifying orthopedic fracture.

Fracture Site	Image Used	Author. Year	CNN Used	Work	Dataset Size	Accuracy	AUC	Winner
Hip (femur neck)	X-ray	Matthew et al. 2019	GooLeNet	Binary classification	805	94%	0.98	
Hip	X-ray	Cheng et al. 2019	DenseNet	Binary classification	3605	91%	0.98	Orthopedist > CNN
Hip	X-ray	Takaaki et al. 2019	VGG-16	Binary classification	3346			CNN > Orthopedist
Hip	X-ray	Yamada et al. 2020	Xception, ImageNet	Binary classification	3123	98%		CNN > Orthopedist
Hip	X-ray	Lee et al. 2020	GoogLeNet-inception v3	Classification	686	86.8%		
Hip	X-ray	Tanzi et al. 2020	InceptionV3, VGG-16, ResNet50	Classification	2453	86% (3 class) 81% (5 class)		
Hip (Atypical fracture)	X-ray	Zdolsek et al. 2021	VGG19, InceptionV3, ResNet	Binary classification	982	91% (ResNet50) 83% (VGG19) 89% (InceptionV3)		
Shoulder (proximal humerus)	X-ray	Chung et al. 2018	ResNet	Binary classification Classification	1891	95%	0.99	Orthopedist > CNN (specialized in the shoulder)
Knee	X-ray	Lind et al. 2021	ResNet-based CNN	Classification	6768		0.87 (Proximal tibia) 0.89 (Patella) 0.89 (Distal femur)	
Ankle	X-ray	Gene et al. 2019	Xception	Binary classification	596	75%		
Ankle (Malleolar)	X-ray	Olczak et al. 2021	ResNet	Classification	5495		0.90	
Ankle (Calcaneal)	CT	Farda et al. 2021	PCANet	Classification, Segmentation	5534	72%		
Wrist	X-ray	Kim et al. 2017	Inception	Binary classification	1389		0.95	
Wrist	X-ray	Thian et al. 2019	ResNet	Binary classification	7356	88.9%	0.90	

Table 1. *Cont.*

Fracture Site	Image Used	Author. Year	CNN Used	Work	Dataset Size	Accuracy	AUC	Winner
Wrist (Scaphoid)	X-ray	Langerhuizen et al. 2020	VGG-16	Binary classification	300	72%	0.77	Orthopedist > CNN
Wrist (Scaphoid)	X-ray	Ozkaya et al. 2020	ResNet50	Binary classification	390		084	Orthopedist > CNN
Vertebra	X-ray	Chen et al. 2021	ImageNet, ResNeXt	Binary classification	1306	73.6%	0.72	Orthopedist > CNN
Vertebra	MRI	Yabu et al. 2021	VGG-16,19, Inception V3, ResNet50	Binary classification	1624		0.95	CNN > Orthopedist

Most of the diagnosis and classification of fractures using deep learning have focused on osteoporotic fractures, and studies on osteoporotic fracture joints with low frequencies are relatively poor [26]. This may be because the dataset for training the CNN model is sufficient because osteoporotic fractures account for a high proportion of the total fracture frequency, and the fracture pattern is relatively standardized, making it suitable for use in fracture classification.

3. Deep Learning for Osteoarthritis and Prediction of Arthroplasty Implants

Osteoarthritis is as familiar to orthopedists as fractures. Therefore, several attempts have been made to diagnose and classify osteoarthritis using deep learning algorithms. Xue et al. trained a CNN model based on VGG-16 with 420 plain hip X-rays [27]. This is one of the earliest studies to apply deep learning methods to the orthopedic field, and the trained model diagnosed hip osteoarthritis with an accuracy of 92.8%. Ureten et al. also presented a model for diagnosing hip osteoarthritis using a similar research design, showing an accuracy of 90.2% [28].

Tiulpin et al. trained a CNN model to classify knee osteoarthritis according to the Kellgren–Lawrence grading scale using a Siamese classification CNN [29]. The model trained using plain knee X-rays showed a multiclass accuracy of 66.7%. In addition, Swiecicki et al. trained a Faster R-CNN using plain and lateral knee X-rays from the Multicenter Osteoarthritis Study dataset [30]. The multiclass accuracy of this model was 71.9%, which showed improved performance compared with the previous study conducted by Tiulpin et al.

Pedoia et al. trained a DenseNet-based CNN based on MRI-T2 images rather than X-ray data, as used in previous studies, and this model showed a high AUC of 0.83 [31]. Kim et al. trained an SE-ResNet-based CNN model using 4366 knee anteroposterior X-rays as a dataset. Furthermore, they trained the model by adding demographic information (age, sex and body mass index), alignment and metabolic data information that can affect knee osteoarthritis, in addition to image information [32]. The diagnostic performance of the image data with additional patient information showed a significantly higher AUC (Table 2).

Advanced osteoarthritis of the hip or knee often requires arthroplasty. Several studies have introduced a model for classifying arthroplasty implants used by patients with deep learning algorithms. Karnuta et al. trained the InceptionV3 network-based CNN model using anteroposterior knee X-rays with nine different implant models inserted [33]. The trained model showed an accuracy of 99% and an AUC of 0.99, classifying the implant models at an almost perfect level. A similar attempt was made at the hip joint. In addition, Borjali et al. created a CNN model trained on 252 plain hip X-rays containing 3 different implant designs, and this model classified implants with 100% accuracy (Figure 2) [34]. Kang et al. also developed a CNN model trained on 170 plain hip X-rays containing 29 different implant designs. This model also showed a high level of performance, with an AUC of 0.99 [35].

Table 2. Summary of diagnostic performance for classifying osteoarthritis.

Location	Image Used	Author. Year	CNN Used	Work	Dataset Size	Accuracy	AUC
Knee	X-ray	Tiulpin et al. 2018	Siamese CNN	Classification	5960	66.7%	
Knee	X-ray	Pedoia et al. 2019	DenseNet	Classification	5042	75%	0.83
Knee	X-ray	Kim et al. 2020	SE-ResNet	Classification	4366	61.6% (with additional information)	0.75 (with additional information)
Knee	X-ray	Swiecicki et al. 2021	Faster R-CNN	Classification	2802	71.9%	
Hip	X-ray	Xue et al. 2017	VGG-16	Binary classification	420	92.8%	
Hip	X-ray	Ureten et al. 2020	VGG-16	Binary classification	434	90.2%	

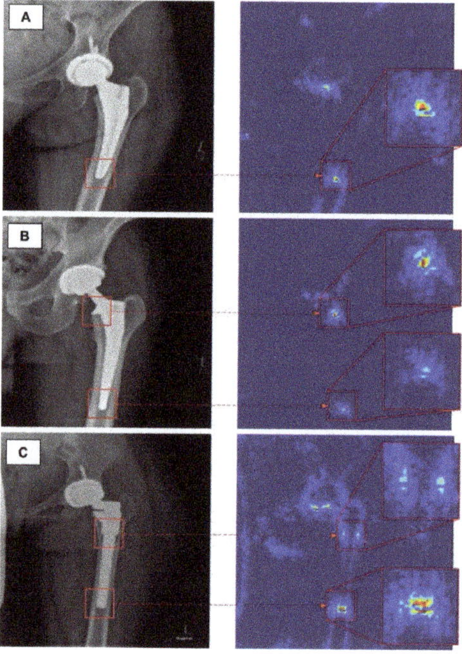

Figure 2. The figure shows how a trained convolutional neural network classifies total hip replacement implants of different designs in A, B and C. Figure obtained from a study performed by Borjali et al. [34].

By contrast, the model classifying shoulder arthroplasty implants showed a relatively low AUC. Urban et al. developed a CNN model trained on 597 plain shoulder X-rays with 16 different implant designs, showing an accuracy of 80% [36]. In addition, Sultan et al. proposed a model for classifying the different designs of four manufacturers using modified ResNet and DenseNet, showing an accuracy of 85.9% [37].

In summary, as in the case of using deep learning for fractures, binary classification of osteoarthritis has a higher accuracy than multiclass classification. In particular, the CNN-based model for specifying arthroplasty implants of the hip or knee shows a high accuracy. This may be because, unlike human bone, the implant design is highly standardized, demonstrating a clear margin on X-rays and providing clear contrast information to the CNN model. However, the classification of shoulder arthroplasty implants shows a low level of accuracy. This may be due to the fact that a shoulder anteroposterior X-ray can show a wider range of positions than an anteroposterior radiograph of the knee or hip.

4. Deep Learning for Joint-Specific Soft Tissue Disease

As for deep learning approaches, an algorithm specialized for detection based on learned images and an algorithm for segmentation by analyzing features have structural differences and have developed into different areas of application [3]. In particular, segmentation has technical difficulties in that it is necessary to preserve spastic information that is easily lost in the outer-layer process of synthesizing the results of the CNN model being trained [38]. Recent studies have attempted to overcome these limitations through techniques such as FCN-based semantic segmentation.

These differences in deep learning algorithms also affect the use of deep learning in the orthopedic field. The deep learning-based studies introduced above are cases of diagnosing and classifying diseases based on X-ray images, and a CNN model specialized for segmentation is not always required [39]. By contrast, for diseases that are diagnosed and classified based on images such as ultrasound or MRI, a satisfactory level of accuracy can be obtained using only a CNN model specialized for segmentation. For example, a CNN model for diagnosing rotator cuff tears is more appropriate for inferring such tears based on the outline of the normal rotator cuff (segmentation) than a method of diagnosis applied by specifying the location where the tear occurred (regional detection).

Therefore, CNN models for diagnosing soft tissue disease in the orthopedic field have mainly been published after 2018, which was when the segmentation technology began to mature. Kim et al. trained a CNN model using a shoulder MRI dataset of 240 patients. The trained model identified the muscle region of the rotator cuff with an accuracy of 99.9% and graded fatty infiltration at a high level [40]. Taghizadeh et al. also conducted a similar study using a shoulder computed tomography of 103 patients as a dataset. The trained CNN model measured fatty infiltration with an accuracy of 91% [41].

Medina et al. introduced a model for segmenting the rotator cuff muscle with 98% accuracy by applying a CNN model trained using the shoulder MRIs of 258 patients [42]. Furthermore, Shim and Chung et al. introduced a model for evaluating the presence of tears and their sizes in the rotator cuff by training a Voxception-ResNet (VRN)-based CNN with 2124 shoulder MRIs. The trained CNN model diagnosed and classified rotator cuff tears with accuracies of 92.5% and 76.5%, respectively [10]. In addition, Lee et al. developed a new deep learning architecture using an integrated positive loss function and a pre-trained encoder. Using this, the location of the rotator cuff tear can be relatively accurately determined, even when imbalanced and noisy ultrasound images are provided [43].

Recent studies suggesting a CNN model for diagnosing meniscal tears, cartilage lesions and anterior cruciate ligament (ACL) ruptures in the knee joint have also been published. Couteaux et al. presented a model that trains a Mask-RCNN with 1828 T2-weighted 2D Fast Spin-Echo images to classify the torn part from the normal area of the meniscus and do so according to the location of the tear [44]. This model diagnosed and classified meniscal tears with an AUC of 0.91. Roblot et al. also proposed a model for diagnosing meniscal tears in a similar way, detecting meniscal tears with an AUC of 0.94 [45].

Chang et al. presented a model for diagnosing complete ACL tears by training a U-Net-based CNN using 320 coronal proton density-weighted 2D Fast Spin-Echo images, demonstrating an AUC of 0.97 [46]. In addition, Flannery et al. trained a modified U-Net-based CNN and evaluated the level of segmentation of the model. The segmentation level suggested by the trained model did not show a statistically significant difference from the ground truth (the value actually suggested by an expert) (Figure 3) [47].

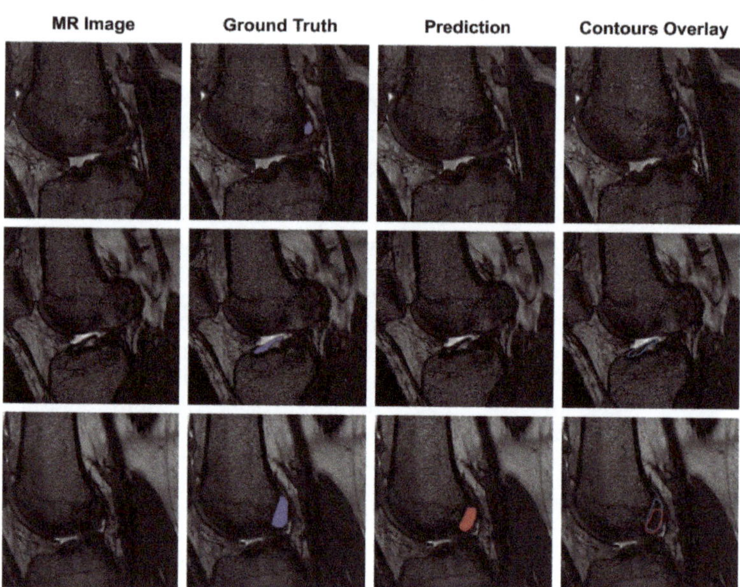

Figure 3. Each row is the same MR slice, and each column is an unsegmented slice (MR Image), an expert measured value (Ground Truth), a trained CNN model predicted value (Prediction) and an overlay of manual and predicted segmentations (Contours Overlay). Figure obtained from a study performed by Flannery et al. [47].

5. Miscellaneous

Concerning bone age, attempts to create a model that automatically predicts a bone's age through the learning of plain X-rays of carpal bones have been conducted since before the first deep learning algorithm was developed. Mahmoodi et al. presented a bone age prediction model with an accuracy of 82% in 2000, using a regression model and a Bayesian estimator [48]. A CNN model using a deep learning algorithm was developed, and it is now possible to predict the bone age with improved accuracy. In addition, Han et al. proposed a model with 97.6% accuracy by training the Inception ResNet v2 model with 5876 hand radiographs [39].

For pediatrics, developmental dysplasia of the hip is one of the most common hip joint disorders in infants and young children, and its diagnosis is difficult owing to the extensive variations in pediatric pelvic anatomy [49]. To create a deep learning algorithm that can diagnose developmental dysplasia of the hip, Zhang et al. trained a CNN model (based on ResNet-101) using 10,219 pelvic anteroposterior radiographs of children. The trained model showed a high AUC of 0.975 [50].

An acute pediatric elbow fracture is also difficult to diagnose, owing to the existence of multiple cartilaginous ossification centers and a highly variable appearance [51]. England et al. trained a CNN using 901 lateral elbow radiographs, and the trained model diagnoses elbow fractures with a high AUC of 0.985 [52].

Central dual-energy X-ray absorptiometry is the reference standard for diagnosing osteoporosis and osteopenia. A CNN model for diagnosing osteopenia and osteoporosis using plain radiography without dual-energy X-ray absorptiometry was recently introduced.

Zhang et al. trained a CNN model with 2564 lumbar X-ray images, and this model showed an AUC of 0.767 and 0.810 for osteoporosis and osteopenia, respectively [53]. Yamamoto et al. trained a CNN with 1131 hip X-rays, and this model diagnosed osteoporosis with an accuracy of 0.885 [54].

For alignment, Pei et al. published an interesting study using a deep learning algorithm to automatically measure the hip-knee-ankle angle. They trained a CNN model with

796 unilateral lower limb X-rays, showing a difference of 0.49° from the ground truth measured directly by orthopedic surgeons [55]. In addition, Rouzrokh and Pouria et al. trained a CNN model with 600 hip anteroposterior and 600 hip lateral X-rays taken after total hip arthroplasty and programmed this model to automatically derive the acetabular component inclination and version. Compared with the ground truth, this model showed a difference of 1.35° for the inclination and 1.39° for the anteversion [56].

Galbusera et al. presented a CNN model trained using biplanar radiographs of the spine. The model automatically calculated the T4-T12 kyphosis, L1-L5 lordosis, Cobb angle of scoliosis, pelvic incidence, sacral slope and pelvic tilt. Among them, the pelvic tilt showed a difference of 2.7° compared with the ground truth, whereas the L1-L5 lordosis showed a difference of 11.5° from the ground truth [56].

Concerning metastasis and infections in the spine, the spine is a joint that receives a high blood supply and is relatively easily exposed to metastasis compared with other joints [57]. Therefore, studies for diagnosing metastatic lesions using deep learning algorithms have mainly focused on the spine. Wang et al. reported that a CNN model trained with sagittal fat-suppressed T2 2D Fast Spin-Echo spine images localized metastatic lesions with a sensitivity of 90% [58]. In addition, Chmelik et al. trained a CNN with sagittal computed tomography images containing 1046 lytic lesions and 1135 sclerotic lesions, and the trained model detected lytic and sclerotic lesions with AUCs of 0.80 and 0.78, respectively [59].

Kim et al. published a CNN model to discriminate between tuberculous and pyogenic spondylitis. They trained the CNN using axial T2-weighted 2D Fast Spin-Echo images, and the trained CNN model divided the two conditions with an AUC of 0.80, with no significant difference from a human reader [60].

As for other applications, in addition to the previously introduced papers, studies using deep learning algorithms in the field of orthopedic surgery have been published. Won et al. introduced a model for grading spinal stenosis by training a Faster R-CNN [61]. Rouzrokh and Pouria et al. attempted to predict postoperative hip dislocation by training a CNN model with 92,584 hip X-rays taken after total hip arthroplasty. The trained model showed an AUC of 76.7% and an accuracy of 49.5% [62].

6. Discussion

Orthopedics, along with dermatology, ophthalmology and cardiology, is the medical field in which research into deep learning algorithms is most actively conducted. Related research has been explosively increasing since 2017, and this trend is expected to continue until the "new winter", when the development of artificial intelligence will reach its limit.

To date, image analysis studies of orthopedic diseases using deep learning have shown excellent results overall. Several studies have reported that in fractures and osteoarthritis, a trained CNN model has a diagnostic accuracy comparable to that of an expert. The studies also presented satisfactory results for the classification of fractures and osteoarthritis. However, the accuracy of multiclass classification did not reach detection, and studies on small joints presented relatively poor results compared with studies on large joints.

Nevertheless, it is expected that this limitation can be overcome for two reasons. First, the CNN model for medical image analysis aims for accurate diagnosis and appropriate classification, and the types of classes required for this purpose are relatively small. When there are few class types, Basha et al. proved that the accuracy can be improved using a CNN model structured as a deeper layer [63]. Therefore, it is expected that the development of a CNN model with deep hyperparameters will increase the accuracy of multiclass classification through medical image analysis. Second, medical images are extremely refined data compared with images used to learn road traffic conditions or climate predictions; that is, researchers can relatively easily obtain appropriate image data without noise, such as different heights of traffic lights or flying birds. This means that even with simple data augmentation such as an affine transformation, an appropriate dataset for training the CNN model can be provided.

Therefore, the authors expect that the development of a CNN model and the accumulation of additional medical images will increase the classification accuracy of fractures and osteoarthritis, which are relatively weak compared with the accuracy of diagnosis. In the same context, it is also expected that the diagnosis and classification of joint-specific soft tissue will be improved, owing to the development of deep learning algorithms advantageous for segmentation. Indeed, there are several recent studies that have completed segmentation at a high level [64,65]. In particular, Hashimoto et al. and others segmented the psoas major muscle through a U-net-based CNN model, and the trained U-net-based CNN model showed an average of 86.6% intersection over union (IoU). U-net is one of the most important semantic segmentation frameworks of CNNs [66] and has the strength of having an architecture that can recognize structural edges. Therefore, U-net is expected to be widely used for segmentation of medical images [67]. Although not in the field of orthopedics, new CNN architectures based on U-Net are continuously being introduced and reporting notable results [68]. Rundo et al. performed prostate zonal segmentation with USE-Net, incorporating Squeeze-and-Excitation blocks (SE) into U-Net [69]. Yeung et al. showed that the model trained with a dual attention-gated CNN (Focus U-Net), which improved the U-Net, segmented the polyp of the colonoscopy image to a satisfactory level [70].

Studies published in the field of orthopedic surgery have thus far been unable to present a CNN model with a higher level of diagnosis and classification than experts. An in-depth discussion is needed as to whether these results are a problem that can be overcome through data accumulation or the development of a better CNN, or whether they are a natural limitation of a CNN model learned from image data.

The authors offer two approaches. First, experts do not solve problems with image data alone. Experts can utilize information other than images, such as the patient's demographic data, the degree of pain, the nature of the disease and a physical examination, which can affect the disease diagnosis and classification. Indeed, Kim et al. reported that a CNN model trained by adding demographic information (age, sex and body mass index), alignment and metabolic data that could affect knee osteoarthritis showed a statistically significantly higher AUC [32]. Therefore, even if an improved CNN model is developed and high-quality image data are accumulated, there is a possibility that the image analysis-based CNN model using a deep learning algorithm will not reach the level of experts.

Second, despite the opinions presented above, the possibility that CNN models will outperform experts in certain fields cannot be excluded, because the CNN model analyzes images from a different point of view than human beings. Among 150 images of scaphoid fractures, Langerhuizen et al. included 23 scaphoid fracture image data that could only be confirmed through an MRI. The trained CNN model showed a lower level of accuracy than orthopedic surgeons, but it detected five of six occult scaphoid fractures that were missed by all human observers [23]. It is therefore necessary to carefully discuss whether an image analysis model using deep learning can outperform experts.

It is clear that the present CNN models have room for improvement. However, this does not undermine the significance of the studies conducted to date. The currently developed CNN model can reduce the task intensity of the expert reader and can be used for the education of non-expert medical workers, such as medical students or specialists during training [71]. In addition, through a developed CNN model, a pediatrician can roughly estimate a patient's bone age using only X-rays without the help of an orthopedic surgeon.

A step away from the fate of clinical doctors and CNN's accuracy battle, there are interesting and more practical studies that give practical help to patients and doctors. Nie et al. converted native medical CT images to higher resolution images through generative adversarial networks (GANs) [72], and this study has the potential to be extended to MRI images [73]. Therefore, it can help a society that has no choice but to use low-quality MRI due to insufficient medical infrastructure or patients who have difficulty using high-quality MRI due to cost problems.

The authors reviewed deep learning approaches for orthopedic diseases applied through image analysis and found some limitations. First, there are no models approved by the Food and Drug Administration, other than a CNN model for predicting the bone age in children and a model for diagnosing wrist fractures [74]. In other medical departments, several models have been approved by the Food and Drug Administration, starting with a deep learning-based model for the automatic diagnosis of diabetic retinopathy in April 2018.

Second, no prospective studies have been conducted [75]. To improve the quality of research and continue applicable studies, a prospective and randomized trial according to the CONSORT-AI guidelines presented in 2020 will be necessary [76].

Third, recently described deep learning methods have mostly been designed to conduct a single task. To be useful in clinical practice, multiple deep learning algorithms will need to evaluate every possible abnormality. Some efforts have been made to overcome these limitations. For example, Grauhan et al. presented a CNN model for diagnosing fractures, joint dislocation and osteoarthritis through plain shoulder radiographs [77].

Finally, there is a need to reduce expert bias on a given dataset. Orthopedic surgeons have traditionally used ultrasound, computed tomography or MRIs to diagnose soft tissue diseases. However, deep learning algorithms often make appropriate judgments beyond human cognition. Kang et al. presented a model for diagnosing SSC tendon tears with a CNN model trained using axillary lateral radiographs, and the learned model showed an appropriate level of accuracy [78]. Thus, orthopedic surgeons may have the freedom to develop CNN models based on their imagination, free from prejudice.

In conclusion, image analysis using deep learning presents a clear milestone in the field of orthopedics and is experiencing explosive growth. The development of a CNN architecture and the accumulation of refined image data are expected to lead to the development of more sophisticated models. However, it is difficult to predict whether a deep learning model that exceeds the capability of experts can be created. Orthopedic surgeons who want to apply a deep learning algorithm to image analysis need to treat data with low prejudice, present research that meets the newly suggested guidelines and focus on developing models that can multitask.

Author Contributions: Conception and design of study, interpretation of data and approval of the version of the manuscript to be published, S.W.C.; interpretation of data, drafting the manuscript and approval of the version of the manuscript to be published, J.L. All authors have read and agreed to the published version of the manuscript.

Funding: This results was supported by "Regional Innovation Strategy (RIS)" through the National Research Foundation of Korea (NRF) funded by the Ministry of Education (MOE). (2021RIS-001(1345341783)).

Institutional Review Board Statement: Not applicable.

Informed Consent Statement: Not applicable.

Acknowledgments: The authors would like to thank the authors for allowing the use of figures in this paper.

Conflicts of Interest: The authors declare no conflict of interest.

References

1. Hubel, D.H.; Wiesel, T.N. Receptive fields, binocular interaction and functional architecture in the cat's visual cortex. *J. Physiol.* **1962**, *160*, 106–154. [CrossRef] [PubMed]
2. Hornik, K.; Stinchcombe, M.; White, H. Multilayer feedforward networks are universal approximators. *Neural Netw.* **1989**, *2*, 359–366. [CrossRef]
3. Zhao, Z.-Q.; Zheng, P.; Xu, S.-T.; Wu, X. Object detection with deep learning: A review. *IEEE Trans. Neural Netw. Learn. Syst.* **2019**, *30*, 3212–3232. [CrossRef] [PubMed]
4. Wang, F.; Casalino, L.P.; Khullar, D. Deep learning in medicine—Promise, progress, and challenges. *JAMA Intern. Med.* **2019**, *179*, 293–294. [CrossRef] [PubMed]
5. Budd, S.; Robinson, E.C.; Kainz, B. A survey on active learning and human-in-the-loop deep learning for medical image analysis. *Med. Image Anal.* **2021**, *71*, 102062. [CrossRef]

6. Fujita, H. AI-based computer-aided diagnosis (AI-CAD): The latest review to read first. *Radiol. Phys. Technol.* **2020**, *13*, 6–19. [CrossRef] [PubMed]
7. Castiglioni, I.; Rundo, L.; Codari, M.; Leo, G.D.; Salvatore, C.; Interlenghi, M.; Gallivanone, F.; Cozzi, A.; D'Amico, N.C.; Sardanelli, F. AI applications to medical images: From machine learning to deep learning. *Phys. Med.* **2021**, *83*, 9–24. [CrossRef]
8. Tjoa, E.; Guan, C. A Survey on Explainable Artificial Intelligence (XAI): Toward Medical XAI. *IEEE Trans. Neural Netw. Learn. Syst.* **2020**, *32*, 4793–4813. [CrossRef] [PubMed]
9. Zhou, B.; Khosla, A.; Lapedriza, A.; Oliva, A.; Torralba, A. Learning deep features for discriminative localization. In Proceedings of the IEEE Conference on Computer Vision and Pattern Recognition, Las Vegas, USA, 30 June 2016; pp. 2921–2929.
10. Shim, E.; Kim, J.Y.; Yoon, J.P.; Ki, S.-Y.; Lho, T.; Kim, Y.; Chung, S.W. Automated rotator cuff tear classification using 3D convolutional neural network. *Sci. Rep.* **2020**, *10*, 1–9. [CrossRef]
11. Singh, A.; Sengupta, S.; Lakshminarayanan, V. Explainable Deep Learning Models in Medical Image Analysis. *J. Imaging* **2020**, *6*, 52. [CrossRef]
12. Kim, B.; Wattenberg, M.; Gilmer, J.; Cai, C.; Wexler, J.; Viegas, F. Interpretability beyond feature attribution: Quantitative testing with concept activation vectors (tcav). In Proceedings of the 35th International Conference on Machine Learning, PMLR, Stockholm, Sweden, 10–15 July 2018; pp. 2668–2677.
13. Gulshan, V.; Peng, L.; Coram, M.; Stumpe, M.C.; Wu, D.; Narayanaswamy, A.; Venugopalan, S.; Widner, K.; Madams, T.; Cuadros, J.; et al. Development and validation of a deep learning algorithm for detection of diabetic retinopathy in retinal fundus photographs. *JAMA* **2016**, *316*, 2402–2410. [CrossRef]
14. Bejnordi, B.E.; Veta, M.; Van Diest, P.J.; Van Ginneken, B.; Karssemeijer, N.; Litjens, G.; van der Laak, J.A.W.M.; The CAMELYON16 Consortium. Diagnostic assessment of deep learning algorithms for detection of lymph node metastases in women with breast cancer. *JAMA* **2017**, *318*, 2199–2210. [CrossRef]
15. Chung, S.W.; Han, S.S.; Lee, J.W.; Oh, K.-S.; Kim, N.R.; Yoon, J.P.; Kim, J.Y.; Moon, S.H.; Kwon, J.; Lee, H.-J.; et al. Automated detection and classification of the proximal humerus fracture by using deep learning algorithm. *Acta Orthop.* **2018**, *89*, 468–473. [CrossRef]
16. Demir, S.; Key, S.; Tuncer, T.; Dogan, S. An exemplar pyramid feature extraction based humerus fracture classification method. *Med. Hypotheses* **2020**, *140*, 109663. [CrossRef]
17. Urakawa, T.; Tanaka, Y.; Goto, S.; Matsuzawa, H.; Watanabe, K.; Endo, N. Detecting intertrochanteric hip fractures with orthopedist-level accuracy using a deep convolutional neural network. *Skelet. Radiol.* **2019**, *48*, 239–244. [CrossRef]
18. Yamada, Y.; Maki, S.; Kishida, S.; Nagai, H.; Arima, J.; Yamakawa, N.; Iijima, Y.; Shiko, Y.; Kawasaki, Y.; Kotani, T.; et al. Automated classification of hip fractures using deep convolutional neural networks with orthopedic surgeon-level accuracy: Ensemble decision-making with antero-posterior and lateral radiographs. *Acta Orthop.* **2020**, *91*, 699–704. [CrossRef]
19. Lee, C.; Jang, J.; Lee, S.; Kim, Y.S.; Jo, H.J.; Kim, Y. Classification of femur fracture in pelvic X-ray images using meta-learned deep neural network. *Sci. Rep.* **2020**, *10*, 1–12. [CrossRef] [PubMed]
20. Lind, A.; Akbarian, E.; Olsson, S.; Nåsell, H.; Sköldenberg, O.; Razavian, A.S.; Gordon, M. Artificial intelligence for the classification of fractures around the knee in adults according to the 2018 AO/OTA classification system. *PLoS ONE* **2021**, *16*, e0248809. [CrossRef] [PubMed]
21. Farda, N.A.; Lai, J.-Y.; Wang, J.-C.; Lee, P.-Y.; Liu, J.-W.; Hsieh, I.-H. Sanders classification of calcaneal fractures in CT images with deep learning and differential data augmentation techniques. *Injury* **2020**, *52*, 616–624. [CrossRef] [PubMed]
22. Ozkaya, E.; Topal, F.E.; Bulut, T.; Gursoy, M.; Ozuysal, M.; Karakaya, Z. Evaluation of an artificial intelligence system for diagnosing scaphoid fracture on direct radiography. *Eur. J. Trauma Emerg. Surg.* **2020**, 1–8. [CrossRef]
23. Langerhuizen, D.W.G.; Bulstra, A.E.J.; Janssen, S.J.; Ring, D.; Kerkhoffs, G.M.M.J.; Jaarsma, R.L.; Doornberg, J.N. Is deep learning on par with human observers for detection of radiographically visible and occult fractures of the scaphoid? *Clin. Orthop. Relat. Res.* **2020**, *478*, 2653–2659. [CrossRef] [PubMed]
24. Chen, H.-Y.; Hsu, B.W.-Y.; Yin, Y.-K.; Lin, F.-H.; Yang, T.-H.; Yang, R.-S.; Lee, C.-K.; Tseng, V.S. Application of deep learning algorithm to detect and visualize vertebral fractures on plain frontal radiographs. *PLoS ONE* **2021**, *16*, e0245992. [CrossRef] [PubMed]
25. Yabu, A.; Hoshino, M.; Tabuchi, H.; Takahashi, S.; Masumoto, H.; Akada, M.; Morita, S.; Maeno, T.; Iwamae, M.; Inose, H.; et al. Using artificial intelligence to diagnose fresh osteoporotic vertebral fractures on magnetic resonance images. *Spine J.* **2021**, *21*, 1652–1658. [CrossRef] [PubMed]
26. Moon, Y.L.; Jung, S.H.; Choi, G.Y. Ecaluation of focal bone mineral density using three-dimensional of Hounsfield units in the proximal humerus. *CiSE.* **2015**, *18*, 86–90.
27. Xue, Y.; Zhang, R.; Deng, Y.; Chen, K.; Jiang, T. A preliminary examination of the diagnostic value of deep learning in hip osteoarthritis. *PLoS ONE* **2017**, *12*, e0178992. [CrossRef]
28. Üreten, K.; Arslan, T.; Gültekin, K.E.; Demir, A.N.D.; Özer, H.F.; Bilgili, Y. Detection of hip osteoarthritis by using plain pelvic radiographs with deep learning methods. *Skelet. Radiol.* **2020**, *49*, 1369–1374. [CrossRef]
29. Tiulpin, A.; Thevenot, J.; Rahtu, E.; Lehenkari, P.; Saarakkala, S. Automatic Knee Osteoarthritis Diagnosis from Plain Radiographs: A Deep Learning-Based Approach. *Sci. Rep.* **2018**, *8*, 1–10. [CrossRef]

30. Swiecicki, A.; Li, N.; O'Donnell, J.; Said, N.; Yang, J.; Mather, R.C.; Jiranek, D.A.; Mazurowski, M.A. Deep learning-based algorithm for assessment of knee osteoarthritis severity in radiographs matches performance of radiologists. *Comput. Biol. Med.* **2021**, *133*, 104334. [CrossRef] [PubMed]
31. Pedoia, V.; Lee, J.; Norman, B.; Link, T.M.; Majumdar, S. Diagnosing osteoarthritis from T2 maps using deep learning: An analysis of the entire Osteoarthritis Initiative baseline cohort. *Osteoarthr. Cartil.* **2019**, *27*, 1002–1010. [CrossRef] [PubMed]
32. Kim, D.H.; Lee, K.J.; Choi, D.; Lee, J.I.; Choi, H.G.; Lee, Y.S. Can Additional Patient Information Improve the Diagnostic Performance of Deep Learning for the Interpretation of Knee Osteoarthritis Severity. *J. Clin. Med.* **2020**, *9*, 3341. [CrossRef]
33. Karnuta, J.M.; Luu, B.C.; Roth, A.L.; Haeberle, H.S.; Chen, A.F.; Iorio, R.; Schaffer, J.L.; Mont, M.A.; Patterson, B.M.; Krebs, V.E.; et al. Artificial Intelligence to Identify Arthroplasty Implants From Radiographs of the Knee. *J. Arthroplast.* **2021**, *36*, 935–940. [CrossRef] [PubMed]
34. Borjali, A.; Chen, A.F.; Muratoglu, O.K.; Morid, M.A.; Varadarajan, K.M. Detecting total hip replacement prosthesis design on plain radiographs using deep convolutional neural network. *J. Orthop. Res.* **2020**, *38*, 1465–1471. [CrossRef] [PubMed]
35. Kang, Y.-J.; Yoo, J.-I.; Cha, Y.-H.; Park, C.H.; Kim, J.-T. Machine learning–based identification of hip arthroplasty designs. *J. Orthop. Transl.* **2020**, *21*, 13–17. [CrossRef] [PubMed]
36. Urban, G.; Porhemmat, S.; Stark, M.; Feeley, B.; Okada, K.; Baldi, P. Classifying shoulder implants in X-ray images using deep learning. *Comput. Struct. Biotechnol. J.* **2020**, *18*, 967–972. [CrossRef] [PubMed]
37. Sultan, H.; Owais, M.; Park, C.; Mahmood, T.; Haider, A.; Park, K.R. Artificial Intelligence-Based Recognition of Different Types of Shoulder Implants in X-ray Scans Based on Dense Residual Ensemble-Network for Personalized Medicine. *J. Pers. Med.* **2021**, *11*, 482. [CrossRef]
38. Guo, Y.; Liu, Y.; Georgiou, T.; Lew, M.S. A review of semantic segmentation using deep neural networks. *Int. J. Multimed. Inf. Retr.* **2018**, *7*, 87–93. [CrossRef]
39. Han, Y.; Wang, G. Skeletal bone age prediction based on a deep residual network with spatial transformer. *Comput. Methods Programs Biomed.* **2020**, *197*, 105754. [CrossRef]
40. Kim, J.Y.; Ro, K.; You, S.; Nam, B.R.; Yook, S.; Park, H.S.; Yoo, J.C.; Park, E.; Cho, K.; Cho, B.H.; et al. Development of an automatic muscle atrophy measuring algorithm to calculate the ratio of supraspinatus in supraspinous fossa using deep learning. *Comput. Methods Programs Biomed.* **2019**, *182*, 105063. [CrossRef] [PubMed]
41. Taghizadeh, E.; Truffer, O.; Becce, F.; Eminian, S.; Gidoin, S.; Terrier, A.; Farron, A.; Büchler, P. Deep learning for the rapid automatic quantification and characterization of rotator cuff muscle degeneration from shoulder CT datasets. *Eur. Radiol.* **2021**, *31*, 181–190. [CrossRef]
42. Medina, G.; Buckless, C.G.; Thomasson, E.; Oh, L.S.; Torriani, M. Deep learning method for segmentation of rotator cuff muscles on MR images. *Skelet. Radiol.* **2021**, *50*, 683–692. [CrossRef]
43. Lee, K.; Kim, J.Y.; Lee, M.H.; Choi, C.-H.; Hwang, J.Y. Imbalanced Loss-Integrated Deep-Learning-Based Ultrasound Image Analysis for Diagnosis of Rotator-Cuff Tear. *Sensors* **2021**, *21*, 2214. [CrossRef]
44. Couteaux, V.; Si-Mohamed, S.; Nempont, O.; Lefevre, T.; Popoff, A.; Pizaine, G.; Villain, N.; Bloch, I.; Cotten, A.; Boussel, L. Automatic knee meniscus tear detection and orientation classification with Mask-RCNN. *Diagn. Interv. Imaging* **2019**, *100*, 235–242. [CrossRef]
45. Roblot, V.; Giret, Y.; Antoun, M.B.; Morillot, C.; Chassin, X.; Cotten, A.; Zerbib, J.; Fournier, L. Artificial intelligence to diagnose meniscus tears on MRI. *Diagn. Interv. Imaging* **2019**, *100*, 243–249. [CrossRef] [PubMed]
46. Chang, P.D.; Wong, T.T.; Rasiej, M.J. Deep Learning for Detection of Complete Anterior Cruciate Ligament Tear. *J. Digit. Imaging* **2019**, *32*, 980–986. [CrossRef] [PubMed]
47. Flannery, S.W.; Kiapour, A.M.; Edgar, D.J.; Murray, M.M.; Fleming, B.C. Automated magnetic resonance image segmentation of the anterior cruciate ligament. *J. Orthop. Res.* **2021**, *39*, 831–840. [CrossRef]
48. Mahmoodi, S.; Sharif, B.S.; Chester, E.G.; Owen, J.P.; Lee, R. Skeletal growth estimation using radiographic image processing and analysis. *IEEE Trans. Inf. Technol. Biomed.* **2000**, *4*, 292–297. [CrossRef]
49. Kyung, B.S.; Lee, S.H.; Jeong, W.K.; Park, S.Y. Disparity between clinical and ultrasound examinations in neonatal hip screening. *CiOS* **2016**, *8*, 203–209. [CrossRef]
50. Zhang, S.-C.; Sun, J.; Liu, C.-B.; Fang, J.-H.; Xie, H.-T.; Ning, B. Clinical application of artificial intelligence-assisted diagnosis using anteroposterior pelvic radiographs in children with developmental dysplasia of the hip. *Bone Jt. J.* **2020**, *102*, 1574–1581. [CrossRef]
51. Rhyou, I.H.; Lee, J.H.; Park, K.J.; Kang, H.S.; Kim, K.W. The ulnar collateral ligament is always torn in the posterolateral elbow dislocation: A suggestion on the new mechanism of dislocation using MRI findings. *CiSE* **2011**, *14*, 193–198. [CrossRef]
52. England, J.R.; Gross, J.S.; White, E.A.; Patel, D.B.; England, J.T.; Cheng, P.M. Detection of Traumatic Pediatric Elbow Joint Effusion Using a Deep Convolutional Neural Network. *Am. J. Roentgenol.* **2018**, *211*, 1361–1368. [CrossRef]
53. Zhang, B.; Yu, K.; Ning, Z.; Wang, K.; Dong, Y.; Liu, X.; Liu, S.; Wang, J.; Zhu, C.; Yu, Q.; et al. Deep learning of lumbar spine X-ray for osteopenia and osteoporosis screening: A multicenter retrospective cohort study. *Bone* **2020**, *140*, 115561. [CrossRef] [PubMed]
54. Yamamoto, N.; Sukegawa, S.; Kitamura, A.; Goto, R.; Noda, T.; Nakano, K.; Takabatake, K.; Kawai, H.; Nagatsuka, H.; Kawasaki, K.; et al. Deep Learning for Osteoporosis Classification Using Hip Radiographs and Patient Clinical Covariates. *Biomolecules* **2020**, *10*, 1534. [CrossRef] [PubMed]

55. Pei, Y.; Yang, W.; Wei, S.; Cai, R.; Li, J.; Guo, S.; Li, Q.; Wang, J.; Li, X. Automated measurement of hip–knee–ankle angle on the unilateral lower limb X-rays using deep learning. *Phys. Eng. Sci. Med.* **2021**, *44*, 53–62. [CrossRef]
56. Rouzrokh, P.; Wyles, C.C.; Philbrick, K.A.; Ramazanian, T.; Weston, A.D.; Cai, J.C.; Taunton, M.J.; Lewallen, D.G.; Berry, D.J.; Erickson, B.J.; et al. A Deep Learning Tool for Automated Radiographic Measurement of Acetabular Component Inclination and Version After Total Hip Arthroplasty. *J. Arthroplast.* **2021**, *36*, 2510–2517.e6. [CrossRef] [PubMed]
57. Lee, B.J.; Kim, S.T.; Yoon, M.G.; Kim, S.S.; Moon, M.S. Chronic osteomyelitis of the lumbar transverse process. *CiOS* **2011**, *3*, 254–257. [CrossRef] [PubMed]
58. Wang, J.; Fang, Z.; Lang, N.; Yuan, H.; Su, M.-Y.; Baldi, P. A multi-resolution approach for spinal metastasis detection using deep Siamese neural networks. *Comput. Biol. Med.* **2017**, *84*, 137–146. [CrossRef]
59. Chmelik, J.; Jakubicek, R.; Walek, P.; Jan, J.; Ourednicek, P.; Lambert, L.; Amadori, E.; Gavelli, G. Deep convolutional neural network-based segmentation and classification of difficult to define metastatic spinal lesions in 3D CT data. *Med. Image Anal.* **2018**, *49*, 76–88. [CrossRef]
60. Kim, K.; Kim, S.; Lee, Y.H.; Lee, S.H.; Lee, H.S.; Kim, S. Performance of the deep convolutional neural network based magnetic resonance image scoring algorithm for differentiating between tuberculous and pyogenic spondylitis. *Sci. Rep.* **2018**, *8*, 13124. [CrossRef]
61. Won, D.; Lee, H.-J.; Lee, S.-J.; Park, S.H. Spinal Stenosis Grading in Magnetic Resonance Imaging Using Deep Convolutional Neural Networks. *Spine* **2020**, *45*, 804–812. [CrossRef] [PubMed]
62. Rouzrokh, P.; Ramazanian, T.; Wyles, C.C.; Philbrick, K.A.; Cai, J.C.; Taunton, M.J.; Kremers, H.M.; Lewallen, D.G.; Erickson, B.J. Deep Learning Artificial Intelligence Model for Assessment of Hip Dislocation Risk Following Primary Total Hip Arthroplasty From Postoperative Radiographs. *J. Arthroplast.* **2021**, *36*, 2197–2203.e3. [CrossRef]
63. Huang, W.; Zhou, F. DA-CapsNet: Dual attention mechanism capsule network. *Sci. Rep.* **2020**, *10*, 11383. [CrossRef] [PubMed]
64. Hashimoto, F.; Kakimoto, A.; Ota, N.; Ito, S.; Nishizawa, S. Automated segmentation of 2D low-dose CT images of the psoas-major muscle using deep convolutional neural networks. *Radiol. Phys. Technol.* **2019**, *12*, 210–215. [CrossRef] [PubMed]
65. Kamiya, N.; Li, J.; Kume, M.; Fujita, H.; Shen, D.; Zheng, G. Fully automatic segmentation of paraspinal muscles from 3D torso CT images via multi-scale iterative random forest classifications. *Int. J. Comput. Assist. Radiol. Surg.* **2018**, *13*, 1697–1706. [CrossRef] [PubMed]
66. Du, G.; Cao, X.; Liang, J.; Chen, X.; Zhan, Y. Medical Image Segmentation based on U-Net: A Review. *J. Imaging Sci. Technol.* **2020**, *64*, 20508. [CrossRef]
67. Hiasa, Y.; Otake, Y.; Takao, M.; Ogawa, T.; Sugano, N.; Sato, Y. Automated Muscle Segmentation from Clinical CT Using Bayesian U-Net for Personalized Musculoskeletal Modeling. *IEEE Trans. Med. Imaging* **2020**, *39*, 1030–1040. [CrossRef]
68. Schlemper, J.; Oktay, O.; Schaap, M.; Heinrich, M.; Kainz, B.; Glocker, B.; Rueckert, D. Attention gated networks: Learning to leverage salient regions in medical images. *Med. Image Anal.* **2019**, *53*, 197–207. [CrossRef] [PubMed]
69. Rundo, L.; Han, C.; Nagano, Y.; Zhang, J.; Hataya, R.; Militello, C.; Tangherloni, A.; Nobile, M.S.; Ferretti, C.; Besozzi, D.; et al. USE-Net: Incorporating Squeeze-and-Excitation blocks into U-Net for prostate zonal segmentation of multi-institutional MRI datasets. *Neurocomputing* **2019**, *365*, 31–43. [CrossRef]
70. Yeung, M.; Sala, E.; Schönlieb, C.B.; Rundo, L. Focus U-Net: A novel dual attention-gated CNN for polyp segmentation during colonoscopy. *Comput. Biol. Med.* **2021**, *137*, 104815. [CrossRef] [PubMed]
71. Kolachalama, V.B.; Garg, P.S. Machine learning and medical education. *NPJ Digit. Med.* **2018**, *1*, 54. [CrossRef]
72. Nie, D.; Trullo, R.; Lian, J.; Petitjean, C.; Ruan, S.; Wang, Q.; Shen, D. Medical image synthesis with context-aware generative adversarial networks. In Proceedings of the International Conference on Medical Image Computing and Computer-Assisted Intervention, Quebec City, QC, Canada, 11–13 September 2017; Springer: Cham, Switzerland; pp. 417–425.
73. Ker, J.; Wang, L.; Rao, J.; Lim, C.T. Deep Learning Applications in Medical Image Analysis. *IEEE Access* **2017**, *6*, 9375–9389. [CrossRef]
74. Benjamens, S.; Dhunnoo, P.; Meskó, B. The state of artificial intelligence-based FDA-approved medical devices and algorithms: An online database. *NPJ Digit. Med.* **2020**, *3*, 118. [CrossRef] [PubMed]
75. Topol, E.J. Welcoming new guidelines for AI clinical research. *Nat. Med.* **2020**, *26*, 1318–1320. [CrossRef]
76. Liu, X.; Rivera, S.C.; Moher, D.; Calvert, M.J.; Denniston, A.K. Reporting guidelines for clinical trial reports for interventions involving artificial intelligence: The CONSORT-AI Extension. *BMJ* **2020**, *370*, m3164. [CrossRef] [PubMed]
77. Grauhan, N.F.; Niehues, S.M.; Gaudin, R.A.; Keller, S.; Vahldiek, J.L.; Adams, L.C.; Bressem, K.K. Deep learning for accurately recognizing common causes of shoulder pain on radiographs. *Skelet. Radiol.* **2021**, *51*, 355–362. [CrossRef]
78. Kang, Y.; Choi, D.; Lee, K.J.; Oh, J.H.; Kim, B.R.; Ahn, J.M. Evaluating subscapularis tendon tears on axillary lateral radiographs using deep learning. *Eur. Radiol.* **2021**, *31*, 9408–9417. [CrossRef] [PubMed]

MDPI
St. Alban-Anlage 66
4052 Basel
Switzerland
Tel. +41 61 683 77 34
Fax +41 61 302 89 18
www.mdpi.com

Applied Sciences Editorial Office
E-mail: applsci@mdpi.com
www.mdpi.com/journal/applsci

www.ingramcontent.com/pod-product-compliance
Lightning Source LLC
LaVergne TN
LVHW071249170726
843515LV00006BA/1360

* 9 7 8 3 0 3 6 5 6 4 8 7 6 *